The Secret Life of Literature

The Secret Life of Literature

Lisa Zunshine

The MIT Press
Cambridge, Massachusetts
London, England

The MIT Press would like to thank the anonymous peer reviewers who provided comments on drafts of this book. The generous work of academic experts is essential for establishing the authority and quality of our publications. We acknowledge with gratitude the contributions of these otherwise uncredited readers.

This book was set in Stone Serif and Stone Sans by Westchester Publishing Services. Printed and bound in the United States of America.

Library of Congress Cataloging-in-Publication Data

Names: Zunshine, Lisa, author.
Title: The secret life of literature / Lisa Zunshine.
Description: Cambridge, Massachusetts : The MIT Press, [2022] | Includes
    bibliographical references and index.
Identifiers: LCCN 2021031221 | ISBN 9780262046336 (hardcover)
Subjects: LCSH: Literature—Psychological aspects. | Cognition in literature. |
    Psychology and literature. | Narration (Rhetoric)—Psychological aspects. |
    Discourse analysis, Literary—Psychological aspects.
Classification: LCC PN56.P93 Z86 2022 | DDC 801/.9—dc23
LC record available at https://lccn.loc.gov/2021031221

10 9 8 7 6 5 4 3 2 1

Contents

List of Illustrations

Preface

The Secret Life of Literature brings together cognitive science and literary history to trace a series of patterns that made their early, modest, appearance in literature at least four thousand years ago and have, since then, grown to become the cornerstone of literary imagination, while remaining largely invisible to readers and critics. It shows how social institutions and political regimes can strengthen or weaken the hold of those patterns and how they present in North American, British, Chinese, Russian, German, and Melanesian, as well as ancient Greek, Roman, and Mesopotamian cultures.

"Cognitive" literary criticism is a relatively new field, yet one already well populated by studies ranging across a variety of genres and cultures.[1] Readers familiar with such studies will notice that this book is organized differently from others. Instead of starting out by reviewing cognitive foundations of my argument, as is often done in such cases, I postpone this review until chapter 3. I do this because I want my readers to be excited about discovering something new about literature, before learning how cognitive psychology and social neuroscience support these discoveries and sharpen their meaning.

A brief road map: The volume is divided into six chapters. Chapter 1 introduces what I call the "secret life of literature," showing how specific patterns of "mindreading" (that is, of the capacity to explain people's behavior as caused by their unobservable mental states, such as thoughts, desires, and intentions) have come to shape our interaction with novels, plays, and narrative poems, as well as with memoirs focusing on imagination and consciousness. The conversation here is more practical than theoretical: I use numerous examples to train the reader to recognize those mindreading patterns in a variety of literary contexts. This chapter also explores the fraught issue of the difference between popular and literary fiction (e.g., Can a computer program distinguish between the two?) and recounts my experience of taking a graduate seminar in creative writing, at my home university, which I did in order to learn if writers themselves are aware of their role in supporting the secret life of literature. (Spoiler alert!) They mostly aren't, and that's a good thing.

Chapter 2 shifts the focus from what the secret life of literature *is* to what it *does*, and the argument becomes more theoretical and historical. I show that writers can intuitively experiment with the real-life relationship between social status and mindreading. Briefly, in real life, the lower one's relative social standing, the more active and perceptive a mindreader they are; in fiction, not necessarily. Here are some highlights of this chapter. If you want to see how characters' mindreading disparity is used in works of literature foregrounding race, go to section 2.6. If you are interested in the work of Mikhail Bakhtin and want to see how this disparity can become a form of heteroglossia in a novel obsessed with social class, turn to section 2.8. To learn what happens to fictional characters' mindreading profiles under totalitarian regimes, take a look at sections 2.10 and 2.11.

Chapters 3, 4, and 5 deal with the history of the secret life of literature, focusing, respectively, on its evolutionary and neurocognitive foundations, its relationship with community-specific mindreading values, and its migration across different genres and national literary traditions. Chapter 3 presents perspectives from social, developmental, clinical, and evolutionary psychology, as well as cognitive neuroscience. Chapter 4 aims to provide a comparative context for some of our unspoken but pervasive beliefs about mindreading. Specifically, it builds on the insight of linguistic anthropologists studying language socialization in Papua New Guinea, that "the similarities and differences between these two practices—thinking about others' internal states and/or talking about them—are often at the heart of culture."[2] We do not often think of literature as expressing a particular mindreading ideology—that is, who gets to talk about people's mental states and who does not and which cultural institutions promote this kind of talk and which suppress it. Once we start thinking about it this way, however, a broad range of practices that we take for granted—for example, readers talking unembarrassedly about characters' intentions; writers using deception, eavesdropping, and shame as recurrent plot devices—appear in new light. Chapter 5 explores the role of those plot devices, particularly lying, in shaping the secret life of literature in ancient China and early-modern Russia.

Chapter 6 turns to children's literature. It follows treatment of mind-reading in stories targeting one- to two-year-olds, three- to seven-year-olds, and nine- to twelve-year-olds, as well as, provisionally, young adult audiences. It inquires, in particular, into the role of tricksters in stories geared toward three- to seven-year-old children, and it looks at the interplay of cognitive and historical factors involved in designating some texts as serious novels and others as "kiddie lit."

A short conclusion speculates about the future of the secret life of literature, imagining travails of an author who decides to write a novel that would break with this pattern. It revisits cultural institutions that would make it hard for the author to do so—hard but not impossible, for, as the preceding chapters will have demonstrated, mindreading ideologies that underwrite the secret life of literature are not cut in stone.

Although this book is a work of literary criticism, it is not intended only for literary critics. I tried to keep it as reader-friendly as possible, by banishing discursive scholarly references to endnotes and not assuming any specialized knowledge on the part of my audience. While working on this project, I shared my research-in-progress not just with literary scholars but also with cognitive and social psychologists, as well as with anthropologists, ethnographers, philosophers, and students of media and communication. My hope is that my argument will continue to be of interest to scholars from those disciplines, as well as to any habitual, or occasional, readers of fiction.

Meanwhile, I gratefully acknowledge the valuable feedback provided by my colleagues from the International Society for the Study of Narrative; the Forum for Cognitive and Affect Studies at the Modern Language Association; the annual conference "Cognitive Futures in the Arts and the Humanities"; the Chinese Association of Cognitive Poetics and Cognitive Literary Studies at China University of Petroleum (Beijing), Hainan Normal University (Haikou), and Guandong University of Foreign Studies (Guangzhou); the Program in Literary Linguistics and Cognitive Literature Studies at the Smolny Institute in Saint Petersburg State University; the Religion, Cognition and Culture Research Unit of the Aarhus Institute for Advanced Studies; the Cambridge Symposium on Cognitive Approaches to Children's Literature; the European Association of Social Psychology; the Center for the Study of the Novel at Stanford University; and the Center for Science and Society at Columbia University.

I am also grateful to friends and colleagues who have, for the past several decades, provided me with invaluable support and whose brilliance and kindness have kept me going: Porter Abbott, Denis Akhapkin, Frederick Luis Aldama, Elaine Auyoung, Michael Austin, Alexandra Berlina, Guillemete Bolens, Fritz Breithaupt, Rhonda Blair, George Butte, Emanuele Castano, Terence Cave, Rita Charon, Mary Crane, Amy Cook, David Richter, Nancy Easterlin, Felipe de Oliveira Fiuza, William Flesch, Monika Fludernik, Thalia R. Goldstein, Paul L. Harris, David Herman, Patrick Colm Hogan, Tony Jackson, Isabel Jaén-Portillo, Suzanne Parker Keen, David Comer Kidd, Karin Kukkonen, Joshua Landy, Haiyan Lee, Howard Mancing, Bruce McConachie, Muqing Xiong, Pascal Nicklas, Keith Oatley, Aaron Ngozi Oforlea, Laura Otis, Alan Palmer, Jim Phelan, Natalie Phillips, Carl Plantinga, Merja Polvinen, Peter Rabinowitz, Alan Richardson, Naomi Rokotnitz, Marie-Laure Ryan, Ralph James Savarese, Bambi Schieffelin, Casey Schoenberger, Nicola Shaughnessy, Julien Jacques Simon, Ellen Spolsky, Gabrielle Starr, Francis Steen, Simon Stern, Peter Stockwell, John Sutton, Mark Turner, Blakey Vermeule, J. Keith Vincent, and Wen Yongchao.

My special thanks are to Michael Holquist (who, sadly, passed away in 2016) and Douglas H. Whalen, my coauthors in a series of experiments carried out at the Haskins Laboratories in New Haven and CUNY Graduate Center and seeking to find out if the "secret life of literature" can be tested empirically. I could not have wished for more creative and encouraging friends and collaborators.

At the MIT Press, I am grateful to Philip Laughlin and his team of anonymous reviewers, whose detailed suggestions for revision have been truly invaluable. In a couple of instances when I could not follow them, the fault is all mine. I am also thankful to Andrew Katz for his thoughtful copyediting of the manuscript.

Some of this book's arguments have first been published elsewhere. I am grateful to the editors of *PMLA*, *Narrative*, and *Eighteenth-Century Fiction* for letting me include material from my essays "The Secret Life of Fiction"

(*PMLA* 130, no. 3 [2015]); "Bakhtin, Theory of Mind, and Pedagogy: Cognitive Construction of Social Class" (*Eighteenth-Century Fiction* 30, no. 1 [2017]); and "What Mary Poppins Knew: Theory of Mind, Children's Literature, History" (*Narrative* 27, no. 1 [2019]).

Last but not least, I thank Joel Kniaz, Etel Sverdlov, and Harry Zunshine.

Figure 1.1
Bradbury Thompson, Tom Sawyer stamp, 1972.

1 The Secret Life of Literature

1.1 What It Looks Like

In a famous scene from an American novel, one twelve-year-old boy is hoodwinking another. The occasion is so iconic that, in 1972, the US Postal Service honored it with a special stamp. Designed by the artist Bradbury Thompson (figure 1.1), the stamp depicts Tom Sawyer pretending to be absorbed in whitewashing a fence, while Ben Rogers watches intently.

Only a minute ago, Ben was playing a game—impersonating a Missouri steamship—but now it has lost all charm for him. All he wants is to take over Tom's chore, and, after appropriate hesitation and negotiation, Tom obliges, quietly exulting in his cleverness: "Tom gave up the brush with reluctance in his face, but alacrity in his heart. And while the late steamer Big Missouri worked and sweated in the sun, the retired artist sat on a barrel in the shade close by, dangled his legs, munched his apple, and planned the slaughter of more innocents. There was no lack of material; boys happened along every little while; they came to jeer, but remained to whitewash."[1]

Ben is sweating in the sun, Tom is sitting in the shade, and Twain is having fun with a biblical reference. His twelve-year-old Herod will soon "slaughter" more "innocents." With macabre logic, Twain describes those innocents as things inanimate. They merely "happen along," as a "material" on which "the retired artist" can work at leisure, dangling his legs and munching an apple.

What underlies these ironic twists—that is, the reason that we understand why the boys are described as being massacred and manipulated—is a series of psychological insights developed by Twain's protagonist. Tom doesn't want his friends to realize that he hates whitewashing the fence. He discovers that if he makes them think that he enjoys it, they'll see it as play instead of work and even pay him for the privilege of doing his chore.

Take another look at those insights. Each of them is structured as a mental state within a mental state within yet another mental state: Tom *doesn't want* his friends to *realize* that he *hates* whitewashing the fence; he *wants* them to *think* that he *enjoys* it. Granted, these are my formulations, but if you try to come up with one of your own, you will discover that, if you want to capture the complexity of the social situation conjured up by Twain, simpler descriptions of mental functioning, such as "he *wants* them to do his work for him" or "they *think* that he *likes* painting the fence " won't do. In fact, they'll misrepresent what's going on, until you find a way to connect them, through another thought or intention. It seems, in other words, that, however you choose to phrase it, you'll need to recursively embed mental states on at least the third level.

Cognitive psychologists and philosophers of mind talk about "mental states" in conjunction with "theory of mind" and "mindreading," which are metaphorical terms[2] used to describe our capacity to see behavior as caused by mental states, such as thoughts, desires, feelings, and intentions.[3] Embedment is yet another metaphor, which comes in handy when we want to talk about complex social dynamics that depend on people's awareness of their own and other people's states. (Although cognitive scientists have several different terms to talk about this kind of awareness, including, for instance, "recursive intention-reading" and "recursive mind-reading,"[4] I prefer the shorter "embedment.") To illustrate the way the term "mental states" will be used throughout this study, here are some examples, with mental states italicized:

- "My last name begins with a *Z*" contains no mental states, embedded or otherwise.[5]
- "I'm *glad* that my last name begins with a *Z* because the teacher may not get to the end of the list today" contains just one mental state: my being happy about being at the end of the class list.
- "I am *afraid* that the teacher *will remember* that she hasn't called on me for a while" contains two embedded mental states: my thinking about what my teacher may be thinking.
- Finally, "I *wonder* if the teacher *realizes* that I'm *hoping* that she won't call on me today because my last name begins with a *Z* and will thus on purpose start at the end of the list" contains three embedded mental states: my thinking about the teacher's thinking about my thinking.

Note that we have to rely on this kind of propositional, or representational, language ("I wonder if she realizes that I'm hoping") to talk about embedded mental states, because it is the tool that we have at our disposal to model the complex intersubjective dynamic of such situations. The actual cognitive processes involved in our experience of those situations may not be structured like embedded representations or may not even "be structured at all."[6] Moreover, mindreading, especially in face-to-face communication, depends on embodied feedback loops (for instance, there may be something about the expression on my teacher's face, as she catches me watching her intently, that may strengthen or weaken my hopes), but these important nuances are left out of our crude linear diagrams.

Later, in chapter 3, I will consider in detail this problem of our limited vocabulary. Here, I want us to focus on something else. Ask yourself, How often, in our daily goings-on, do we thus embed mental states on the third and fourth levels? Or, to put it differently, how often do we find ourselves involved in social situations that would require this kind of language to describe them? Although it certainly happens—I am thinking now about faculty meetings, fraught family get-togethers, and love triangles—a majority of our routine social interactions probably don't require such complex embedments. For instance, I see my neighbor coming out of his house and strolling toward his car, and I assume that he *wants* to go somewhere; or I see my son pulling out a box of pencils, and I assume that he *intends* to draw. (I may not be consciously aware of my assumptions, yet they may influence my subsequent course of action.)

So, on the one hand, yes, "human collaborative activity and cooperative communication both rest on . . . recursive intention-reading."[7] But, on the other hand, thinking about thinking about thinking (third-level embedment) "occurs in interpersonal cognition in real life less frequently" than, for instance, thinking about thinking (second-level embedment). The former, as the psychologist Patricia Miller et al. put it, "has a lower ecological plausibility."[8]

Hence an important difference between our daily mindreading and our experience of reading literature. Literature creates intersubjective situations of a kind that can be described as depending on "complex embedments of mental states" at a much greater frequency than it happens in our daily life. Specifically—and this is what I call the secret life of literature—to make sense of what's going on in novels, plays, and narrative poems, as well as in memoirs focused on imagination and consciousness, we constantly embed mental states on at least the third level. The key word here is "constantly," for neither literary critics nor lay readers appreciate the true scale of this phenomenon. To put it starkly, literature, *as we know it today* (this is an important point that I will keep emphasizing) cannot function on lower than the third level of embedment. As such, it differs, for example, from expository nonfiction, such as newspaper articles and textbooks,[9] which may contain occasional forays into the third level but can also subsist, quite happily, on just the first and second levels.

Literature, of course, is a capacious concept, and it encompasses many more genres than I just listed. To give just a few examples, it includes personal essays by writers ranging from Sei Shonagon and Michel de Montaigne to Wole Soyinka and Joan Didion; mirrors for princes, from Augustine's *The City of God* to Machiavelli's *The Prince*; and political speeches, from Lincoln's "The Gettysburg Address" to King's "I Have a Dream." While these texts range widely in their frequency of complex embedment (and there are, among them, some pretty spectacular embedders), they mostly do not depend on it to the same high degree as do novels, plays, narrative poems, and memoirs concerned with consciousness.

In the latter, embedded mental states can be found on the level of individual sentences, paragraphs/stanzas, and whole chapters/acts.[10] They can belong to characters, narrators, (implied) authors, and readers, in a vast variety of combinations.[11] In "Tom *wants* his friends to *think* that he *enjoys* his chore," the third-level embedment involves the novel's characters. But at the same time, yet another complex embedment arises from an intricate give-and-take between the narrator and his audience.[12] The narrator *expects* that his readers will *appreciate* his mischievous *intention*, as he likens Tom, in the same breath, to King Herod and to a retired artist. Again, this is my formulation, but if you try to explain how this passage achieves its ironic effect, you are likely to find yourself speculating about how the author might have been intuitively anticipating his readers' thinking.[13]

It would be wrong to assume, however, that we factor mental states of the implied author and reader into any complex embedment. Of course, we can say, "the implied author *wants* us to *know* that Tom *wants* his friends to *think* that he *enjoys* his chore," and call it a case of fifth-level embedment instead of third, but those extra levels are redundant because they don't contribute anything to our understanding of the passage. In contrast, the references to King Herod and to a retired artist are the kind of "communicative event"[14] that necessitates a recognition of a particular intentionality behind it.[15]

When we read, we do not spell it out to ourselves the way I just did. Indeed, in spite of the language that I may use to describe it—such as "we are aware" or "the author wants us to know"—most of it doesn't rise to the level of conscious awareness. Nevertheless, on some level we must be keeping track of those complex intentionalities (which is a term I will use interchangeably with "mental states" to avoid sounding repetitive), because, otherwise, how would we explain to ourselves, say, Twain's evocation of the Massacre of Innocents in a scene that had nothing to do with infanticide? To recognize an allusion or to appreciate a metaphor is to acknowledge an intention.[16]

Throughout this book, I use the term "implied mental states" to refer to thoughts and feelings of characters, narrators, authors, and readers that are thus *not* spelled out but are nevertheless integral to our making sense of what we read. But, of course, a work of fiction may also contain complex embedments of mental states that are explicitly spelled out by the author. For instance, think of the time when Tom first encounters Becky Thatcher and starts showing off by engaging in various "dangerous gymnastic performances." Becky observes him for a while, then throws him a flower and disappears inside her house. Tom keeps up his antics for some time, because he *hopes* that she is still *aware* of his *interest* in her. Or, as Twain puts it, explicitly describing Tom's embedded thoughts, "Tom comforted himself a little with the *hope* that she had been near some window, meantime, and been *aware* of his *attentions*."[17]

Here is another explicitly spelled-out complex embedment. When Aunt Polly punishes Tom for breaking a sugar bowl and then finds out that it was Sid who broke it, she can't bring herself to confess that she has been in the wrong—for "discipline forbade that"—and goes "about her affairs with a troubled heart," while Tom, perfectly aware of her remorse, is quietly exalting in it. He *knows* that his aunt *is yearning* for his *forgiveness* (third-level embedment), and he *enjoys* knowing that she is yearning for his forgiveness (fourth-level). Or, as Twain puts it, "He knew that in her heart his aunt was on her knees to him, and he was morosely gratified by the consciousness of it."[18]

What is crucial about these third- and fourth-level embedments is that they do not just occasionally happen along. Instead, any given paragraph contains multiple complex embedments, sometimes implied, sometimes explicitly spelled out, sometimes a combination of the two. As I am writing this and leafing through *Tom Sawyer*, I reach almost at random for a complex embedment here and a complex embedment there; but in pretty much every case, I can turn to a group of sentences preceding or following any passage that I just quoted for you, and it will contain another implied or explicitly spelled-out complex embedment.

I started this chapter with a picture of a postal stamp, so before moving on, let us briefly circle back to the visual. Do visual representations, such as paintings and movies, also depend on complex embedment of mental states? A short answer to this question is that they do—feature movies more consistently than paintings—and in ways specific to their contexts.

For instance, Bradbury Thompson's portrayal of Tom and Ben is brimming with intentionalities: that of the artist (who apparently decided to portray the boys younger than they are in the story); that of the particular beholder (for I am aware, as I am looking at this stamp, of trying to square the artist's vision with my own perceptions of Tom Sawyer, formed years ago, in a different language, and then layered with the later, "American" impressions); and, of course, that of the characters themselves (i.e., Tom *wants* Ben to *think* that he is too *absorbed* in his task to even notice him). As we take in this stamp, we may toggle between different constellations of complex embedments. This "secret life" of visual images deserves its own study, but for now we return to literature.

1.2   A Dime a Dozen

Sometimes, shortly after I'd given a talk about embedded mental states, I would receive emails from members of my audience, something to the effect of "Have you noticed this embedment in such and such work of fiction?" On the one hand, such letters make me happy: they show that the senders are now aware of this phenomenon and want to share their new awareness. On the other hand, a part of me is wondering if it means that I failed to get across one of my key points, which is that third-level embedments in literature are nothing to write home about: they are a dime a dozen. True, their frequency increases dramatically with the advent of certain genres, such as ninth-century Chinese tales of romance, eleventh-century Japanese novels, sixteenth-century Spanish novels, and eighteenth-century English novels. But even the earliest works of literature available to us, such as *The*

Epic of Gilgamesh (ca. 2100 BC), already feature some complex embedments. And, generally (although with some fascinating exceptions, which I address later), when it comes to a work of fiction written within the past three hundred years, to discover a third-level embedment in it is roughly as exciting as to discover a noun.

Along the same lines, I suggest to students interested in "cognitive" approaches to literature that merely locating a series of complex embedments in this or that text does not constitute literary analysis. The question is not whether such embedments are there—for they are pretty much guaranteed to be there—but what effect they have on our interaction with the text. For instance, if complex embedments involve mental states of characters *and* are explicitly spelled out, then the text in question is more likely to be considered "popular fiction." In contrast, "literary fiction" of the kind that may end up on a college syllabus tends to include embedded mental states of narrators and implied authors and readers (in addition to mental states of characters) *and* to imply mental states (in addition to or instead of explicitly spelling them out). So thinking about different types of complex embedment allows us to understand something new about the distinction between "low-brow" and "high-brow" literature.

It also alerts us to cultural contexts that sustain those distinctions. For instance, when students encounter a work of fiction in a college literature course, they tend to work harder on reading implied complex embedments into it and expect to be rewarded for doing so (more about that in chapter 4). In contrast, when readers are faced with a text that they judged a priori as "having lower literary merit"—as, for instance, may be the case with readers prejudiced against science fiction—they may "exert less inference effort"[19] in situations that require supplying mentalistic explanations of characters' behavior. This is not to say that our intuitions about embedded mental states are solely determined by the context in which we read a given text but that we are sensitive to *both* such contexts and the cues supplied by the text.

Here is another way in which paying attention to complex embedments opens up new venues in literary analysis. It turns out that some fictional characters are consistently portrayed as more capable of embedding mental states on a high (i.e., third and fourth) level than are others. What factors inform the intuitive decision, on the part of the author, to make one character more "sociocognitively complex" than another? More often than not, the decision seems to be influenced by the character's social status, which is figured out along the lines of class, gender, or race. A "cognitive" approach thus builds on and complements the rich literary-critical tradition of exploring the role of class, gender, and race considerations in the construction of fictional subjectivity.

Then there is also the issue of the history of complex embedment in literature. What combinations of cognitive/cultural/historical contingencies make it more or less likely that a particular literary tradition would be characterized by a commitment to complex embedments? In some ways, this is the most difficult question one can ask, and we may never come up with a definitive answer. Yet exploring this issue is important, if only because it forces us to become aware of a broader range of historical factors than we usually settle for, in our critical studies.

So the question is not "Are there any complex embedments in this text?"— because, almost always, there are—but "What work do they do?" and "How has it come to be that way?" These are the questions that I encourage my students to ask and that I, myself, ask in the chapters that follow. But, first, I want to give you a range of examples of complex embedments, to show what forms they take in different texts. I hope that, after reading this chapter, you, too, will be struck by this phenomenon: apparently so essential to our interaction with literature yet so invisible, flying, mostly, under the critical radar.

1.3 Explicitly Spelled-Out Mental States

The majority of complex embedments in literature are implied rather than explicitly spelled out. In fact, some texts contain next to zero explicitly articulated embedments. Still, many do. In this section, I present examples of such *explicit* embedments, starting with works of fiction published recently and then moving back in time and ending with *The Epic of Gilgamesh*. I do not analyze any of the passages—that will come in later sections. I just pile them up to give you some idea of the range of literary texts that depend on explicit complex embedments. (In what follows, emphasis is mine throughout, unless stated otherwise.)

In Elena Ferrante's *The Story of the Lost Child*, the protagonist says to a friend, "I'm laughing out of despair, because I've never been so offended, because I *feel humiliated* in a way that I *don't know* if you can *imagine*."[20] In Sally Rooney's *Conversations with Friends*, Frances reports her thoughts as she observes a man raising both eyebrows at another man in response to something another woman has said: "I *thought* it was cowardly of Philip to look at Andrew, whom I *knew* he didn't even *like*, and it made me uncomfortable."[21] In Rachel Cusk's *Transit*, a woman gets up to leave after a lunch with her friend, while "darting frequent glances" at her: "It was as if she was *trying to intercept my vision* of her before I could *read* anything into what I saw."[22] In Tsitsi Dangarembga's *Nervous Conditions*, the narrator doesn't want to think through the implications of the "treacherous mazes" of her thoughts about the inescapability of female victimization: "I didn't want to reach the end of those mazes, because there, I *knew*, I would find myself and I was *afraid* I would not *recognize* myself after taking so many confusing directions."[23] In Jokha Al Harthi's *Celestial Bodies*, a young woman named Khawla cannot understand why her sister, Asma, does not realize that the religious texts that she is so fond of bore other people to death: "Khawla was *astonished* at how *oblivious* Asma seemed to the *awful boredom* these ancient books induced."[24] In Ben Lerner's *Leaving the Atocha Station*, the protagonist thinks that his communication with his Spanish friend, Theresa, is becoming a travesty: "I saw her reflected in my eyes, saw that she knew, or was coming to know, that what interest I held for her, all of it, was virtual, that my appeal for her had little to do with my actual writing or speech, and while she was happy to let me believe she believed in my profundity, on some level she was aware that she was merely encountering herself."[25]

I chose my next example on a lark. The sentence that contains the spelled-out embedment *is* the story in its entirety, Joy Williams's "The Museum." Here it is: "We were not *interested* the way we *thought* we would be *interested*."[26]

With my next example, I want to show you that science fiction writers do not shun complex embedment. This, in response to the assumption that I encounter often enough (and that has served as the impetus for the study quoted earlier),[27] which is that works of science fiction get by without representing complex mental states.[28] In Philip K. Dick's *Do Androids Dream of Electric Sheep?*, characters can preprogram their feelings on a special "mood organ." This allows the author both to depict complex emotions arising during a marital quarrel and to comment on the presumably mechanical nature of their emotional life.

So here is Rick Deckard trying to decide how he wants to make himself feel during an unpleasant conversation with his wife: "At his console he hesitated between dialing for a thalamic suppression (which would abolish his mood of rage) or a thalamic stimulant (which would make him irked enough to win the argument)." Rick *wonders* if he *wants* to quell or to ratchet up his *feeling of rage*. Moreover, his wife is watching him closely, ready to "dial the maximum" on her mood organ if he dials "for greater venom" on his, that is, *intending* to become even *angrier* in response to his *anger*.[29]

To run a bit ahead of myself, this scene also contains some implied embedments, though, perhaps, to appreciate them one has to be rereading the novel. Repeat readers may enjoy the irony of the situation in which the character whose job it is to hunt down and kill androids—those, presumably, not capable of feeling genuine emotions—himself uses a mood organ. As Ralph James Savarese puts it, "here, the technological apparatus is active and animate; the human hero, passive and inanimate. . . . Even his feelings aren't strictly organic."[30] Were we to spell out the underlying embedment (which, again, is *not* something we consciously do when we read), we may say that the author wants his readers to be aware of the muddled thinking behind the discrimination between those who "truly" experience emotions and those who "choose" to experience them.

Back to explicit embedments. In Shirley Jackson's short story "The Beautiful Stranger," an unhappily married suburban wife has a sudden revelation that her emotionally abusive husband is gone and in his place there is a "beautiful stranger." This new man, moreover, knows that she is afraid that her husband may return and is thus not surprised when she looks up at him for reassurance that he is not her husband: "She was *aware* from his smile that he had *perceived* her *doubts*, and yet he was so clearly a stranger that, seeing him, she had no need of speaking."[31]

In E. M. Forster's *Howards End*, Margaret Schlegel's fiancé, Henry Wilcox, is revealed to have had an affair, ten years before, with a woman who is now Leonard Bast's common-law wife, Jacky. Margaret's sister, Helen, fresh from the ruckus at the Wilcox's garden party, at which Henry and Jacky have accidentally come face-to-face, flies back to London and forces her brother, Tibby, to consider a baffling dilemma that he'd rather not consider: "Ought Margaret to *know* what Helen *knew* the Basts to *know*?"[32]

These were all examples from relatively recent literary past. Let us now start moving further back in time. In Lev Tolstoy's *Anna Karenina* (1877), Alexei Karenin is made to listen to Anna's delirious speech while she, as everybody believes, is dying: "Alexei Alexandrovich's inner disturbance kept growing, and now reached such a degree that he ceased to struggle with it; he suddenly *felt* that what he *had considered* an inner *disturbance* was, on the contrary, a blissful state of soul, which suddenly gave him a new, previously unknown happiness."[33]

In Jane Austen's *Pride and Prejudice* (1813), Miss Bingley talks to Elizabeth about Mr. Wickham's regiment in Mr. Darcy's hearing, because she *hopes that* Elizabeth will be *embarrassed imagining* Mr. Darcy *thinking* about the Bennett girls' involvement with that regiment: "[She] had . . . *intended* to discompose Elizabeth, by bringing forward the idea of a man to whom she *believed* her *partial*, to make her betray a *sensibility* which might injure her in Darcy's *opinion*, and perhaps to remind the latter of all the follies and absurdities by which some part of her family were connected with that corps."[34]

In Cao Xueqin's novel *Dream of the Red Chamber*, also known as *The Story of the Stone* (ca. 1750–1760s), Dai-yu explains to Bao-yu why she is angry at him for having earlier tried to prevent their cousin Xiang-yun from making fun of her: "But what about that look you gave Yun? Just what did you mean by that? I *think* I *know* what you *meant*. You meant to warn her that she would cheapen herself by joking with me as an equal."[35]

In Shakespeare's sonnet 42, the speaker is constructing a complicated argument in order to console himself for the heartbreaking discovery that his mistress and his friend are having an affair. The "loving offenders," he proposes, are actually doing it for his sake: they want to prove their devotion to him. The young man wants to love what the speaker loves: "Thou *dost love* her, because thou *know'st* I *love* her." Just so, his mistress allows herself to be loved by the young man because she, too, *wants* to be *loved* by a man whom the speaker *loves*: "And for my sake even so doth she abuse me, / Suffering my friend for my sake to approve her."

The ending of the poem ("But here's the joy; my friend and I are one; / Sweet flattery! then she loves but me alone") is open to two different interpretations. Either the speaker is *happy* that his mistress *wants* to find new ways of expressing her *love* for him, or the speaker *is aware* that he is trying to make himself *feel better* by *thinking* that his mistress *wants* to find new ways of expressing her *love* for him. It is a choice between self-flattery and self-awareness, and it is a complex embedment of mental states either way.[36]

In Nizami Ganjavi's twelfth-century narrative poem *The Story of Layla and Majnun*, Kais doesn't feel jealous of other boys in school who stare "at Layla open-mouthed" or, if the school is closed, "roam the alleyways and the passages between the market-stalls, all in the hope of catching a tiny glimpse of her dimpled face," because he *knows* that they *don't love* her as much as he *loves* her: "Naturally, Kais knew that the other boys desired [Layla], but he also *knew* that they *could not desire* her as much as he *did*, and so their antics did not perturb him in the least."[37]

Let us go yet further back, to the ninth century's *Book of Exeter*. In the Old English poem "The Wanderer," the speaker *wonders* why he is not more *depressed* when he *thinks* about death:

> Indeed I *cannot think*
> why my spirit
> *does not darken*
> when I *ponder* on the whole
> life of men throughout the world,
> How they suddenly
> left the floor (hall),
> the proud thanes.[38]

In Petronius's *Satyricon* (first century AD), Lichas wants to sleep with Encolpius to make up for Encolpius's currently sleeping with Lichas's long-term mistress, Tryphaena. Encolpius is not interested, and Lichas arranges for Tryphaena to fall for Encolpius's slave and lover, Giton. Lichas hopes that the jealous Encolpius will want to make Tryphaena angry by taking up with him and thus takes "the trouble to draw [his] attention" to Tryphaena's relationship with Giton. The plan works well. As Encolpius reports, "Therefore I was the more ready to treat him nicely, and he was delighted beyond measure—being of course quite sure that my lady's ill-treatment of me would kindle my disgust, and that in my anger I should feel more kindly disposed to him."[39]

In *The Odyssey* (eighth century BC), one of Penelope's suitors, Eurymachus, wants to assure her that her son, Telemachus, mustn't be afraid of him: "To this Eurymachus son of Polybus answered: . . . 'Telemachus is much the dearest friend I have, and has nothing to fear from the hands of us suitors. Of course, if death comes to him from the gods, he cannot escape it.' He said this to quiet her, but in reality he was plotting against Telemachus."[40] Eurymachus *wants* Penelope to *stop worrying* about the suitors' *intentions* vis-à-vis Telemachus. (Homer, of course, hastens to explain to us that Eurymachus is lying, but our conversation about lying and literary history will have to wait until a later chapter.)

In *The Epic of Gilgamesh* (ca. 2100 BC), the king of the city of Shurrupak, named Utnapishtim, is told by God Ea to tear down his house and build a boat that would allow him and his family to survive a great flood that is about to kill everybody else. Utnapishtim then asks Ea, very reasonably, how he should explain his actions to other people in Shurrupak: "Then Ea opened his mouth and said to me, his servant, 'Tell them this: I have learnt that Enlil is wrathful against me, I dare no longer walk in his land nor live in his city; I will go down to the Gulf to dwell with Ea my lord. But on you he will rain down abundance, rare fish and shy wild-fowl, a rich harvest-tide. In the evening the rider of the storm will bring you wheat in torrents.'"[41]

Ea *wants* the people to *believe* that another god, Enlil, is *angry* at Utnapishtim and that by going down to the Gulf, Utnapishtim hopes to escape Enlil's wrath. There is plenty of cruel irony in the picture of abundance about to rain on the city that Ea expects Utnapishtim to plant in the heads of his doomed compatriots. By the time Utnapishtim is telling this story to Gilgamesh, the giant flood has already taken place, so he must be aware of this irony. That is, he *knows* that Ea *wanted* to make sure that the citizens of Shurrupak wouldn't *get alarmed* at the sight of his boat and try to do something to escape the coming disaster, just as he knows what Ea's fanciful talk of "rich harvest-tide" and "wheat in torrents" truly portended. But to talk about irony and implicit realizations, we must go to the next section.

1.4   Implied Mental States: Dramatic Irony and Beyond

In ancient Mesopotamia, Ea was associated with wisdom, magic, and mischief—a trickster figure. Indeed, trickster tales, with their plots of deception, may have been the earliest fictional contexts for implied complex embedments. And so must have been drama, for what is "dramatic irony" but a cultural shortcut for implicitly acknowledging a particular mindreading dynamic? The audience knows that a character doesn't know. And what is it that the poor character is so fatally unaware of? More often than not, it has something to do with the intentions of another character, of a deity, or, even, with the character's own intentions, which have been rendered calamitously obscure to them. This state of affairs doesn't have to be explicitly spelled out, yet the audiences must be aware of it (an awareness that necessitates embedding complex mental states!) in order to make sense of what is going on.

Because dramatic irony is thus a prototypical implied third-level embedment, I start this section with several examples from plays and then segue to narrative poems, short stories, and novels. The trickster tales will have to wait until a later chapter, dealing with the history of complex embedments.

Note, too, that, when talking about drama, I focus on playscripts and not performance. The latter, of course, brings in more and different complex embedments than are present in the script. For instance, the social psychologist Tiziano Furlanetto and his colleagues have found that when an actor's "gaze and action [do] not signal the same intention," observers engage in a stronger "spontaneous perspective-taking," which suggests that, "in presence of ambiguous behavioral intention, people are more likely take the other's perspective to try to understand the action." Thus, if we imagine a character onstage who, in the middle of a complex social interaction with someone else, would start reaching for an object without looking at it, that action alone would complicate our perception of their intentions vis-à-vis others.[42] As Furlanetto et al. put it, "observing a person grasping without looking may thus be perceived as ambiguous. What is he planning to do? Why is he not looking at the object he is reaching for?"[43]

This is just one small example of the role of embodiment in modulating and complicating an audience's perception of actors' embedded intentionality.[44] In general, exploration of embedded mental states that emerge when actors widen and explore the space between their characters' words and their body language deserves a separate study. It is not my aim here to undertake such a study, so we return to mental states implied by texts alone.

In Shakespeare's *Othello*, Iago *wants* Othello *to think* that Desdemona *is in love* with Cassio. In *Romeo and Juliette*, Romeo *does not know* that Juliette is not dead but merely *wants* some people to *think* that she is dead. In *Measure for Measure*, Duke Vincentio *wants* Isabella to *think* that he *doesn't believe* her story about Angelo's "intemperate lust."[45] In *Twelfth Night*, Maria, Sir Toby Belch, and Fabian *want* Malvolio to *think* that Olivia *loves* him.

These are all act- and scene-level implied embedments. For a quick example of a sentence-level implied embedment in drama, consider a scene from *Twelfth Night*, in which Malvolio first courts Olivia and then exits the stage in full anticipation of the "greatness" that will soon be "thrust upon" him. Once he is gone, Shakespeare has another character, Fabian, make a "nod to the audience,"[46] which starts off a complex embedment involving the audience, the author, and the characters. When Fabian says, "If this were played upon a stage now, I could condemn it as an improbable fiction,"[47] the author slyly tells his spectators that he *knows* that they *know* that the characters *don't know* that they are upon a stage now.

More scene-level complex embedments: In George Etherege's *The Man of Mode* (1676), Dorimant *knows* that his new mistress, though believing herself injured by him, will nevertheless help him to deceive his old mistress (who also happens to be her friend) because she is *afraid* that the old mistress will *realize* that the new mistress has been lying to her.[48] In Oliver Goldsmith's *She Stoops to Conquer* (1773), Hastings *doesn't want* his friend Marlow to *know* that he is *mortified* that Marlow gave the jewels that Hastings had earlier entrusted him with, to Mrs. Hardcastle for safekeeping.

In Wang Shifu's play *The Story of the Western Wing* (thirteenth century), Oriole's maid, Crimson, encourages student Zhang to pursue her young mistress, because she *thinks* that she *knows* Oriole's true *feelings* about the attractive young man. As Zhang becomes too importunate, however, and Oriole responds with indignation, Crimson *realizes* that she must have been *wrong* in *assuming* that Oriole *cares less* about her honor than her love for Zhang.

In *Layla and Majnun*, when the main protagonists, still children, are basking "in the glow of each other's love," the narrator asks us—that is, *wants* us—to *imagine* what other people around them *may be thinking* about their *feelings*: "Did others realize what had happened between Kais and his Layla? Did they see the stolen looks, the furtive glances that passed between them? Could they read the signs and crack the codes of secret love that bound their hearts together? Who knew about them and how much was known? Until one day, in the market, a voice was heard to say, 'Kais and Layla are in love. Have you not heard?'"[49]

In Austen's *Emma* (1815), Frank Churchill *wants* onlookers to *think* that he is *interested* in Emma in order to conceal his engagement with Jane Fairfax. In Forster's *Howards End*, all throughout the novel, that is, "throughout Margaret's various conversation with the Wilcoxes, her marriage to Henry Wilcox, and her sister's involvement with the Basts," readers *know* that Margaret *doesn't know* (while the Wilcoxes do know) that the late Mrs. Wilcox had *wanted* her to inherit Howards End.[50]

In the opening of Anton Chekhov's short story "Rothschild's Fiddle" (1894), we learn that the "town was small, worse than a village, and populated almost only by old people, who died so rarely that it was quite annoying."[51] One can't help wondering what kind of person would consider it so patently obvious that old people should hurry up and die. To map out our implicit reaction explicitly, the narrator *wants* the reader to *wonder* who would *want* old people to die and why.

In Lu Xun's "A Madman's Diary" (1918), the protagonist finds it infinitely amusing that a doctor, whom his older brother brought in to consult, says that he'll be "better" if he rests "quietly for a few days." Because he thinks that the doctor is "the executioner in disguise" and that what he and the brother really want is to eat him, resting quietly for a few days will only fatten him up and thus give them "more to eat." So he laughs uproariously and watches them turn pale, "awed" by his "courage and integrity": "I could not help roaring with laughter, I was so amused. I knew that in this laughter were courage and integrity. Both the old man and my brother turned pale, awed by my courage and integrity."[52] The reader *knows*, however, that the protagonist *doesn't realize* that the reason that the two men turn pale is that they *think* that his laughter is a sure sign of his insanity. Or, to put it differently, the implied author *wants* the readers to *realize* that the mad protagonist *misinterprets* the body language of his visitors.

When Maggie, the protagonist of Hannah Pittard's novel *Listen to Me* (2016), finds an empty bottle of champagne in the recycling bin, her heart sinks. Her husband, Mark, has apparently tossed "without ceremony" the bottle left over from their last anniversary, which she has been saving. She wonders, next, if this is a test and if Mark is "measuring her steadiness"— for she has been going through a rough time lately—"by relieving her of an ultimately trivial trinket." If so, she decides, she "would pass his test with flying colors." That is, she *wants* him to *think* (were he to see the bottle, now placed "at the very top" of the bin) that she *is not overly sentimental* about it. Except that (dramatic irony!) Maggie has just torn off and saved as a keepsake "a sliver of the pink foil," which means that she is actually failing the test that she imagines Mark has set up for her.[53] Or, to spell out this embedment explicitly, the implied author *wants* the reader to *realize* that Maggie *is fooling herself.* (And so is, for that matter, Mark, but the implied embedments involved in his self-deception are constructed on other occasions.)

The first sentence of Zadie Smith's *On Beauty* (2005), "One may as well begin with Jerome's emails to his father," overflows with embedded intentionality.[54] The implied author *wants* her readers *to know* that the action will be filtered through the *consciousness* of a reflective narrator. And there is more, of course. Those who are familiar with the opening of *Howards End,* "One may as well begin with Helen's letters to her sister," will sense yet another set of intentions in Smith's first sentence.[55] The author wants her readers to know that the action will be filtered through the consciousness of a reflective narrator—*and* that she means her novel to be a meditation on Forster's novel. There are no direct references to mental states in the sentence about Jerome's emails to his father, yet its impact on the reader is directly bound to its embedded intentionality.[56]

I don't think we notice it, though. Were I to articulate my feelings upon first opening Smith's novel, I would say that I experienced a pleasing jolt of recognition and something that could be expressed in words as, "Oh, so it's that kind of book!" It is when I try to slow down and figure out what kind of mental work goes into "Oh, so it's that kind of book!" that I end up considering the embedded intentions of the author.

Let us now revisit Williams's one-sentence story "The Museum": "We were not interested the way we thought we would be interested."[57] I used it in the previous section as an example of explicitly spelled-out embedments in literature, but its affective punch may actually reside with its implied embedments. "The Museum's" protagonists watch closely their emotional responses, especially when they find themselves in a cultural context that is expected to elicit a particular kind of response. The story thus draws the reader's attention to a specific sensibility: one predicated on self-awareness yet not always happy about the burden of such awareness. This may be the reason why at least one reviewer characterized "The Museum" as "rueful."[58]

As with explicit embedments, discussed earlier, I want to see here how far back into literary history I can reach to find examples of implied complex embedment, especially those involving implied authors and readers. Let's start with a novel written in the second century AD, Apuleius's *The Golden Ass*. When one of its characters, goddess Venus, learns that her son, Cupid, has ignored her order to humiliate and destroy Psyche (of whose beauty Venus was jealous) and instead married Psyche and that they are now expecting a baby, she rushes into the bedroom where Cupid lies and begins "roaring with all the strength in her":

> Pretty classy goings-on, huh? A nice way to make your family look good! . . . I was in a fight to the finish with a girl, and now I have to put up with her as my daughter-in-law? And what's more, you worthless, disgusting hound, you assume that you're the only one fit to breed, as if I'm too old to have a baby. This is just to let you know: I am going to have another son, much better than you, and to humiliate you even more I'm going to adopt one of the slaves born in my house, sign everything over to him: those wings and that torch, and that bow, and your actual arrows—all the tools of my trade, which I didn't give you to use like this.

It's totally up to me, because there was no money set aside from your father's estate to buy you this equipment.[59]

Venus wants Cupid to know that she is extremely angry. What Venus doesn't know, however, is that, just now, Cupid has abandoned Psyche for not trusting him and following the advice of her envious sisters and that Psyche is desperate to win back Cupid's love. (Were Venus to know all this, she might try attacking Psyche while the girl is lonely and vulnerable, instead of simply venting her anger at her son.)

Those are straightforward enough embedments, but they are not what makes the passage hilarious. What makes it hilarious is the interplay of mental states of the implied author and the reader. As the novel's recent translator Sarah Ruden puts it, Apuleius "exquisitely [manages] the tension between the high and low, the inside and outside points of view."[60] The goddess of love, beauty, fertility, and prosperity comes across as garrulous, jealous, feeling her age, and penny-pinching. Apuleius knows that we don't expect Venus to sound like this, and we *know* that he *knows* that we *didn't expect* this.

The comic effect of Venus's speech—if, that is, we find it funny!—stems from this embedded awareness. This point is worth emphasizing because to phenomenologically "get" the joke, readers must swiftly process embedded mental states. My map here thus seeks to capture something that readers actually interpretively do, "rather than being an analytical account of the semantics of the text, divorced from the reader."[61]

As always with complex passages, there are often several ways to map them out. I just suggested one—"we know that Apuleius knows that we didn't expect Venus to sound like this"—but a different mapping is also possible. Readers may or may not remember that, within the novel, the story of Cupid and Psyche is narrated by an old crone who keeps house for pirates and who wants to soothe and entertain a young woman kidnapped by those pirates. So if we do remember it, we can say that "Apuleius uses the old crone as his framing device because he *wants* a narrator *incapable of imagining a* Venus who would *feel* differently from herself under these circumstances."

Let us revisit *Gilgamesh*, which is considered to be one the earliest surviving works of literature. When we learn that Ea wants Utnapishtim to tell the people of Shurrupak that the same god who is angry at Utnapishtim will bless them with good fortune ("But on you he will rain down abundance, rare fish and shy wild-fowl, a rich harvest-tide. In the evening the rider of the storm will bring you wheat in torrents"), it is difficult not to think that Ea is having a joke at the expense of the Shurrupakians.[62] Not only does he *want* those people to *think* that they are *beloved* by a god (when exactly the opposite is the case), but he also chooses a very particular vocabulary to convey his lie. How we read the complex give-and-take of intentions implied by his evocation of rain, tide, and torrents depends on whether we perceive Ea as being sadistic, philosophical, or just true to his trickster "nature." But whichever way we view him, we seem to assume that the narrator of Gilgamesh *wants* to draw the audience's *attention* to Ea's *desire* to comment on what is to come. This is to say that to judge the ethics of the situation, we have to be intuitively aware of the underlying mental states and hence the irony implied by the disjunction between what is stated and what is intended.

1.5 Who Are "We"? Historical Speculations and Empirical Studies

Do we have any evidence that *Gilgamesh*'s early audiences were also aware of this ironic disjunction? While we may never be able to know for sure, thinking about this question raises important issues. One such issue is the identity of the "we" whose reaction to *Gilgamesh* I seem to be quietly presenting as normative and then comparing with that of its early listeners/ readers. To put it broadly, are some readers more aware of implied mental states in literature than are others? Does it take a particular training in reading and interpreting works of literature to become a part of this enlightened "we" community?

I can tell you right away that I do not have definitive answers to these questions. What I do have is a series of considerations that bear upon them, directly or indirectly: some based on historical analysis of patterns of reading, others on ethnographical studies of indigenous performance genres, yet others on lab experiments conducted by interdisciplinary teams of literary critics and cognitive scientists. I will share these considerations with you, and then we can see if they add up to any kind of provisional answer.

Let us start with studies that suggest that expert readers of literature become sensitized to certain features of literary texts, including various types of implied intentionality. The social psychologists David Comer Kidd and Emanuele Castano, who study effects of reading fiction on theory of

mind, have shown that long-term exposure to literary fiction makes readers less willing to settle for unambiguous interpretation of mental states
and more eager to look for cues of intentionality.[63] This may mean that the
more literature one reads, the more implied mental states one is prepared
to see in what one is reading.

Moreover, the literary critic and neuroscientist Natalie Phillips has found
that professors of literature did not make good subjects in fMRI experiments
that focused on reading for pleasure, because they found it hard to stop
close reading, that is, analyzing what they read. (At least that was what happened when the text in question was deemed worthy of close reading; one
wonders if those professors would have a similarly hard time refraining from
analyzing a work of science fiction.) In contrast, graduate students qualified
for participation in such experiments because they still seemed to be able
to read classic literature for pleasure, even though they were en route to
becoming professional close readers. (Note that Phillips does not explicitly
equate close reading with uncovering new embedded mental states,[64] while I
believe that close reading typically involves such uncovering.[65]) So one take-
home message from her experiments is that if you spend a good portion of
your life not just reading literature but also thinking about it, your experience of reading (which, as I have argued elsewhere, necessarily involves
mindreading)[66] becomes different from that of people who either don't read
literature or read it but don't think about it professionally.

What we should not do, however, based on these studies, is to overstate
that difference and treat it as a *constant*. Consider the work of the cognitive
literary critic Andrew Elfenbein, who brings together empirical studies of
what readers do today and historical reconstructions of the eighteenth- and
nineteenth-century practices of reading. Elfenbein emphasizes continuity
between different kinds of reading, reminding us that "many readers . . . are
neither novices nor experts but somewhere in between" and that even expert
readers are routinely faced with pressures and distractions that lead them
to engage in merely "good enough processing, which occurs when [they]
process what they have read just enough to make sense of it."[67] What this
means is that we should not assume that experienced readers of literature
(even, perhaps, the professors from Phillips's experiment) would *always* see
more implied embedments in the text than would less experienced readers.

This may be a good time to bring up the difference between reading
and interpretation. Although I would dearly love to believe that my maps

of mental states reflect something obvious and hence do not require any special interpretive effort, the truth is that some of them do, and I am not always the best judge of the extent to which such maps depend on my having taken extra time to think about the passage under consideration. To put it bluntly, if I am giving you the results of my interpretive effort while claiming that this is really an effortless reading of the kind that just about any expert reader would produce under the circumstances, then I am inflating the difference between expert reading and novice reading. To quote Elfenbein again, "full comprehension and reading do not co-occur, which is why literary scholars should hesitate more than they do to make 'reading' synonymous with 'interpretation.'"[68]

That said, there is some room between "full comprehension" and "good enough" comprehension. While neither expert nor novice readers may immediately and fully comprehend a rich variety of complex embedments structuring a given passage, they may grasp enough of some of those embedments' meaning to carry them through. In fact, Elfenbein's discussion of automatic processes involved in reading literature provides a useful framework for thinking about fictional mindreading. If we adopt his model (which itself is based on the work of the psychologist Agnes Moors), we can characterize embedment of complex mental states as "top-down automatic processing: processes that have become automatic as a result of training and repetitive practice." As Elfenbein explains, "[Such] processes are usually unconscious, but they are not inaccessible to consciousness. They can become conscious when attention is directed to them. . . . Some of these processes include comprehending (understanding the meaning of what is read) and situation model building (integrating what has been read with general world knowledge, cognitive and emotional inferences, predictions, and evaluations)."[69]

More often than not, such "understanding the meaning of what is read" and "situation model building" involve attributing mental states to fictional characters, narrators, and the author. And, just as in real life, much of this fictional mindreading does not rise to the level of awareness, except when we consciously direct our attention to it.

Expert readers of literature may, indeed, develop, "through long practice, a set of strategies for understanding imaginative literature,"[70] which means that sometimes they would indeed be more attuned to intentionality cues in the text than would be less experienced readers. That said, automatic ascription of mental states to observed behavior is something that, arguably, all readers do. Literary scholars may thus regularly shift "between strategies common to many readers" and those that they have developed as professionals.[71] While in the latter mode, they are more likely to be aware of a richer set of counterintuitive implied embedments—and thus begin to interpret the text—than they are when in the former mode, when, for instance, they may be distracted or in a hurry or uninterested in what they are reading or having decided that this particular text does not deserve much attention.

So where does all this leave us in respect to the initial question of whether *Gilgamesh*'s earliest audiences might have been as aware of its ironic interplay of intentionalities as "we" can be today. I suspect that, *even then,* some members of the audience—those whose experience with imaginary intentionalities was, for whatever reason, more extensive than that of other people—were particularly eager, on some occasions, to intuit more implied mental states in the text. To *them*, Ea's promise to the people of Shurrupak, that soon they will be happily drowning in the torrents of fish and fowl and wheat, might have, sometimes, felt more "dramatically ironic" than it did to others.

But, one may argue, surely, given the variety of artifacts that experiment with nuances of intentionality today (all those novels and movies!), surely, our culture, as a whole, must be more attuned to implied embedded mental states than would be a culture not exposed to such an abundance. I agree with this argument on the condition that we humbly acknowledge that we often have no clue what performative and literary genres may be thriving (or had thrived, thinking back to *Gilgamesh*) in a culture different from ours.[72] So, as long as we keep our potential ignorance in mind, I would say that, yes, a community with a long and rich tradition of representing mental states might be more open to intuiting complex mental states in a given cultural artifact, though, even within that community, some people would still be more eager to look for cues of intentionality than would be others.

So here is one way to think about a hypothetical community of readers—those "we" and "us" and "ours"—which I regularly evoke in this book when I describe embedded intentions along the lines of, "the narrator of *Gilgamesh* wants to draw *our* attention to Ea's intentions, as Ea refers to the flood that will soon destroy the unsuspecting Shurrupakians, as a downpour of abundance." My hope is that such first-person plural pronouns designate readers who are paying "good enough" attention to what they are reading, which is to say that they are neither terribly distracted nor engaged in the act of professional literary interpretation. Perhaps I don't always manage to hit that sweet spot (i.e., that "good enough" state of the reading mind) in my mapping of mental states, but this is to what I aspire.

Moreover, given the pragmatics of when and where I wrote and you are reading this book, it is reasonable to assume that these "good-enough" readers have had some exposure to cultural artifacts—such as novels, plays, television series, and movies—that call for attribution of mental states to a broad variety of actual and imagined entities. It remains open to debate whether such exposure makes them (us) better mindreaders in their daily life.[73] Still, it provides them with some training in teasing out hidden intentionalities of characters, authors, and implied audiences, a training that comes in handy in a culture that (mostly) values thinking and talking about one's own and others' mental states. (I have more to say about this in chapter 4, in which I talk about cultures that may not encourage such conversations.)

1.6   Studying "Us" in the Lab

Now I will tell you about a different kind of attempt to figure out if there is such a thing as a collective of readers when it comes to the processing of complex embedment of mental states. To see if there is any evidence that different readers are likely to agree on their estimate of the level of embedment in a given passage, a team of cognitive scientists and literary scholars, headed by Douglas H. Whalen, (late) Michael Holquist, and myself, ran a series of experiments at Yale University, Haskins Laboratories in New Haven, and the CUNY Graduate Center. The first set of experiments presented participants with short vignettes, crafted specifically for the occasion and featuring different levels of embedment; the second, with an excerpt from an actual work of literature, Harper Lee's novel *To Kill a Mockingbird*.

To give some idea of what our artificial vignettes looked like, here are two of them, one in which each sentence contains a second-level embedment and another in which each sentence contains a fourth-level embedment:

* [*second level*] I am not even sure why Stephanie wants to go the movies with Alice and me. She hates the kinds of movies that we like. I remember the last time Alice wanted us to see this retrospective of silent films. We both thought that Stephanie wouldn't enjoy it at all. She went along

and sat through the whole four-hour thing, but we could tell that she was bored. I think I need to figure out how to talk to the two of them about this problem.

- [*fourth level*] I think my daughter begins to find it a bit irksome that when we visit my aunt she has to be very careful about choosing topics of conversation that won't offend. She knows, for example, that my aunt can't stand it if we suggest that it's not a good idea for her to live alone. We also have to keep in mind not to argue with her about her conviction that she can remember her doctor appointments without ever writing anything down.[74]

Altogether, we had eighty-four vignettes (387 sentences) ranging in their level of embedment from zero to five. When it comes to results, the "great majority of responses were within one level (94.2%), but differences did account for 25.54% of the judgments."[75] This is to say that the participants' judgments were in perfect agreement with the experimenters' judgments in 74.5 percent of cases, while the greatest difference involved one level of embedment in either direction. (That is, in 10.5 percent of cases, participants judged the vignettes to have one more mental state than did the experimenters, and in 9 percent of cases, they judged the vignettes to have one fewer mental states.) In contrast, the participants judged the vignettes to have two more mental states than did the experimenters only in 2 percent of cases, and they judged them to have two fewer mental states only in 0.78 percent of cases; and the numbers went even further down with the difference of three levels, to 0.17 percent and 0.21 percent, respectively.

What we found in the second set of experiments was that an excerpt from *To Kill a Mockingbird*, which featured twelve sentences (three consecutive paragraphs), yielded a lower rate of agreement. Specifically, in half the cases, the participants' judgments were essentially the same as the experimenters', while in three cases, they were above the experimenters' judgments, and in three cases, below.[76] I will not discuss here the setup of our studies, because we already have done it, extensively, elsewhere.[77] I will focus only on three take-home lessons that are relevant for us now.

First, as we realized in the process of devising our vignettes, elements of style such as metaphors, alliterations, and allusions bring in mental states.[78] A single metaphor, even a subdued one, can inadvertently change the tone of a whole vignette, evoking a mental state in a reader, even if that mental state may be too subtle to describe in a propositional format.[79] To adapt Patrick Colm Hogan's argument from a related context, such a mental state may not be "strongly activated." It will be, "rather, 'primed' in the cognitive sense of the term. Thus [it will be] partially activated in such a way as to affect the orientation of thought and feeling without entailing precise, reasoned consequences."[80] While we did our best to control for this "priming" factor in our synthetic vignettes by draining them of anything that could be seen as a sign of style, we were forcefully reminded that it is style rather than straightforward propositional statements (such as "I know that she knows that I know") that may generate complex embedments in literary texts.

Second, during one of the sessions in which we introduced our subjects to the concept of counting embedded mental states on the level of an individual sentence (i.e., the unit level on which we eventually settled), we discovered something similar to what Natalie Phillips later observed in her fMRI studies of pleasure reading. It became clear to us that when our subjects happened to be expert readers—such as graduate students in English and comparative literature—they sometimes saw more mental states in a given sentence/paragraph than did lay readers.[81] This observation made a lot of sense if you would consider that people who apply to graduate programs in literary studies may already have higher-than-average interest in intricate social situations that call for attribution of complex mental states[82] and that they may become even more so after spending years dissecting and interpreting mental states of literary characters, authors, and other scholars.[83]

Still, even with those complicating factors (that is, mental states introduced by various elusive elements of style and the difference in our subjects' expertise), "the broad agreement about the levels of embedment in individual sentences demonstrated by our experiment [showed] that sentence-level embedment of mental states is a real phenomenon that can be reliably assessed in a laboratory setting."[84] Studying "us" in the lab is, thus, a legitimate endeavor, especially if one clearly differentiates between one's expectations, in the case of artificial vignettes, and excerpts from literature.

For here is the third take-home lesson from our experiments. It was very encouraging to learn that both researchers and subjects could be trained to judge levels of embedment quickly. It was also heartening to see that their subsequent judgments—in the case of vignettes—displayed a sizable agreement. That said, disagreements—which were especially pronounced in the case of *To Kill a Mockingbird*—turned out to be illuminating in their own right. In fact, one of the conclusions of our last study was that, particularly when it comes to individual sentences in works of literature, high agreement rates on their levels of embedment should not be expected. While disagreements may have multiple causes, including flaws in the design of experiments, one clear cause must be the "necessary complicating role of large-scale (i.e., paragraph, chapter, and cross-chapter) embedments of mental states in the perception of the sentences," while another may have to do with the role of personal responses to literature.[85]

To illustrate how such complications work, I will now turn to a novel that our last study mentioned only briefly: E. M. Forster's *Howards End*. Specifically, I will show that a seemingly clear-cut sentence carries the potential for expanding, contracting, and otherwise changing its levels of embedment. This happens because, far from being a one-shot game, the sentence is part of the dynamic mindreading ecology of the novel—what the philosophers of mind Hanne De Jaegher and Ezequiel Di Paolo would call "the ongoing engagement" between the text and its readers.[86]

1.7   Are Embedments *in* the Text?

"Ought Margaret to know what Helen knew the Basts to know?"

What could be more straightforward than this example of explicitly spelled-out embedded mental states from Forster's *Howards End*? Yet this straightforwardness is treacherous. The sentence is a Trojan horse harboring implied mental states that rush at us as soon as we move in for a closer look.

Until now, I avoided providing much context for my examples of spelled-out embedments. I did so because I wanted to first clearly lay out the terms of my discussion: "here are the explicits, and here are the implieds." But reality is messier than this neat division may imply: explicitly spelled-out embedments are often integrated with implied ones. Sometimes the relationship between the two is complementary, but, just as often, the implied embedments subvert the explicit ones.

For instance, taken on its own, "Ought Margaret to know what Helen knew the Basts to know?" seems to present a social dilemma and invite a discussion of what to do next. However, if we look at its context, we realize that this sentence actually does something very different. It mocks drawn-out conversations about relationships and *refuses* to discuss the social dilemma in question. It thus makes it possible for us to pass over the question about what Margaret ought to know, filing it away, as it were, as a bit of a tedious joke.

To see how it works, let us expand the quote:

> [Tibby Schlegel] had never been interested in human beings, for which one must blame him, but he had had rather too much of them at Wickham Place. Just as some people cease to attend when books are mentioned, so Tibby's attention wandered when "personal relations" came under discussion. Ought Margaret to know what Helen knew the Basts to know? Similar questions had vexed him from infancy, and at Oxford he had learned to say that the importance of human beings has been vastly overrated by specialists. The epigram, with its faint whiff of the eighties, meant nothing. But he might have let it off now if his sister had not been ceaselessly beautiful.[87]

This is a very complex passage, and, as is often the case with such, I expect, as I map out its implied embedments, that my understanding of what is going on may not coincide with yours. Still, different as your understanding may be, I encourage you to take note of the mental states involved. For my argument depends not on the unique perceptiveness of my interpretation but on the *complexity of embedments* expected from a reader to make sense of this passage.

The narrator *anticipates* that readers will *dislike* Tibby for not being *interested* in personal relations. The narrator *wants* his readers to *imagine* what it might have *felt* like for Tibby to grow up in a household where such relations were constantly discussed. By doing so, the narrator *wants* us to *recognize* Tibby's *aversion* to such discussions as a self-defense mechanism, even as he lets us *suspect* that he himself may still *consider* Tibby's *supercilious thinking* unsympathetic.

Planted in the middle of these implied embedments, "Ought Margaret to know what Helen knew the Basts to know?" acquires rather unflattering overtones. Instead of signaling social complexity, it signals impatience with overthinking "personal relations." To put it differently, instead of taking the content of the phrase at its face value and engaging earnestly with the question of whether Margaret ought to know and so on, we may now dismiss this content, because, as Tibby has shown us, one way of dealing with this dilemma is to say, "Who gives a hoot?" The implied embedments thus undercut the explicit one.

"Ought Margaret to know what Helen knew the Basts to know?" can be said to be a case of "free indirect discourse," which is yet another term that, similar to "dramatic irony," functions as a cultural shortcut designating a specific mindreading dynamic. To use our present vocabulary (instead of a more traditional literary-critical one)[88] to describe this dynamic, we can say that free indirect discourse occurs when the implied author wants the reader to distrust information that the text *seems* to be treating as true. Thus, while it may *seem* that whether Margaret knows what Helen knows is a real concern, on closer inspection, it turns out to be just another example of the type of question that Tibby doesn't like thinking about.

But guess what? The question whether Margaret knows and so on may be a red herring, but it won't be put to rest. Far from being confined to the immediate environs of one paragraph, some of its implied complex embedments continue to unspool throughout the novel. For instance, we learned, in Forster's previous chapter, that Margaret already knows that her fiancé, Henry Wilcox, had had an affair, ten years ago, with the woman who was to become Leonard Bast's common-law wife and that Margaret has already forgiven Henry. This means that Helen's present worries are misplaced. That is, Helen *doesn't know* that Margaret *already knows* "what Helen knew the Basts to know."

But wait, there is more: Margaret had written a note to Helen—before she realized that Helen may already know about the affair—and that note was driven by Margaret's *wish* (anticipating and mirroring Helen's present wish!) to preserve Helen from *knowing* what Margaret *knew*. Over that note, Forster explains, Margaret "took less trouble than she might have done; but her head was aching, and she could not stop to pick her words."[89] Ironically, upon receiving that less-than-carefully-worded letter, Helen can't make any sense of it and thus assumes that it is the doing of Henry Wilcox (as she puts it to Tibby, "He makes Meg write")[90] and that Henry wants to prevent Margaret from learning the truth about his past. For readers who remember Henry's role in suppressing his late wife's will (which would have Margaret inherit the Wilcoxes' country house, Howards End), Helen's assumption may ring less mistaken than it is.

In other words, every time you change the context for the original straightforward embedment, "Ought Margaret to know what Helen knew the Basts to know," your interpretation shifts,[91] and every one of these interpretive shifts (i.e., on the level of individual sentences, paragraphs, or chapters) unfolds as yet another complex embedment of mental states involving characters, readers, and the implied author.

Let me pause here, before you start feeling like Tibby and ask exasperatedly, "Ought readers to know what Zunshine considers the implied author to intend?"—a question that may really mean, as we have seen, "Who gives a hoot?"—and decide how many hoots we should give about any of this. There are two points I want to make here. The first is a simple assertion: to read *Howards End* is to embed complex mental states incessantly, whether you are aware of it or not. Yours may not be the same as mine, but that difference in content is less important than the structural similarity: the fact that neither of us can make sense of the text without constantly embedding *some* complex mental states.

The second point is that complex embedments are not merely *in* the text, ready to affect the same way whoever opens the book.[92] Instead, they emerge as a specific reader *acts* on what they are reading by intuitively choosing a context in which to make sense of a potential embedment. Thus, while one reader may indeed focus on the question of whether Margaret ought to know what Helen knew the Basts to know, another reader (or the same reader on a different occasion) may adopt some of Tibby's indifferent perspective on personal relations and pass over that question; while yet another may particularly respond to the implied author's attitude toward Tibby's superciliousness regarding Helen's concern, and so forth. I suspect that were we to bring *Howards End* to the lab and ask our subjects to count the levels of embedment involved in this paragraph, the numbers that they would report would remain generally high—that is, between three and five—but there would be quite a bit of fluctuation within that range, given that the content and configuration of embedments would differ from one reader to another.

Here is, then, one way to describe the experience of reading literature. As we read, we construct contexts in which to make sense of potential embedments, and then we use the information that we derived from those embedments to construct contexts for subsequent embedments. We can think of this process of continuous and contingent construction as "participatory sense-making"—to borrow the term that De Jaegher and Di Paolo use to characterize mindreading involved in daily social interactions.[93] This means that by the time readers arrive to "Ought Margaret to know what Helen knew the Basts to know," they already "have a history of interaction" with the novel,[94] that is, they have already been primed, by preceding embedments, to treat some contexts as more relevant than others. Some of this priming has been planted (so to speak) by the author, but some has not.[95]

Let us take a closer look at the aspects of priming that are less predictable and thus fall outside the range of responses that I seek to capture with my hopeful "we." So far I have described the construction of contexts for complex embedments as a forward-oriented process: with past embedments influencing embedments-to-come. But the "participatory sense-making" can move backward as well as forward. Something that we just read may trigger a complex embedment that hails from a preceding chapter, an embedment that has been lying dormant until now.

For instance, perhaps I did not pay much attention, the first time around, to the nuances of the narrator's view of Tibby's attitude toward his sister's dilemma, yet, later, as I come across another social situation involving Tibby's perspective of other people's emotions, I may find myself realizing that the narrator has been feeling less than charitable toward this character for a while and thus retroactively revise the meaning of "Ought Margaret to know . . ." The cognitive literary scholar Anezka Kuzmičová describes this reverse sense-making in terms of "probes" that illuminate this or that aspect of our past reading experience. As she explains, "in reading long-form narrative . . . the number of verbal probes that can guide one's grasp of the preceding text . . . is endlessly [high]. In light of this insight, it seems a mystery how any two people can ever come close to converging in their subjective experience of a story or novel."[96]

Here is something to deepen this mystery yet further. Kuzmičová observes that "for many leisure readers" (a group that comes closest to my ideal of "good-enough" readers), "the added value of narrative lies . . . in momentarily becoming conscious of one's self and one's problems in specific ways that may be less readily available otherwise." These "persona realizations inform consciousness" in a variety of ways. "Often enough they may come in the form of propositional thought ('Oh my, this character is acting just like me'). Just as often, however, they may assume the form of mental imagery," feeding on "personal memories triggered by the narrative." To psychologists, such associations are known to be "much more common in literary narrative compared to other types of reading materials." Their frequency "directly affects the pleasure taken in reading. . . . It is in this sense that literature affords a unique form of self-consciousness, in which you focus on yourself and yet you do not, because the story you are reading is really about others."[97]

To give you an example of this kind of unique (no *we* here—only *me*!) experience of embedding, imagine that I am reading the "Ought Margaret to know" passage at a particular juncture in my life at which I may feel a sharp pang of recognition by thinking of myself as a beleaguered Tibby surrounded by overbearing Helens. I would thus be more likely to construct an embedment in which I would give more weight to the nuances of that paragraph that portray Tibby with sympathy and compassion. The problem is that were that process to take place in a lab and were someone to ask me to map out the embedments of mental states present in this paragraph, I would not include any of my personal reflections into my report and instead would come up with something along the lines of, "The narrator *wants* his readers to *imagine* what it might have *felt* like for Tibby to grow up in a household where such relations were constantly discussed. By doing so, the narrator *wants* us to *recognize* Tibby's *aversion* to such discussions as a healthy boundary-setting reaction."

What does it mean for the experiment—for its accuracy, reliability, replicability, and so on—that my map would effectively bury that very important aspect of self-consciousness, in which I focus on myself and yet I do not, because the story I am reading is really about others?[98] At the very least, it means that we have to remember that, as any other literary-critical tool, our maps of embedment may conceal as much as they reveal about readers' interactions with the text.

1.8   Enactive Embedments

Each instance of reading thus has its own unfolding history dependent on a uniquely situated reader—a particular person at this exact point in their life.[99] There is no predicting how a literary text will meet a reader at a given time; in what direction the probes "of consciousness will be thrust";[100] which contexts will have more traction and which will have less; and, ultimately, what specific sequence of complex embedments the text and the reader will *jointly* create.[101]

This perspective on reading is congenial with the so-called enactive school of cognitive science, which emphasizes that mind is always constituted by "organism-environment interactions."[102] In particular, cognitive scientis s committed to the enactive paradigm caution against thinking of mindreac-ing as a form of problem-solving: one "detached individual trying to figue out the other."[103] To be fair, cognitive literary critics, such as myself, have never approached mindreading as problem-solving. As I have emphasized on numerous occasions, mindreading takes place away from conscious access: it is too fast and intuitive and enmeshed with body language to be thought of along the formal lines of "figuring out the other." Still, there are some occasions on which we would do well to heed that warning, and one such occasion is studying embedments in the lab.

For think again about the first part of our experiment, in which we pre-sented our subjects with context-free synthetic constructs such as "I am not even sure why Stephanie wants to go the movies with Alice and me." Our expectations certainly conformed to the model of "detached individuals try-ing to figure out the other." There was a "correct" answer associated with each sentence; the vignettes were to be decoded, and our subjects were the decoders.

In contrast, reading a work of literature does not reduce the text anc the reader to those roles. Instead, reading can be described as a form o social interaction between the two autonomous agents (i.e., the text and its reader) that unfolds as they settle on a particular sequence of contexts for complex embedments.[104] Unlike decoding, this enactive process of cocre-ation is less predictable, less likely to yield high agreement rates, and harder to study in the lab.[105]

Still, harder does not mean impossible, and there are some unexpected bonus points along the way. For instance, when my colleagues and I were working on tallying the data collected from our subjects, I noticed that cognitive scientists began to sound like literary critics, that, in fact, we all began to sound like participants in a literature seminar, avidly discussing motivations of characters, narrators, and readers in order to figure out why this or that sentence was assigned this or that level of embedment. As far as interdisciplinary projects go, a study of embedded mental states may thus be a particularly gratifying experience for a literary scholar, because it builds on the strengths of each participating discipline (e.g., I had to defer to my colleagues from the Haskins Laboratories for their expertise in brain-imaging techniques and statistical analysis) without losing sight of the complexity of the issues under consideration.[106]

Elfenbein has pointed out that scholars of literature have long been prone to "quick condemnation" of each other's work as "reductive," so it is "not surprising" that, given what they *think* psychologists are doing, they now accuse them "of the same perceived sin."[107] What my experience suggests is that one way to put that stale prejudice to rest is to develop a collaborative project with one's colleagues from cognitive science: to hear them talk your language and to attempt to understand theirs.[108] Then even those aspects of the project that would seem to point toward shortcomings of studying literature in the lab (such as a failure to obtain high agreement rates on the level of embedment in passages from a novel) may yield important insights about the participatory nature of literary mindreading.

1.9 Sitting Ducks

It seems that in literature certain types of explicitly spelled-out complex embedment—specifically those involving characters' assertions about their own and other characters' mental states—may function as sitting ducks. Just like "Ought Margaret to know what Helen knew the Basts to know?" they may be set up to be upended by their contexts. So when we come across a sentence that reads suspiciously like one of my awkward mindreading maps—for example, "he thinks that she knows that he knows"—we may want to be on the lookout for implied embedments that would subvert these explicit ones.

Here are some examples of such subversion.

Recall Shirley Jackson's "The Beautiful Stranger." While its protagonist is happily interpreting the man's smile as an indication that he knows that she has been worried that he may be her husband, after all ("She was *aware* from his smile that he had *perceived* her *doubts*"),[109] readers may have a reason to doubt her insight. Although it is possible that the husband has been (say) abducted by aliens and somewhat imperfectly replicated, which means that the smiling man *is* a beautiful stranger,[110] another explanation is that the protagonist has gone insane. If this is what we think is happening, then, even while we're following the wife's train of (embedded) thoughts, we are also aware that the implied author *wants* us to *consider* that she is *misinterpreting* the meaning of her husband's smile.

My second example comes from Lara Vapnyar's *Still Here* (2016), which tells several interlocking stories of Russian immigrants in the United States

during the dot-com bubble of the late 1990s. At one point in the novel, a
recently-separated-from-her-husband woman named Vika meets an attractive stranger. He tries to start a conversation with her, but she rejects hi
overture outright because he strikes her as a variation on "type of Husband."
And, as we learn through a spelled-out embedment, she thinks that, righ
now, she needs something different: "A Husband knew her the way she didn'
want to be known, at her worst, her ugliest, her most embarrassing. . . . Wha
she needed was a Lover."[111]

Though not quite reaching, in its bluntness, the parodic level of Forster'
explicit embedment, Vapnyar's "a Husband *knew* her the way she *didn't want*
to be *known*" still stands out in the sea of implied embedments surrounding it. These embedments include Vika reflecting with anguish on having
disappointed a terminally ill patient in her care, who had hoped for an
emotionally honest response from her; Vika deciding to visit the Metropoli-
tan Museum of Art because she "truly enjoys" it and not because she cares
whether other people think that she likes art; and readers beginning to suspect that Vika doesn't realize how much she misses her estranged husband

And so, perhaps, the real reason that Vika rejects the stranger is not that
he is a "type of Husband" but that he is not *her* husband.[112] Of course, it
will take many chapters before she becomes aware of that. Right now she is
denied that intuition. All she has at her disposal is an explicit embedment,
which is compact and expressive and almost aphoristic ("a Husband knew
her the way she didn't want to be known") and, as such, provides Vika with
a convincing (perhaps too convincing!) explanation of her own behavior.

Muriel Spark's *The Girls of Slender Means* (1963) takes place in bombed-out
London in the early summer of 1945. The novel centers on a group of young
women living in the dormitory-style "May of Teck Club" and on their male
admirers. In the following passage, the explicit embedments describe one of
these men's awareness of his thoughts, while the implied embedments make
us wonder whether this self-awareness truly differentiates him from another,
much less sympathetic character: "The Colonel seemed to be in love with
the entire club, Selina being the centre and practical focus of his feelings in
this respect. This was a common effect of the May of Teck Club on its male
visitors, and Nicholas was enamoured of the entity in only one exceptional
way, that it stirred his poetic sense to a point of exasperation, for at the same
time he discerned with irony the process of his own thoughts, how he was
imposing upon this society an image incomprehensible to itself."[113]

Nicholas thinks of the girls from the May of Teck Club as beautiful and pathetic in their communal poverty and, moreover, glorying in their economic hardship. In their heroic penury, they are emblematic of war-torn England at its best. Of course, the girls themselves experience their poverty as a temporary evil that they can't wait to overcome. Not altogether blind to their perspective, Nicholas is *aware* that the girls *wouldn't recognize* themselves in his *vision* of them, or, as Spark puts it, Nicholas "*discerned* with irony" that "he was *imposing* upon this society an image *incomprehensible* to itself."

Those are the spelled-out mental states. But then Spark also seems to *want* us to *suspect* a certain affinity between the *feelings* of Nicholas and an American colonel (especially since they both sleep with the same girl, Selina). This is not a pleasant comparison, for, the colonel is obtuse and philistine, while Nicholas is sensitive and sophisticated. Still, the way I see it, Spark won't let her readers off this hook. We wonder uncomfortably— that is, she *wants* us to *wonder*—if Nicholas's *self-awareness* is enough to mark him as different from other men who are "*in love* with the entire club" (as opposed to being attracted to one particular person) and thus displace onto it their sexual or political fantasies.

Hence a word of caution to a fictional character: thinking that you know well your own, or someone else's, mind—especially if you are spelling it out as a complex embedment—may not bode well for you.

1.10 Why Maps of Mental States Are Ugly

Bitter is the fate of the literary critic who has selflessly dedicated herself to pursuing the secret life of literature. Droning on that "the implied author *wants* the implied reader to *understand* that this character *doesn't know* . . ." does not endear her to her readers. And who can blame them? There is nothing appealing about such mindreading "maps." They are boring, repetitive, almost grotesque, and often hard to follow. They look pathetic next to the texts that they claim to represent. There is even a vague feeling of violence being inflicted on the elegant originals. The originals recover well (they have seen it all), but the critic may be stuck with the reputation of a plodding pedant.

Recent work by Max Van Duijn, Ineke Sluiter, and Arie Verhagen may help to explain why mindreading maps look so off-putting next to original texts. As these scholars suggest, by the end of the second act of *Othello,* "the audience has to understand that Iago *intends* that Cassio *believes* that Desdemona *intends* that Othello *believes* that Cassio *did not intend* to disturb the peace."[114] This looks, to me, like a very complex embedment (in fact, I would make it simpler, by scaling this map at least one level down), but imagine Shakespeare actually making Iago step forward and regale his audience with an aside in which he would say something along the lines of "I want Cassio to believe that Desdemona intends that Othello . . ." and so forth.

Better to not imagine it. Not only would it sound unbearably stilted, but also, after a while, it would become "hard or even impossible for a reader or hearer to make the right inferences" about the characters' intentions.[115] For, as the cognitive evolutionary psychologist Robin Dunbar and his colleagues have shown, "fifth-order intentionality" (fifth-level embedment of mental states) represents "a real upper limit for most people," that is, the level at which their understanding of the situation worsens dramatically.[116] (Works of literature, I should add, do not often go the fifth level and higher. Extremely intricate social nuances can be conveyed on the third and fourth level of embedment. Even for such authors as Henry James—who, one may assume, would soar freely in the fifth-level empyrean—there is plenty to do on the third level. Shakespeare, Jane Austen, Muriel Spark, and Penelope Fitzgerald; Pushkin, Dostoevsky, Tolstoy, and Tatiana Tolstaya; Apuleius and Heliodorus, Cao Xueqin and Murasaki Shikibu, ply most of their unhumble trade on the seemingly humble third level.)

And so Shakespeare does not make Iago step forward and spell out his intentions as a sequence of embedded mental states. Instead, as Van Duijn and his coauthors explain, "narrative takes over"; that is, readers and viewers have at their disposal a number of "strategies characteristic of (literary) narrative discourse that support [their] ability to keep track of the [mental states] of characters." These strategies supply "support and scaffolding for readers' abilities to process [embedded mental states] by providing cues that prompt them to construct a fictional social network using mainly the same socio-cognitive skills as in real-life interaction."[117]

To construct a map, we strip off this vital scaffolding. While embedded mental states in their natural environment are often implied, distributed over a paragraph or a scene, and embodied, we spell them out and force them into sentence-like propositions. "He thinks that she thinks that he wants X"; "she remembers that she used to think that were X to happen, she would feel Y." But who in their right mind would enjoy reading that kind of stuff? If a work of fiction is a living, breathing body, then a map of embedded mental states is a skeleton, with all the appeal and charm of a skeleton.

There is, thus, a good reason why writers themselves don't let those bones stick out. "He thinks that she thinks that he wants X" may be what's going on, but they do not put it that way. If they do, then, as we have seen with the "Ought Margaret to know what Helen knew the Basts to know" example, it may be a joke, a parody, or a comment on someone's lack of interest in social subtleties.

In the section that follows, I consider a fascinating case of the difference between the skeleton and the body. It shows that thinking on at least the third level of embedment is essential to the writing process, even if writers do not articulate it consciously to themselves. It so happened that this author (i.e., Patricia Highsmith) articulated it, but one can hardly hope to find many such examples in print. (Although, as I show immediately after, there *are* ways of finding other cultural contexts in which writers can be observed working through these issues.)

1.11   Do Writers Themselves Make Such Maps?

When the idea for a novel about a passionate love affair, between a gorgeous older woman and a young woman struggling to make it on her own in New York, occurred to Highsmith, she jotted in her diary the following description of the first meeting between the protagonists: "I see her the same instant she sees me, and instantly, I love her. Instantly, I am terrified, because I know she knows I am terrified and that I love her. Though there are seven girls between us, I know, she knows, she will come to me and have me wait on her."[118] I *know* she *knows* I am *terrified*. I *know* she *knows* I *love* her. This is good enough for a map, so that the writer herself knows what's going on in the scene, but it won't do for an actual novel. Here is how this scene looks in Highsmith's *The Price of Salt* (1952):

> Their eyes met at the same instant, Therese glancing up from a box she was open-ing, and the woman just turning her head so she looked directly at Therese. She was tall and fair, her long figure graceful in the loose fur coat that she held open

with a hand on her waist. Her eyes were gray, colorless, yet dominant as light or fire, and caught by them, Therese could not look away. She heard the customer in front of her repeat a question, and Therese stood there, mute. The woman was looking at Therese, too, with a preoccupied expression as if half her mind were on whatever it was she meant to buy here, and though there were a number of salesgirls between them, Therese felt sure the woman would come to her. Then Therese saw her walk slowly toward the counter, heard her heart stumble to catch up with the moment it had let pass, and felt her face grow hot as the woman came nearer and nearer.[119]

If we map out this paragraph, we may come up with several third-level embedments. Some of them may even be similar to "I know she knows I love her" from Highsmith's diary. But unlike those explicit embedments the ones in *The Price of Salt* are implied. That is, they may still supply the underlying bone structure for the first encounter between Carol and Therese but they are not anymore visible to the naked eye.

No wonder my own maps of embedded mental states—structured as strings of mentalizing verbs, such as "think" or "believe"—are destined to be clunky and off-putting. Although reading literature means reading mental states,[120] it seems we can only enjoy those mental states in context. Just as we, apparently, cannot absorb vitamins when we take them in the form of pills, "pure" mental states do nothing for us, except, after a very short while, irritate us. Highsmith's desiccated embedments, "I *know* she *knows* I am *terrified*. I *know* she *knows* I *love* her," may have a poetic ring to them. Still, they accrue a certain interest and cultural value (as, when a literary critic, such as myself, is thrilled to discover them in the writer's diary) only because she has already seduced us with the text in which these mental states are *implied*.

Moreover, something else happened in the process of building up from the bare bones of "I know she knows I love her." Other embedments came into being, those involving not just the main characters but the implied author and the implied reader and arising from the style of the narrative and its historical contexts. Observe, for instance, that, while Therese feels helplessly "caught" by the "light or fire" of Carol's eyes, Carol, too, is powerfully compelled to come "nearer and nearer." If one remembers that this dance of fatally attracted butterflies is taking place in 1952, one wonders if Highsmith wanted her audience to fear that her story would fall into the predictable 1950s pattern of depicting a lesbian love relationship as doomed.[121]

It is an interesting question at what point in the second part of the twentieth century the expectation of that particular doom faded. Or, to put it in terms of our present discussion, at what point has it become possible *not to think* that the author *expects* that the reader would *assume* that a story about a love affair between two women cannot end well?

Our awareness of historical contexts is thus yet another factor to consider when we ask if implied mental states are already "in" a given text or are intuited into it by some readers but not others. I suggested earlier that complex embedments arise as social situations built by a text are filtered through the unique consciousness of a particular reader. But, as the case of *The Price of Salt* shows, specific historical circumstances and their attendant ideologies also influence what kinds of implied mental states would be read into a text. A given reader's awareness of the author's stylistic choices—here, reference to "fire" that "catches" the hapless prospective lover—may alert them to intentionality behind the scene. However, their construction of the meaning of that intentionality—Is this a common poetic trope or a sign of danger? Is the protagonists' relationship doomed because of their sexual orientation?—would reflect, among other things, their position in a particular historical moment.

1.12   What Do Writers Actually Say When They Talk about the Secret Life of Literature?

Several years ago, I enrolled in a graduate seminar in my university's MFA program. My goal was to see if writers are aware of the "secret life of literature," that is, if they are aware of the extent to which their texts depend on the constant embedment of complex mental states. That meant paying close attention both to our workshop discussions and to my own writing process, for, like other students in that class, I had to come up with two original short stories and have others comment on them.

Here is what I found, in brief. It is impossible to write fiction while thinking about embedding mental states, because the state of mind in which one puts oneself as a creative writer is different from that of a literary critic. But here are two important caveats.

First, even though I do not think *consciously* of embedding complex mental states when I am writing fiction, I, nevertheless, keep coming up with social situations that call for such embedments. So one way to rephrase what I said earlier is to say that a creative writer puts oneself into a state of mind in which one produces complex embedments without being aware of doing so.

Second, after the first draft is done and I start revising it, thinking consciously of ways to add yet another mental state to this or that social situation becomes helpful, to some degree. It seems that, in the process of
revision, a writer begins to think like a critic or, at least, *more* like a critic
than they did before.

As to whether writers talk about embedding complex mental states during their workshop discussions, the dynamic is similar. They are not familiar with this vocabulary and thus do not use these terms. Nevertheless,
when they comment on each other's drafts, their suggestions for improvement tend toward making social situations present in the original more
emotionally complex, which, of course, depends on cultivating complex
embedments of mental states.

While some of those suggestions center on characters, many involve various states of awareness between the implied reader and the implied author.
Again, the "implied reader" and the "implied author" are not the terms
writers use. They talk instead about texts, protagonists, narrators, authors,
and readers. Thus, they may say, "The protagonist doesn't know it, but are
we supposed to think that the text knows it?"[122] or "Even if the narrator is
unsure what the story is about, the reader must sense that the author knows
what the story is about, what it's doing."[123] Or, to quote from a workshop
participant's written response to one of my stories, "The simplest way I can
think of for this would be to utilize a third-person perspective so that the
narrator could give us insight that the current narrator wasn't willing to. Or
you could leave it in first person and just use the asides of the narrator to
also give us possible suspicions that might be fleeting in her mind that she
refuses to give much thought to."[124]

In other words, when writers are writing and revising/talking about their
craft, they operate on a high level of embedment, even if they are not aware
of it. Indeed, if my own experience is to be trusted, *consciously* focusing on
embedding mental states is detrimental to all of these processes, although
it is significantly more detrimental during the initial writing stages. The
"secret life of literature" must remain secret even to the people who make
literature happen.

Let me now show you how an author may use a feedback received from
their peers to make a given social situation more emotionally complex by
bringing in more embedments. Here are two excerpts—an original and a

revision—from one of my stories written for the workshop. The story features a middle-aged protagonist thinking back to a time when she was nineteen and she and her best friend, "Julia," were in love with the same man, "Zhenia." The man eventually chose Julia, and the protagonist remembers asking Julia about what the two of them did together: "I don't wish to know where Julia and Zhenia go together and what they do. But, of course, I keep asking, and she tells me."

There is already one complex embedment here. The protagonist is aware that she can't stop herself from doing something that, she knows, will make her feel bad. But look what happened after I followed the advice given to me by the workshop participant who suggested highlighting the difference between the past and the present narrator, so that the asides of the present narrator can "give us possible suspicions that might be fleeting in her mind that she refuses to give much thought to":

> I didn't wish to know where Julia and Zhenia went together or what they did. But, of course, I kept asking, and she kept telling. Today, I think it is odd that Julia didn't seem to realize that it was painful for me to listen to those stories. But, perhaps, she did realize it, which was yet another sign that she had already given up on our friendship. I can say that now, knowing how quickly we were about to grow apart, in spite of my desperate attempts to hold on to her. At the time, however, I interpreted her behavior differently. It made sense to me that she would not think that I might be hurt by Zhenia's choice. After all, I didn't consider myself lovable either.

To map some of the new complex embedments structuring this passage, the older narrator *thinks* it is odd that Julia *did not think* that her friend would *feel bad* hearing her stories; the older narrator *wonders* if Julia *did know* that her friend would *feel bad* hearing her stories; the older narrator is *aware* that her younger self *was not willing* to *consider* that Julia *did not care* about their friendship anymore; the older narrator is *aware* that her younger self *believed* that no one could *love* her; and so forth.

I chose an excerpt from the revised version of my story that contains mostly spelled-out mental states in order to make this discussion more manageable. Initially I had wanted to give you a passage that contained no explicit mental states—only implied ones (which are often more interesting)—but then I realized that doing so would require supplying much more information about the story's plot. Because explicit embedments often present on the level of individual sentences, while implied ones may function on the level of paragraphs, chapters, and plots, explicit embedments are easier to demonstrate.

Here, then, are two key takeaway messages from my experience of taking an MFA course. First, the process of generating complex embedments without being aware of it, while writing, provides a useful insight into our reading practices. For there, too, mentalizing takes place mostly away from conscious access. The "felt experience of reading," the cognitive literary scholar Elaine Auyong reminds us, is "distinct from the mental acts underlying it."[125] To make sense of what we read, we constantly process complex embedments, yet if we pause and take a stock of doing so, the pleasure of reading may evaporate.

Second, we can now come back to the main claim of this book—which is that literature as we know it today cannot exist without embedding mental states on at least the third level—and add the following. Readers for whom this secret life of literature is *most fully present* (even if they do not think about it in those terms) are writers in the process of writing and revising. I dedicated several preceding sections of this chapter to figuring out if some readers are more immediately attuned to complex embedments in literature than others are. While that question mostly remains open, we can confidently say that there is at least one group highly attuned to such embedments, and these are writers when they are writing.

This view finds support in the work of Robin Dunbar, who has suggested that the reason that "good writers [are] so rare" is that they have to constantly keep in mind a higher-order intentionality than do readers.[126] Dunbar and I differ in one respect: I think that, both in our daily life and while reading literature, we operate on a somewhat lower level of embedment that he thinks we do. Thus, he writes that "in everyday social life, we probably don't work at much beyond the third order most of the time,"[127] while I would say (along with Patricia H. Miller et al.) that we don't work at much beyond the *second* level and rise to the third level only occasionally.[128] Similarly, Dunbar observes that writers "are among the very small proportion of individuals who can successfully cope with sixth and seventh order intentionality,"[129] while I think that literature can get plenty complex on the third level and expect that writers do not have to reach to such highs as sixth and seventh very often.

But those nuances notwithstanding, I find Dunbar's argument that writers have to process more higher-level embedments than do readers congenial to my argument that people who are most appreciative of high levels of embedment in literature are those in the process of creating those embedments. As Dunbar puts it,

> When the audience ponders Shakespeare's Othello, for example, they are obliged to work at fourth order intentional levels. . . . [But whatever level of intentionality they are working on], Shakespeare himself is being forced to work at one level of intentionality higher, because he must intend that we (the audience) believe that Iago intends . . . , etc. . . . In effect, a successful story-teller has to be able to work at the very limits of normal adult competence in social cognition. The significance of this is perhaps best reflected in the contrast with the fact that, in everyday social life, we probably don't work at much beyond third order most of the time. . . . The need to be able to work at one or more orders . . . higher than the reader means that the story-teller has to be a rather unusual individual: they are among the very small proportion of individuals who can successfully cope with sixth and seventh order intentionality.[130]

Dunbar and I thus focus on different manifestations of the same phenomenon. He says that literature *sometimes* operates on the sixth and seventh level of intentionality; I say that it *constantly* operates on at least the third. As you can see, these claims are complementary rather than mutually exclusive. The bottom line is that we both think that literature turns up the volume on something fundamental to our everyday social functioning (i.e., mindreading) and that people who operate the dial are the ones who immediately feel the difference.

1.13   Bodies without Minds

"Constant" sounds a whole lot like "universal," which has a bad rap in literary studies, so let us face this issue squarely here. When I say that literature as we know it today cannot function without constantly embedding mental states on at least the third level, do I claim that the secret life of literature is, in effect, "universal"? And, if so, do I also claim that there are no exceptions to this unspoken "rule"?

To start with the second question first: of course, there are exceptions. (For instance, in the next chapter, I will show that some socialist-realist novels published in the Soviet Union and East Germany operated on a lower-than-third level of embedment.) This said, before we pronounce a particular text an exception, we'd better take a good look to make sure that it actually is. In my experience, works of literature that leap to people's minds when they start searching their mental databases for exceptions, typically do not turn out to be such, upon closer inspection.

Patrick Colm Hogan provides a useful framework for thinking about exceptions in his work on "literary universals."[131] Perhaps, one day, what I call the secret life of literature will indeed be considered a literary universal, on the terms that he outlines, but I don't think we are there yet. At this point, it is still an empirical issue. This is to say that we'd do well to keep our mind open and continue checking for this pattern as we study literature from different cultural traditions and historical periods. I expect that social contexts of complex embedment would differ from one author, text, genre, and culture to another, and I think that sensitivity to those contexts is a more interesting and immediate research challenge than the adjudication of the question of universality.

Meanwhile, let us look at some texts that often figure as candidates for exception. What happens, for instance, when writers craft stories that contain no explicit references to mental states, for instance, when their characters seem to come across as lacking "psychology," "interiority," and "depth" or else live in a dystopian society that eschews any discussion of emotional life? More often than not, such stories still contain numerous complex embedments of mental states, but they are all implied. This is to say that readers have to do all of the heavy lifting associated with reading intentions into the behavior of characters and/or into stylistic choices of the author.

And readers do step up to that plate—for otherwise they wouldn't be able to make sense of what is happening in the story *or* appreciate its tone. Yet, ironically, even while they do that, they may continue to take at face value the text's claims (so to speak) to "mindlessness." Consider Evgeny Zamyatin's novel *We* (1921), set in a dystopian future where feelings are jettisoned for mathematical formulas. *We* has apparently fooled enough readers in several languages, because, when I give talks about complex embedment, it is one of the two novels (the other one being Alain Robbe-Grillet's *Jealousy*, which I will discuss later) almost inevitably brought up during the question-and-answer period as an example of a work of fiction that contains no mental states, much less any embedded ones.

Yet *We* constantly prompts us to construct embedded mental states to make sense of what is going on. Look at the first meeting of its protagonists, D-503 and I-330, narrated by D-503: "All this without smiling, I'd ever say with certain reverence (perhaps she knows that I'm a builder of the "Integral"). But I'm not sure—in her eyes or eyebrows—there is some strange irritating X, and I can't quite catch it, can't assign it a numerical expression."[132]

There is a whole constellation of complex embedments here. For instance, D-503 *wonders* if I-330 is *impressed* because she *knows* what he does. Also, he is *irritated* that he *can't fathom* her exact *attitude*. Moreover, the implied reader *understands* that D-503 *doesn't realize* that he's *falling in love* with I-330. The fact that we don't notice any of these and even may end up thinking of Zamyatin's novel as devoid of mental states testifies to the unreflective speed with which we attribute thoughts and feelings when we encounter behavior (more about this in chapter 5, on the history of complex embedment in literature).

Here is another example. It was suggested to me by a colleague sympathetic to the idea that, in eighteenth- and nineteenth-century Europe, writers heavily relied on complex embedment—what with all those thick courtship novels focused on characters' feelings! Modernists, too: just think of Proust's and Woolf's obsession with the multiply storied consciousness. But surely (so my sympathetic colleague thought), latter-day postmodernist authors have outgrown all that preoccupation with psychology. Take Cormac McCarthy's *Blood Meridian: Or the Evening Redness in the West* (1985). Its characters are notorious for their lack of interiority, which means we do not need to embed mental states as we follow their actions.

To see if this supposition is true, consider the opening of the novel. *Blood Meridian* tells the story of a nameless teenager, "the Kid," who joins a gang of scalp hunters terrorizing the border between the United States and Mexico in 1849–1850. We start by learning about the birth and upbringing of "the Kid":

See the child. He is pale and thin, he wears a thin and ragged linen shirt. He stokes the scullery fire. Outside lie dark turned fields with rags of snow and darker woods beyond that harbor yet a few last wolves. His folks are known for hewers of wood and drawers of water but in truth his father has been a schoolmaster. He lies in drink, he quotes from poets whose names are now lost. The boy crouches by the fire and watches him.

Night of your birth. Thirty-three. The Leonids they were called. God how the stars did fall. I looked for blackness, holes in the heavens. The Dipper stove.

The mother dead these fourteen years did incubate in her own bosom the creature who would carry her off. The father never speaks her name, the child does not know it. He has a sister in the world that he will not see again. He watches,

pale and unwashed. He can neither read nor write and in him broods already a
taste for mindless violence. All history present in that visage, the child the father
of the man.[133]

Looking at these three paragraphs, you can see why this novel may strike
some readers as not featuring any thoughts and feelings. This is a far cry
from, say, Marcel Proust's *Remembrance of Things Past*, in which a typical
sentence embeds explicit mental states, as in, "Sometimes when, after
kissing me, she opened the door to go, I longed to call her back and say
to her 'Kiss me just once more,' but I knew that then she would at once
look displeased, for the concession which she made to my wretchedness
and agitation in coming up to give me this kiss of peace always annoyed
my father, who thought such rituals absurd."[134] On the other hand, even
though McCarthy's "Kid" doesn't seem to be able—in stark contrast to the
little boy in Proust—to consider other people's feelings, McCarthy's prose
achieves its uncanny effect by embedding mental states of the mysterious
narrator, the implied author, and the reader.

For there is a very peculiar narratorial consciousness at work in these
early paragraphs. McCarthy's narrator inserts himself in the story ("I looked
for blackness, holes in the heaven") and starts making the case, as it were,
against the Kid. First, by being born, the Kid murdered his own mother,
though, admittedly, she was complicit in the crime. She "did," after all,
"incubate in her own bosom the creature who would carry her off." There
is another victim, too. The mother's death destroyed her husband, a former
schoolteacher, a weak soul, who now "lies in drink," quoting from poets
"whose names are now lost." The child "watches" his father—the word
"watches" is repeated twice. He even "crouches" as he "watches": a little
predator, in whom there "broods already a taste for mindless violence." The
puzzling opening sentence now makes sense, too. "See the child," ladies
and gentlemen of the jury, see the defendant on the stand.

He has known all along how it would turn out—the "I" of the second
paragraph—the narrator who watched the heaven on the night the Kid
was born. God-like he is, but also accomplished, in ways that only certain
sophisticated readers would appreciate. He wants those readers to know
that, unlike other riffraff populating the story, *he* recognizes the unintelligi-
ble sounds issuing from the drunk father as bits of forgotten poems. He also
can cite from the poet whose name has not been forgotten—Wordsworth—
and he does so, appropriately, to support his point: "the child the father of
the man."

Thus, already in the first paragraphs of the novel, McCarthy *wants* his readers to *know* that the story will be told by a narrator who *is determined* to aggrandize himself and to condemn the Kid. Of course, we don't put it this way to ourselves, but to the extent to which we are aware of the strange tone of the opening, starting with "See the child," we are embedding the implied author's intentions. (To quote again one of the MFA workshop's participants, "fiction is a cohesive intentional work.")[135]

What it all adds up to is that *Blood Meridian* embeds complex mental states just as *Remembrance of Things Past* does, even if, in direct contrast to Proust's novel, *Blood Meridian* contains almost no explicit references to mental states. We embed implied intentions of the narrator and the author to make sense of the novel's *tone*—the crucial component of McCarthy's poetic prose.

1.14 Minds without Bodies

But if some stories pretend to be "mindless" and thus make us work harder at reading mental states into their characters' body language, the opposite—that is, stories in which mental states are spelled out but there are no bodies behind them—is also possible. Consider Daniel Defoe's *Robinson Crusoe* (1719), whose protagonist regularly ponders intentions of "Providence," an entity that has landed him on a desert island: "These reflections made me very sensible of the goodness of Providence to me, and very thankful for my present condition, with all its hardships and misfortunes; and this part also I cannot but recommend to the reflection of those who are apt, in their misery, to say, 'Is any affliction like mine?' Let them consider how much worse the cases of some people are, and their case might have been, if Providence had thought fit."[136] As Crusoe imagines people who complain about their affliction, he *wants* them to *consider* that had Providence *thought* fit to land them in an even worse situation than they are currently in, it could have easily done so.

Crusoe is not alone thinking about various "secret intimations" of the "invisible intelligence."[137] Other fictional instances of such "intelligences" range in form from the karmic destiny of Cao's *Dream of the Red Chamber* to "Aubrey McFate" of Nabokov's *Lolita*. What such nebulous entities have in common is their apparent capacity for intentions and attitudes, which characters and readers try to fathom, all the while generating embedded mental states.

Here, for instance, is Mrs. Plinth, a well-heeled provincial lady from Edith Wharton's short story "Xingu" (1916). Mrs. Plinth can't help feeling keenly that the heavenly power that has made her rich intended for her the honor of hosting distinguished visitors, an honor currently usurped by another, less worthy lady, Mrs. Ballinger: "An all-round sense of duty, roughly adaptable to various ends, was, in her opinion, all that Providence exacted of the more humbly stationed; but the power which had predestined Mrs. Plinth to keep footmen clearly intended her to maintain an equally specialized staff of responsibilities. It was the more to be regretted that Mrs. Ballinger, whose obligations to society were bounded by the narrow scope of two parlour-maids, should have been so tenacious of the right to entertain [the current special guest]." [38]

Mrs. Plinth resents that Mrs. Ballinger refuses to acknowledge the intention of Providence, which wanted Mrs. Plinth to host distinguished visitors. Providence, apparently, is as invested in Mrs. Plinth's social success as it is willing to let some people, including Robinson Crusoe, to get away relatively scot-free, while smiting others. We may have come a long way from Apuleius's Venus and Cupid: divine entities that guide fictional characters have, nowadays, shed their bodies. But their social minds are as keen and active as ever, plotting and picking favorites among mortals.

1.15 One Body, Many Mental States

Let us stay with Robinson Crusoe a bit longer. If you want to know how many fictional characters one needs to start generating complex embedments, the answer seems to be just one. A single character can embed enough mental states to sustain a three-hundred-page novel, as does Crusoe, who spends twenty-three out of his twenty-eight years on a desert island with nobody to talk to. (Friday joins him only at the tail end of his confinement.) His loneliness does not prevent him, however, from engaging in introspective musings such as this one:

> From this moment I began to conclude in my mind that it was possible for me to be more happy in this forsaken, solitary condition than it was probable I should ever have been in any other particular state in the world; and with this thought I was going to give thanks to God for bringing me to this place.
>
> I know not what it was, but something shocked my mind at that thought, and I durst not speak the words. "How canst thou become such a hypocrite," said I, even audibly, "to pretend to be thankful for a condition which, however thou mayest endeavour to be contented with, thou wouldst rather pray heartily to be delivered from?" [139]

This passage is typical for Defoe's novel, which demonstrates on every page ample narrative possibilities of the embedded consciousness of a solitary protagonist.[140] Crusoe *imagines* that he can be *grateful* to God for bringing him to a place where he can be *happier* than anywhere else in the world. But then he is *shocked* that he *would pretend* to be *grateful* for a condition that he would, in fact, *prefer* to escape. He accuses himself of becoming a hypocrite—"hypocrisy" being yet another cultural shorthand for a complex embedment, for a hypocrite *wants* to make others *think* that he or she has *beliefs* and moral standards that he or she, in fact, *does not have.*

Unlike Crusoe, the speaker of William Wordsworth's poem "Lines Composed a Few Miles above Tintern Abbey, On Revisiting the Banks of the Wye during a Tour" (1798) is not alone: accompanying him on his "tour" is his sister, Dorothy. Still, for most of the poem, he is thinking about the relationship among his various selves situated at different points in time, watching himself, for instance, to form impressions that, he knows, will influence him for years to come:

> And now, with gleams of half-extinguished thought,
> With many recognitions dim and faint,
> And somewhat of a sad perplexity,
> The picture of the mind revives again:
> While here I stand, not only with the sense
> Of present pleasure, but with pleasing thoughts
> That in this moment there is life and food
> For future years.[141]

The speaker *imagines* his future self being made *happy* by *remembering* how *happy* he was here (by remembering, that is, his "present pleasure"). David Herman has described this literary dynamic as "distributed temporality,"[142] and we can see this interplay among the mental states of past, present, and future selves throughout "Tintern Abbey." Embedments arising out of a temporally distributed self can be encountered in any work of literature, but they may be particularly common in memoirs (be they prose or poetry, such as Nabokov's *Speak, Memory* or Wordsworth's *Prelude*) concerned with imagination and consciousness,

1.16 Many Bodies, One Shared Mental State

But if a single character can be a source of mental states embedded on the third and fourth level, the opposite is also true. A large group of characters

can share a single mental state—thus forming what the cognitive narratologist Alan Palmer calls an "intermental unit."[143] Such an intermental unit
can then be embedded within other mental states the same way as a mental
state of just one character can be embedded within other mental states.

To illustrate this, here is another, typically self-reflexive sentiment of Crusoe, who begins by contemplating his own feelings and then turns to the
thoughts of an intermental unit comprising, perhaps, millions of people:
"But it is never too late to be wise; and I cannot but advise all considering
men, whose lives are attended with such extraordinary incidents as mine,
or even though not so extraordinary, not to slight such secret intimations
of Providence, let them come from what invisible intelligence they will."[14]
Crusoe is thinking about the thoughts of, if not the whole of humankind
then a large part of it. He *wants* "all considering men" to *pay attention* to the
intentions of Providence. This is as large a group of people as they come—a
massive "intermental unit"—all sharing one mental state, which is embedded, in its turn, within the thoughts of the protagonist.

We also may want to take a look into the novelistic construction of
crowds and ask how writers get around the problem of representing a large
number of minds—fifty, a hundred, a thousand—numbers that would
instantaneously take us outside our zone of cognitive comfort were we to
try to imagine the mental states of those people one by one. It seems that
authors can deal with this challenge in several ways. They may portray a
crowd through two or three distinct personalities—the spokespeople who
capture various points of view held by the multitude. Or they may depict
a crowd as being of "one mind," shouting or grumbling in unison. This, in
turn, makes it possible for this unified "mob mind" to interact with two or
three other distinct individuals, who respond to the mob's concerns, so that
the cumulative number of embedded mental states still stays within the
comfortable range of four.[145]

Think, for instance, about the preelection scene in George Eliot's *Middlemarch* (1871), which starts with Mr. Brooke, who is running for Parlia-
ment, giving a short speech in front of a large crowd of potential voters. In
response to his claptrap, first one heckler and then another ("the invisible
Punch") make fun of him. Mr. Brooke, however, misunderstands their reactions, thinking that the second heckler intends to ridicule the first, until
"a hail of eggs" directed at him and his effigy makes the crowd's feelings
abundantly clear:

"That reminds me," [Mr. Brooke] went on, thrusting a hand into his side-pocket, with an easy air, "if I wanted a precedent, you know—but we never want a precedent for the right thing—but there is Chatham, now; I can't say I should have supported Chatham, or Pitt, the younger Pitt—he was not a man of ideas, and we want ideas, you know."

"Blast your ideas! we want the Bill," said a loud rough voice from the crowd below.

Immediately the invisible Punch, who had hitherto followed Mr. Brooke, repeated, "Blast your ideas! we want the Bill." The laugh was louder than ever, and for the first time Mr. Brooke being himself silent, heard distinctly the mocking echo. But it seemed to ridicule his interrupter, and in that light was encouraging; so he replied with amenity—

"There is something in what you say, my good friend" . . . here an unpleasant egg broke on Mr. Brooke's shoulder.[146]

Readers may walk away with an impression that a multitude of "weavers and tanners of Middlemarch" have expressed their opinions about Mr. Brooke's candidature,[147] when all we really have here are two distinct (if invisible) spokespersons and Mr. Brooke's initially mistaken view of their attitude toward each other. Once the crowd's minds have thus been compressed to a manageable number, we are ready to process the scene's complex embedments of mental states and consider its meaning. For instance, as Eliot's biographer Nancy Henry explains, a "crowd of [Middlemarch voters] detects and mocks the insincerity of Mr. Brooke's commitment to reform."[148] Or, to put it in terms of our present discussion, this crowd *knows* that Mr. Brooke only *wants* them to *think* that he *cares* about reform.[149]

1.17 Downgrade This!

As we are nearing the end of this chapter, let us revisit the issue of "simplifying" our descriptions of mental functioning, first brought up in the section on *Tom Sawyer*. For, I can still imagine a reader who thinks that it just *may* be possible to make sense of scenes that, as I claim, embed mental states on at least the third level while staying on the first or second level. To see what that would look like—that is, what downgrading the levels of embedment does to a story—let us take a look at three examples from classical Roman, Greek, and Japanese literature.

In Apuleius's *The Golden Ass* (second century AD), a young widow learns that her beloved husband was treacherously murdered during a boar hunt by the man who had long wanted her himself. Unaware that she knows about his perfidy, that man is now pressing the widow for marriage. She "pretend[s] to be won over" and suggests that they have a clandestine affair, "just until the year travels the full length of its remaining days," at which point they would wed. She wants him to believe that she is eager to sleep with him yet is ashamed that people would think it unseemly for a new widow. So he agrees to come to her house late at night, muffled "from head to foot and bereft of [his] escort," thus leaving himself vulnerable to her gory revenge.[150]

Let us see how much of this episode's meaning is retained if we insist on scaling down its levels of embedment:

- "The widow is *eager* to sleep with the man who killed her husband." This is one mental state, and you can decide for yourself how accurately it describes what is going on.

- "The man *thinks* that the widow is *eager* to sleep with him." That's two embedded mental states, and this configuration is still wrong, because it reflects only the limited perspective of the doomed character.

- "The widow *wants* the man *to think* that she *wants* to sleep with him" or "The widow *wants* the man *to think* that she *is afraid* of what people will say if she becomes his mistress so early into her bereavement." Once we start operating on the third level, we, finally, begin to capture the complexity of the situation.

In Heliodorus's *An Ethiopian Romance* (third century AD), an Egyptian priest, Calasiris, tells to his acquaintance Cnemon the story of the first meeting of the protagonists, Chariclea and Theagenes. During a public celebration at the altar of Apollo, Theagenes is supposed to receive a torch from a priestess (Chariclea) with which to light the altar piled with animal sacrifices. The surrounding crowd includes Chariclea's adopted father, Charicles, who is, however, too busy right now to observe his daughter closely:

> At first they stood in silent amazement, and then, very slowly, she handed him the torch. He received it, and they fixed each other with a rigid gaze, as if they had sometime known one another or had seen each other before and were now calling each other to mind. Then they gave each other a slight, and furtive smile, marked only by the spreading of the eyes. Then, as if ashamed of what they had done, they blushed, and again, when the passion, as I think, suffused their hearts, they turned pale. In a single moment . . . their countenances betrayed a thousand shades of feeling; their various changes of color and expression revealed the

commotion of their souls. These emotions escaped the crowd, as was natural, for each was preoccupied with his own duties; they escaped Charicles also, who was busy reciting the traditional prayer and invocation. But I occupied myself with nothing else than observing these young people.[151]

Calasiris *knows* that Charicles *doesn't know* that Chariclea and Theagenes are *falling in love* with each other. We may not articulate this to ourselves as we read the novel. But later, when Calasiris hatches a plot to help the young people elope together, it makes sense to us because it hinges on Calasiris's *knowing* that Charicles *doesn't know* that Chariclea *loves* Theagenes. Get rid of one of those levels of embedment and the elopement plot falls apart.

In Murasaki Shikibu's *The Tale of Genji* (eleventh century AD), shortly after Genji's mother's death, the emperor sends a messenger to the boy's grandmother, inviting her and Genji to the palace. Upon receiving the grieving emperor's letter, the grandmother talks to the messenger about what it means for her to have outlived her only daughter:

"Now that I know how painful it is to live long," she said, "I am ashamed to imagine what that pine must think of me, and for that reason especially I would not dare to frequent his Majesty's Seat. It's very good indeed of him to favor me with these repeated invitations, but I am afraid that I could not possibly bring myself to go. His son, on the other hand, seems eager to do so, although I am not sure just how much he understands, and while it saddens me that he should feel that way, I cannot blame him. Please let his Majesty know these, my inmost thoughts."[152]

Observe that, while declining the emperor's invitation, Genji's grandmother quotes from a poem, *Kokin rokujo* 3057, in which, as the translator, Royall Tyler, explains, "the poet laments feeling even older than the pine of Takasago, a common [lyrical] exemplar of longevity: 'No, I shall let no one know that I live on: I am ashamed to imagine what the Takasago pine must think of me.'"[153] The bereaved mother knows that the emperor will be pained by her refusal to visit him, and she *wants* him to *understand how she feels.* By evoking the poem (which is itself a third-level embedment of mental states: "I am *ashamed* to *imagine* what that pine must *think* of me"), she makes him aware of a somewhat unexpected nuance of her grief: shame. If the emperor considers that even a tree would reproach her for outliving her child, he would surely understand that she doesn't want to be seen by others, especially in a place to which people go with the purpose of being seen, such as the emperor's palace.

Try conveying any of this through lower-level embedments. "Genji's grandmother *is thinking* about a poem" (one mental state) or "Genji's grandmother *wants* the emperor to *recall* a famous poem" (second-level embedment) or "Genji's grandmother *wants* the emperor to *pity* her" (also second-level embedment) all distort the meaning of what is going on. Until we start thinking on at least the third level—for instance, "Genji's grandmother *wants* the emperor to *understand* that she is too *depressed* to make an effort to be seen by others"—our reading of the passage remains tone-deaf.

1.18   Can a Computer Program Tell the Difference between "Popular Fiction" and "Literature"?

Can one design a computer program that will count levels of embedment in a given sentence, paragraph, or chapter? The possibility of such a program has been mentioned to me on several occasions, with cautious enthusiasm by computer scientists and with dread by my colleagues from literary studies. I would be excited to see software for counting mental states in fiction because I suspect that it will fail and that its failure will be as illuminating as was the failure of various artificial intelligence projects in the 1950s–1970s.

The latter, as you may remember, alerted scientists to the unprecedented complexity of evolved human cognition. The machines could not replicate cognitive processes that came so easily to people that they hadn't even been aware of them. Just so, by failing to register embedded mental states in literature, a computer program would illuminate cognitive processes that make reading literature possible and that we take completely for granted, such as a constant attribution of embedded mental states to characters, implied authors and readers, and narrators.

It will be particularly instructive if, in this case, the failure turns out to be selective. For I believe that a computer may be able to count embedments in some texts but not in others. That is, it may succeed with works of fiction that embed mental states of their characters *and* describe these mental states explicitly but not with those that embed implied mental states of characters, narrators, implied authors, and readers.

Consider this passage from John Irving's novel *The 158-Pound Marriage* (1974): "'I am going to get a lover,' she said, 'and I'm going to let you know about it. I want you to be embarrassed when you make love to me wondering if I am bored, if *he* does it better. I want you to imagine what I say that

I can't say to you, and what *he* has to say that you don't know.'"[154] I believe that one can indeed design a computer program that will do well with this novel. Make it pick such words as "want," "embarrassed," "wonder," "bored," and "imagine," and you will have a fairly accurate map of a given passage's embedment. "I *want* you to be *embarrassed* because you *wonder* if I *am bored*"—that's fourth-level embedment, and a computer may just be able to perform this calculation.

Now picture software faced with a sentence from Cao Xueqin's *Dream of the Red Chamber*, in which its female protagonist, Lin Dai-yu, reflects on her winsome cousin, Xue Bao-Chai: "And now suddenly this Xue Bao-chai had appeared on the scene—a young lady who, though very little older than Dai-yu, possessed a grown-up beauty and aplomb in which all agreed Dai-yu was her inferior."[155] What's going on in this sentence? Here is one way to spell out the mental states that we infer as we make sense of it: the narrator *wants* his readers to *realize* that Dai-yu *feels distressed* because she *is certain* that everyone around her *considers* her inferior to Bao-chai. That's at least four embedded mental states, but to articulate them, we have to take in subtle cues, such as the unhappy tone with which Dai-yu refers to her cousin. She calls her "a Xue Bao-Chai" (一個薛寶釵) or "this Xue Bao-chai" in David Hawkes's translation. The use of the pronoun "this" or "a" (yīgè) before a personal name is particularly important here, because it reflects Dai-yu's anguished sense of propriety. She can't say anything harsh or vulgar, so a vaguely dismissive "this" becomes an expression of her irritation and jealousy.

If we look for explicit references to mental states that this sentence contains, we notice the word rendered by the translator as "agreed" (wèi, 謂).[156] This word may describe an attitude of some people around Dai-yu, but the meaning of the passage does not reside with it. Instead, as we've seen, that meaning is expressed through embedded mental states implied but not stated by the text.

What will a computer do in this case? It may pick up on the word "agreed," but, as we have already seen, that word contributes little to the complex embedment present in the sentence. The problem is that a computer program cannot register *implied* mental states, much less figure out context-specific relationships that organize these mental states into embedments. Because in *Dream of the Red Chamber*, any word—including "a" and "this"—can create an implied embedded mental state, only a human mind, with its infinite sensitivity to contexts, can follow it.

But what about passages from *Dream* that spell out embedded mental states of its characters? After all, Dai-yu's diatribe about the look that Bao-yu gave to Xiang-yun, which I quoted in section 1.3 ("But what about that look you gave Yun? Just what did you mean by that? I *think* I *know* what you *meant*. You meant to warn her that she would cheapen herself by joking with me as an equal"), is not terribly different from Irving's "I want you to be embarrassed when you make love to me wondering if I am bored."

The difference between the two is that Cao's novel (as, indeed, other texts that we tend to put on our course syllabi) does this only occasionally. In contrast, *The 158-Pound Marriage* or, for that matter, Dan Brown's *Da Vinci Code*, Stephenie Meyer's *Twilight Saga*, or Danielle Steel's *Against All Odds* do it constantly. Computers will have a ball counting mental states in the fly-by-night favorites that spell out mental states of their characters and do not demand that their readers process implied mental states of narrators and implied authors.

Several strains of research in social and developmental psychology may bear on these issues. For instance, the social psychologist Emanuele Castano and his colleagues, working with theory of mind and fiction, suggest that "life-time exposure to literary fiction positively predicts attributional complexity, while exposure to popular fiction negatively predicts it."[157] (Psychologists use the term "attributional complexity" to describe motivation to seek complex explanations for human behavior, explanations that include though are not limited to, mental states.)[158] Although Castano et al. are careful to observe that "literary and popular fiction foster different socio cognitive processes and cognitive styles, all of which are important,"[159] the distinction between the two has been central to their research projects for a while.[160] Moreover, following up on Kidd and Castano's earlier studies, the cognitive neuroscientist Iris van Kuijk and her colleagues have suggested that, compared "to popular fiction, reading literary fiction might encourage participants to process the meaning of words, sentences and their relationships more deeply and that might produce [theory of mind] differences."[161]

Literary critics may take issue with the term "popular," on several counts. For instance, they may object to the cognitive scientists' identification of popular fiction with "character-based" stories and the consequent exclusion of science fiction from the domain of the literary.[162] They are also aware of the slipperiness of the term, because historically, it is known to have covered a broad range of texts, some of them straddling "the cat-egories of literary, genre, and popular."[163] Nevertheless, when we compare patterns of embedment in Irving, Meyer, Brown, and Steel with patterns of embedment that we encounter, say, in Cao, Tatiana Tolstaya, and Zadie Smith, the difference seems to be quite obvious. However you choose to call them—popular, genre fiction, mainstream, lowbrow—novels by Meyer, Brown, and Steel spoon-feed complex embedments to their readers, which must have an effect on those readers' theory of mind that is different from the texts that require them to work at constructing them.

For instance, there are intriguing studies by developmental psycholo-gists who have found that adding explicit references to thoughts, feelings, and intentions of characters in stories for young children does not promote their understanding of mental states.[164] I will discuss those studies in chap-ter 6 (i.e., on children's literature), but, for now, I just want you to note that, even at a young age, *actively figuring out* implied mental states based on context seems to result in different sociocognitive outcomes than merely *being told* what this or that character thinks.

1.19 Conclusion: Close (Mind)Reading

If you are a teacher of literature, you may have noticed by now, particularly with the Cormac McCarthy example but also with the excerpts from Mark Twain, Cao Xueqin, E. M. Forster, and Patricia Highsmith, that the process of identifying embedded mental states in literature looks a lot like close reading—a "fundamental practice" of literary analysis, which consists of "examining closely the language of a literary work or a section of it."[165] The reason that an inquiry into embedded mental states may end up as a close reading is that close reading is often an explication of mental states, those of characters, narrators, authors, readers, and other critics.

We do not think about it in these terms, but it is worth paying attention to. Next time you are developing a close reading with your students, pause and take a closer look at the embedment of mental states that you perform along the way. Conversely, think about passages that you tend to select for this kind of exercise. See if they tend to "promise" (something that experi-enced instructors learn to perceive at a glance) a discussion that is likely to embed complex mental states.

Of course, as Jonathan Culler observes, "there are all sorts of ways of achieving closeness in reading."[166] These range from memorialization, translation
into a foreign language, and inquiry into how culture shapes the meaning
of the text to looking for "conflicts or tensions" within the text, which can
be manifested by "ambiguous words, undecidable syntax, incompatibilities
between what a text says and what it does, incompatibilities between the
literal and the figurative, . . . and so on."[167] Note, however, how integral attribution of complex mental states is to nearly all of those endeavors. Consider,
for instance, translation as a (somewhat less popular, today) form of close
reading.[168] Central to translating is figuring out what the author *meant* by
this or that choice of word in the source language—and hence, which word
would convey the author's *intention* most accurately in a target language.[169]

Here is one way to think about the sociocognitive role played by all those
various practices of achieving "closeness in reading." It is as if it were not
enough, for some of us, to merely process texts that continuously embed
complex mental states. If we happen to live in what I have dubbed elsewhere
a "culture of greedy mindreaders," we may also join special communities for
doing so.[170] Those communities reward their members (e.g., students, critics)
who are adept at discerning complex intentionalities present in literature,
prizing, in particular, intentionalities that are unexpected and yet plausible.

Far from being an isolated phenomenon, the omnipresence of complex
embedment in literature is thus supported by a variety of cultural practices. Those practices seem to emerge in response to specific historical circumstances (as did the close scrutiny of textual "conflicts or tensions" in
literary studies) and thus are not typically thought of as bound with the
intricacies of our social cognition. In the chapters that follow, I will bring
the two together. That is, I will show that the cognitive and the historical
are inextricably connected in the case of complex embedment and that to
understand why a particular work of fiction embeds mental states the way
it does, we have to inquire into the political and cultural history of its creation and reception.

2 Mindreading and Social Status

Until now, I have talked about what complex embedments *are*, that is, what they may look like in novels, plays, and narrative poems, especially if they don't even seem to be there. In this chapter, I focus on what complex embedments *do*, that is, how writers use them to shape readers' perception of their characters. For, as it turns out, characters may differ in their ability to embed their own and others' thoughts and feelings. How do writers (intuitively) decide who should be more capable of complex embedment and who should be less so and what it may mean in the context of their stories? To answer these questions, we start with the real-life dynamics of mindreading and see what makes us better at figuring out the mental states of other people. "Better" in real life is not exactly the same as "more complex" in fiction, but the underlying cause is, curiously, similar, and it has to do with one's social status.

2.1 Confessions of a Bad Mindreader

How good are we at reading other people's minds? Clearly, not great. Our "misinterpretations about the intentions of others often provoke responses that are themselves misinterpreted, leading the interaction into a spiraling [dynamic] likely to engender a general breakdown."[1] Cultural traditions, social stereotypes, and professional occupations all play roles in hindering the way we understand each other's intentions.[2] Some cognitive anthropologists even go so far as to say that while "human society" may "rest on a bedrock" of mindreading, mindreading is "not a particularly useful tool for predicting and interpreting" people's behavior, because it "typically misattributes the mental states of others."[3]

Yet the misreading of mental states of others may not always be the main culprit. What makes it worse is that, in complex social situations, we do not read other people's minds in isolation from our own.

Let us say, for instance, that I am angry at someone for what I perceive as a personal slight. While it may seem that I attribute a certain nefarious intention to them, the actual mindreading dynamic may be more complicated. For what makes me angry may be not merely my perception of what *they* are thinking. Instead, it is my expectation about what *I* ought to think in response to what I perceive they are thinking. This may work out differently on different occasions, but what many of those miserable occasions have in common is my *assessment* of my possible responses to what I experience as their intentions.

And that assessment can be wrong.

This is to say that, mistaken as we may be in our unreflective attributions of mental states to others, we can be even more off the mark when we consciously reflect on our own thoughts and feelings. As the cognitive anthropologist Dan Sperber puts it, "even in the case of seemingly conscious choices, our true motives may be unconscious and not even open to introspection; the reason we give in good faith may, in many cases, be little more than rationalizations after the fact."[4]

Moreover, we do not have neat little storage facilities in our minds where our "true motives" are held and that we could access if only we could somehow tear through the mist and debris that surround them. Instead, we construct our motives similarly to how we construct memories: ad hoc, grabbing what seems to be emotionally "good to wear" right now. In the words of the cognitive literary scholar Patrick Colm Hogan (citations removed), "We often think that we simply and directly know our own motives, the causes of our emotional responses and behaviors. But considerable research has shown that this is not the case. . . . [People] tend to experience their affective feelings as reactions to whatever happens to be in focus at the time. . . . [If] the person is unable to specify either the origin or the target of affect he or she is experiencing, then this affect can attach itself to anything that is present at the moment."[5]

So, for instance, when I huff and puff in response to what someone said or did, something in me is constructing a chain of reasoning along the lines of, "I am the kind of person who would experience the Y kind of emotional reaction to X. They are saying/doing X. Don't they know how I am bound to respond to this?" From here, it is a very short step to reading into their actions a range of disagreeable motives, from thoughtlessness to the intention to aggravate me.

At least this is how it seems to work with me when I am at my worst. I do not claim that this emotional pattern applies to everyone, or even to me all the time. But were I to generalize from this private experience, I would say that reading other people's minds in complex social situations is often bound up with reading our own minds. This means, given how strikingly uninsightful we are when it comes to our motives, that misreading other people's minds may also be bound up with misreading our own minds, in fact, sometimes predicated on it.

We misread other people's minds alongside misreading our own or even *because* we misread our own. One wonders why evolution couldn't come up with something better than this hapless "mindreading" adaptation . . .

2.2   How to Become a Better Mindreader

But wait! Becoming better mindreaders is within our grasp. All we have to do is to take a demotion in our social hierarchy. Studies have shown that people in weaker social positions engage in more active and perceptive mindreading than do people in stronger social positions. It works even when we know that it's just a game: "when one is given the role of subordinate in an experimental situation, one becomes better at assessing the feelings of others, and conversely, when the same person is attributed the role of leader, one becomes less good."[6]

The scholar of Icelandic sagas William Ian Miller may add to this insight that to become a better mindreader, one may want to place oneself in a society in which "margins for error [are] smaller." For instance, "blood-feuding people" of medieval Iceland "had to be practically wiser and more cunning than we are now, if only because . . . the stakes [were] higher for them in routine social interactions and transactions":

> [Life] hung in the balance more often for them than it does for us in the free West, considerably more so. You had to be pretty good at discerning motives in others, reading their inner states—better than we safe souls are, for sure. I marvel at the unfathomable complacency that can allow someone to walk down the sidewalk intently texting a message and thinking that if he bumps into someone or forces them unknowingly to give way, that he will not have to account for

himself, secure that he will not suffer a much deserved beating to help him regain a modicum of manners, to assist him in the project of avoiding giving unwarranted offence to others.[7]

By giving oblivious texters a second chance, modern liberal democracies may be blunting the edge of their mindreading prowess. Though one also wonders if mindreading prowess purchased at the price of the constant threat of "beatings" may not be a rather stressful proposition.

There are plenty of commonsense reasons why it would be vitally important for someone in inferior social position to be attuned to the intentions of people above them. What I want to add to those is a possible psychological reason based in the dynamic that I described in the previous section, which is that reading and misreading other people's minds is bound with reading and misreading our own. Recall my fraught chain of reasoning— "I am the kind of person who would experience the Y kind of emotional reaction to X. They are saying/doing X. Don't they know how I am bound to respond to this?"—and think what happens when the "they" in question are of higher social status than I am. How likely is it that I would expect "them" to care about my feelings and persevere with my high-and-mighty "I am the kind of person who . . ."?

But if I don't expect them to care about my feelings and I don't bother anticipating my emotional response to their lack of caring, then I effectively remove my mental states from the equation. This may make me less blinded to their actual intentions and thus turn me into a "more active and perceptive" mindreader.

Again, as in the case of theory of mind being sharpened by the anticipation of a beating, "more active and perceptive mindreading" does not necessarily imply a happy or even healthy mindreader. There is plenty of research in the social sciences about negative effects of low socioeconomic status on one's well-being.[8] Of course, "weaker social position" is a relative concept, and it does not always imply a low socioeconomic standing. For instance, I have a lower social status than the dean of my college, which means that, in a meeting with her, I would be reading her mind more assiduously and perhaps more accurately than she would be mine. (Indeed I have had my share of faculty meetings in which we all sit around the table and try to figure out what our dean *really* meant by this or that oblique promise.) Yet as a tenured faculty member at a research university, I am not exactly an underprivileged type. Still let us not lose sight of the fact that when heightened mindreading ability reflects one's current weaker social position, there must always be some degree of stress involved.

Consider, too, that those who are in superior social position may assert and "exert their status precisely by refusing to read mental states of others."[9] Mindreading obtuseness can function similarly to strategic ignorance: "it is the interlocutor who has or pretends to have the *less* broadly knowledgeable understanding of interpretive practice who will define the terms of the exchange."[10] The powerful, writes Rebecca Solnit, "swathe themselves in obviousness in order to avoid the pain of others and their own relationship to that pain. There's a large category of acts hidden from people with standing: the more you are, the less you know."[11]

On a more personal (and, hopefully, less insidious) level, I can think of other situations in which one may refuse to read minds of others to assert one's power over them. For instance, I am aware of not wanting to look too closely into what my grade-school son and his friends may be thinking when I prevent them from doing something that they want to do, because I think that it is dangerous or inconvenient or that we don't have enough time. By choosing to be a bad mindreader, I construct myself as a figure of parental authority, not a happy or optimal stance but one that may get me through a busy afternoon.[12]

Incidentally, what an "American White Middle Class (WMC)"[13] parent may guiltily characterize as "bad" mindreading, a Western Samoan parent may consider as a prosocial pedagogical measure. For instance, as the linguistic anthropologist Elinor Ochs explains, in Samoa, it is the responsibility of a lower-ranking person (e.g., child) to make their perspective clear to a higher-ranking person (e.g., adult). Ochs does not talk about mindreading as such, focusing instead on utterance interpretation, but the status-sensitive dynamic of "perspective-taking" that she describes maps well onto our present distinction between high- and low-status mindreaders:

> In [a highly stratified] Samoan society, sib and parental caregivers work hard to get children, even before the age of two years, to take the perspective of others. This demeanor is a fundamental component of showing respect, a most necessary competence in Samoan daily life. . . .
>
> In Samoan interactions the extent to which parties are expected to assume the perspective of another in assigning a meaning to an utterance of another varies with social rank. In speaking to those of lower rank, higher ranking persons are not expected to do a great deal of perspective-taking to make sense out of their

own utterances or to make sense of the utterance of a lower ranking interlocu or. Higher ranking persons, then, are not expected to clarify and simplify for lower ranking persons. For example, caregivers are not expected to simplify their speech in talking to young children. . . . And exactly the reverse is expected of lower ranking persons. Lower ranking persons take on more of the burden of clarifying their own utterances and the utterances of higher ranking interlocutors.[14]

All this said, would you want to become a better mindreader? If a blunted interest in other people's intentions denotes your higher social standing, shouldn't you be grateful for this status quo and not aspire to a greater mindreading perspicacity?

But here is something else to consider. Better mindreading may be associated with relative powerlessness and social stress, yet it also can be experienced as—and, indeed, become—a source of power on its own. Consider Héctor Tobar's meditation on growing up, unbeknownst to him, in the same community with James Earl Ray, the future killer of Martin Luther King Jr.: "Whereas Ray denied any commonality with the black people around him, I believe I have no choice but to study the white people around me, and to understand them as part of my American story—even men and women who hate and slander my people. Like many other Latino residents of this country, I derive a sense of power from observing the lives of people who cannot see the full measure of my humanity."[15]

While Ray (arguably) maintained his social superiority by refusing to read the minds of the Black people around him, Tobar (arguably) conformed to his lower social standing as a Latino by making an extra effort to "understand" the white people who refused to see him as fully human. Yet Tobar felt empowered by his interest in their intentionality, and, in the long run, his commitment to understanding and describing other people's complex subjectivity has fueled his career as an acclaimed writer, while Ray's white-supremacism-driven lack of interest in mindreading turned him into an outcast and a murderer.

Hence, one way of looking at these two outcomes is to register the role of status-sensitive mindreading in the perpetuation of oppression and discrimination. Another is to note the availability of professions (e.g., writer, lawyer, psychologist, manager) that require strenuous mindreading efforts and, as such, may serve as means of elevating one's social standing. I will address the subject of institutional venues that reward active mindreading in chapter 4 of this book; here, I merely want to point out that a "better" mindreading skill is not an unmixed blessing in a postmodern industrial society marked by racism and inequality.

2.3   How It Works in Literature: Two Models

In any given work of literature, some characters may carry on complex mindreading reflections, whether explicitly spelled out or not, while others settle for simpler ones. In deciding (not necessarily consciously) which will do which, writers may end up correlating their characters' social status and their mindreading ability. (This "may" is important, because writers may also end up *not* correlating the two: the pattern I am describing here is far from universal.) There are two ways of doing so. The first—let us call it the first model—is that writers follow the real-life dynamic and make characters of lower social standing capable of embedding more complex mental states than those above them are. The second—let us call it the second model—has writers invert that dynamic, making those who are high in the social hierarchy also high in the mindreading hierarchy.

Here is what these two models do not predict:

- They do not guarantee that the high-embedding character will be correct in their attributions of mental states. For instance, Jane Austen's Emma embeds complex mental states regularly as she plots her love matches, yet, just as regularly, she is wrong.

- They do not define characters' ethics. As the cognitive literary critic Blakey Vermeule has shown, crafty villains can be "masterminds" carrying on triple or even quadruple mental embedments.[16]

- They do not map neatly, or at all, onto the familiar literary-critical distinction between "round" and "flat" characters.[17]

- Finally, they do not say anything about the aesthetic value of the text. A work of fiction can follow either model, or it may not. Indeed, in some texts, such as Jennifer Egan's *A Visit from the Goon Squad*, social hierarchies are fluid—as it were, *intersectional*—so it is not clear at any given point which aspect of a character's social standing (gender, race, clout, or salary) ought to be considered as predictive of their relative capacity for complex embedment.

Here is something that only the second model can predict. When a writer portrays people in weaker social position as less capable of complex embedment than people in stronger social position, it may be indicative of a particular ideological agenda on their part. They may be anxious about their own position in the class hierarchy or wanting to please a particular segment of their readership who would prefer to see social inferiors who "know their place." Agendas vary. We may speculate about them (as I do in the sections that follow) and never learn the truth. Still, when a writer seems to have opted for the second model—the one that inverts the real-life correlation between low social standing and more active mindreading—it alerts us to a possible point of tension bound to a specific historical moment. This is one of many occasions on which historically minded literary scholars and cognitive literary theorists may benefit from each others' insights.

Before I turn to a series of case studies representing either of the two preceding models, I want to remind you that the pattern that I am discussing here is far from universal. Writers may *not* foreground the difference between their characters' capacity for contemplating complex mental states, or, if they do end up foregrounding this difference, they may *not* correlate it with characters' social standing. This is to say that factors other than social status (along the lines of class, race, or gender) may influence the author's intuitive decision to make one character more sociocognitively complex than another.

Consider Lev Tolstoy's *War and Peace* (1869), which tells the history of several Russian aristocratic families against the background of the Napoleonic Wars. Its characters include Napoleon Bonaparte as well as Russian Field Marshal Mikhail Kutuzov, whose decision to let Napoleon occupy the abandoned Moscow, in September 1812, led to the eventual demise of the French army. Tolstoy makes both Napoleon and Kutuzov contemplate Moscow just as it is about to be taken over the by French, but if Napoleon's thought processes run along the lines of, "a city occupied by an enemy is like a girl who lost her innocence,"[18] Kutuzov is thinking about the complex social dynamics engendered by the place's vulnerability. Thus, he is aware that, when other Russian generals feel compelled to keep talking about defending Moscow, they do it not because they believe that it can be done (i.e., just like him, they know that it "cannot be defended") but because, for them, this kind of talk creates a fine "pretext for quarrel and intrigue."[19] Tolstoy's portrayal of Kutuzov as significantly more sociocognitively complex than Napoleon reflects not the difference in these characters' social standing (which would be hard to define) but the author's patriotism and his hatred of "the Corsican monster."[20]

I do not want you to think, based on this example, that Tolstoy *never* correlates his characters' social standing with their ability to embed complex mental states. The question of whether he does is an empirical one and can be explored, if a critic is so inclined. I just want you to observe that the intuitive decision, on the part of an author, to make some characters more sociocognitively complex than others may be influenced by a wide spectrum of factors, ranging from personal political preferences to conventions of the genre (e.g., a sympathetic double agent in a spy thriller may be expected to embed mental states on a higher level than her counterpart from an opposing side does). The two models that I discuss here by no means exhaust the scope of possibilities open to a writer, although they do provide a fascinating glimpse into literature's experimentation with real-life social dynamics.

2.4 The First Model: Reflecting the Real-Life Mindreading Dynamic in *Mansfield Park*

The protagonist of Jane Austen's novel *Mansfield Park* (1814) is female, young (merely a child when she first enters the house of her rich relatives), and poor—a charity case with no obvious claims to beauty or intelligence. To survive and thrive in social circumstances stacked against her so thoroughly, she has to be particularly attuned to other people's wishes and intentions, and so she is. Again and again, the "little" Fanny Price is placed on the top of the mindreading chain, in direct inversion of her social position vis-à-vis her relatives and acquaintances.

One of several ways in which Austen accomplishes this inversion is to first present us with a seemingly complete scene, outlining everyone's embedded feelings—which seem complex enough, for the time being—and then superimpose Fanny's mind on top of that scene. For instance, when Fanny's cousins and their guests—the golden youth of Mansfield Park—embark on their ill-conceived theatrical production, we learn that Julia Bertram is jealous of her sister Maria, who is clearly preferred by Henry Crawford; that Maria ignores Julia's feelings; and that Julia hopes that Maria's fiancé, Mr. Rushworth, will become aware of the impropriety of her behavior and expose her to public humiliation: "[Julia] was not superior to the hope of some distressing end to the attentions which were still carrying on there, some punishment to Maria for conduct so shameful towards herself as well as towards Mr. Rushworth. . . . Maria felt her triumph, and pursued her purpose, careless of Julia; and Julia could never see Maria distinguished by Henry Crawford without trusting that it would create jealousy, and bring a public disturbance at last."

To this mix of second- and third-level embedments, Austen then adds Fanny's awareness of Julia's feelings, while also making sure that there is no reciprocal awareness (and hence comparable complexity) on Julia's side: "Fanny saw and pitied much of this in Julia; but there was no outward fellowship between them. Julia made no communication, and Fanny took no liberties. They were two solitary sufferers, or connected only by Fanny's consciousness."[21]

Fanny's consciousness is indeed the place where various characters get "connected" or, to put it differently, where many of the novel's fourth-level embedments take shape. To spell one of them out (an exercise that typically results in painfully pedestrian prose, for, in the original text, those high-level embedments are often implied rather than laid out in their full propositional glory), we can say that Fanny *knows* that Julia *is miserable* because Julia *knows* that Henry *likes* Maria. We can further say that Fanny *intuits* that Julia *hopes* that Mr. Rushworth *will realize* that Maria's behavior *makes* people around them *think* that he is a fool and revenge himself on her and that, though otherwise compassionate toward Julia, she can't quite find it in herself to empathize with this particular hope of her cousin's.

Change of scenery. Maria marries Mr. Rushworth and reconciles with Julia, and both sisters leave Mansfield Park. The passage that I am looking at now takes place after Henry Crawford proposes to Fanny, is rejected, and decides to convince her to reconsider. During a quiet evening in a Mansfield drawing room, after Fanny, her aunt Lady Bertram, her cousin Edmund, and Henry have been talking together for some time, Henry turns to Fanny to inquire more closely about her involuntary response (i.e., a shake of the head) to something that he just said. Edmund, who approves of Henry's courtship, wants to make it easier for Henry to talk to Fanny privately. Accordingly, he takes up a newspaper and removes himself from the general conversation. Lady Bertram, he knows, won't be in Henry's way because she rarely thinks of anything other than the convenience of her favorite pug.

Once again, Fanny's perspective is added *after* the scene has been set, for, much as she wants to come across as focused solely on her needlework, she can see what Edmund is doing with that newspaper: "[As] Edmund perceived, by [Henry's] drawing in a chair, and sitting down close by her, that it was to be a very thorough attack, that looks and undertones were to be well tried, he sank as quietly as possible into a corner, turned his back, and took up a newspaper, very sincerely wishing that dear little Fanny might be persuaded into explaining away that shake of the head to the satisfaction of her ardent lover. . . . Fanny . . . grieved to the heart to see Edmund's arrangements."[22]

Fanny's capacity for complex embedment contrasts starkly with that of Lady Bertram, seated right next to her (who seems incapable of embedding thoughts and feelings above the second level), but also with that of the two young men. Henry wants to know what Fanny disapproves of. Edmund knows that Henry wants to know what Fanny disapproves of. Fanny, however, knows that Edmund knows that Henry wants to know what Fanny disapproves of. To put it starkly, in terms of embedded intentionalities, Henry has intentions regarding Fanny; Edmund is aware of Henry's intentions regarding Fanny, but Fanny is aware of Edmund's intentions regarding Henry's intentions regarding herself. Here, as on many other occasions, "the dear little Fanny" is one or two mental states ahead of whichever Bertram or Crawford happens to be at hand.

In scenes that do not immediately involve Fanny, characters' ability and willingness to imagine other people's mental states is recalibrated to reflect their immediate power relations. For instance, excited about the theatrical production, Tom Bertram chooses not to understand Edmund's warning that Maria is about to dishonor their family (i.e., by developing a relationship with Henry Crawford while about to be married to Mr. Rushworth). When Edmund invites Tom to *consider* their mutual *awareness* of Maria's growing *disregard* for her fiancé's *feelings* (as he puts it, "to attempt [private theatrics] would be imprudent, I think, with regard to Maria, whose situation is a very delicate one, considering everything, extremely delicate"),[23] Tom ignores that invitation and insists that the play will entertain their mother.

To quote Eve Kosofsky Sedgwick, Tom "pretends to have the *less* broadly knowledgeable understanding of interpretive practice," yet he is the one who will "define the terms" of their conversation. As the older brother operating within the system of primogeniture, he can afford to be obtuse when it suits him, while Edmund must keep honing his younger brother's skill of being convincing without giving offense. On a gentleman's estate, mindreading hierarchies reflect the social pecking order.

2.5   The First Model in Pre-Revolutionary China and Russia

We now turn to authors from very different cultural traditions. In the eighteenth-century Chinese classic *Dream of the Red Chamber*, by Cao Xueqin (ca. 1750–1760), girls and young women typically embed mental states on a higher level than rich men and older rich women do.[24] Moreover, although these female characters are beautiful, accomplished, and pampered by their families, they are powerless. Their fates are decided by their elders, who cannot—and will not—read their emotions and, consequently, doom their young charges to lives of misery or to early deaths.

The striking mindreading skills of Cao's young women stand out in the long history of the literary response to social stratification in premodern China. As Haiyan Lee observes, "[In societies] structured by kinship sociality . . . theory of mind is certainly present and useful but not always prized in social life and does not animate expressive culture to the same extent [as it does] in modern commercial societies structured by stranger sociality, cosmopolitanism, and social mobility. . . . The hierarchical structures of [kinship sociality] place a greater premium on theory of mind for subordinates than for the powerful, hence attaching a tinge of opprobrium to its exercise."[25] When subordination follows the lines of gender, mindreading acumen—configured as cunning— follows closely: "Women in a patriarchal and patrilineal society, especially young daughters-in-law, are structurally motivated to be inward-looking, to adopt a calculating, fawning, and defensive mentality, and to orient their action around the intentions of the more powerful (senior, male) members of the kin group."[26]

Fawning, defensive, and calculating underlings, female or male, do not make for sympathetic fictional characters, which is why such personages tend to "ply shady trades as go-betweens, procuresses, litigation masters, soothsayers, brokers, and garden-variety hangers-on who prey on the honest and unsuspecting." Yet, as Lee argues, "[In some] exceptional circumstances . . . mind-reading becomes an asset and the consummate practitioner is admired and celebrated as a cultural hero. Most of these circumstances involve forces of good combatting forces of evil, as in warfare or criminal investigation. More rarely, theory of mind is mobilized to emplot romantic courtship."[27]

In other words, we can read the literary history of premodern China as punctuated by the appearance of works that valorize a character's capacity for complex embedment of mental states. Those works include warfare chronicles (such as Luo Guanzhong's fourteenth-century *The Romance of the Three Kingdoms*) and detective novels (such as the eighteenth-century case studies of Judge Dee), as well as the bildungsroman-courtship-novel extraordinaire *Dream of the Red Chamber*. Although some of *Dream*'s young women (most obviously, Wang Xi-feng) still come across as defensive and calculating, most are true to the ideal that the middle-aged Cao set out to bring back to life, after finding himself one day, in low spirits, "thinking about the female companions" of his youth: "As I went over them one by one, examining and comparing them in my mind's eye, it suddenly came over me that those slips of girls—which is all they were then—were in every way, both morally and intellectually, superior to the 'grave and mustachioed signior' I am now supposed to have become."[28]

And so, in direct contrast to the young women of, for instance, the anonymous late sixteenth-century classic *The Plum in the Golden Vase*, whose sharpened capacity for high-level embedment of mental states makes them cheats, liars, and hypocrites,[29] the cognitive complexity of the girls from *Dream* manifests itself in their admirable social sophistication and poetic sensibility. Far from damaging their personalities, their subordinate status lends poignancy to their moral and intellectual superiority.

Let us cross national boundaries again. If we look at Russian literature before the 1760s (that is, before Russian writers became exposed to western European models, a topic that I discuss at some length in chapter 5), we see something very similar to what Lee describes as the association of such complexity with "pipsqueaks," that is, with socially insignificant personages who, nevertheless, manage to create problems for "gentlemen."

There is, for instance, Frol, from the anonymous *The Tale of Frol Skobeev* (1680–1720), a social nonentity who rises to wealth and nobility by thinking one step (i.e., one mental state) ahead of various aristocratic figures who come his way. Frol is a pettifogger (remember Lee's observation that a social nonentity may use his mindreading skills to become a "litigation master"?), who tricks the only daughter of a rich courtier into sleeping with him (by crossdressing as a woman) and then elopes with her. When the distraught parents find out what has happened, they first want to prosecute the rogue but then relent and start showering the young couple with land and money, all the while cursing their "thief" and "knave" of a son-in-law.[30]

They relent because Frol knows how to manipulate their feelings. When they send a servant to inquire about the health of their child, Frol asks his wife to pretend to be sick and tells the servant, "See for yourself, my friend, how she's doing: that's what parental wrath does—they scold and curse her from afar, and here she is, dying."[31] Frol *wants* his parents-in-law to *think* that their *anger* is killing their daughter, a stratagem that quickly cools their wrath and sets Frol on the way to prosperity.

Critics consider *The Tale of Frol Skobeev* an early example of Russian picaresque.[32] Viewed in the context of the present argument, this characterization raises the intriguing possibility of a cognitivist reading of the literary figure of the picaro.[33] From Mateo Alemán's *Guzmán de Alfarache* (1599–1604) to Daniel Defoe's *Moll Flanders* (1724), picaros use their superior mindreading skills to flatter, bully, cheat, and steal their way to economic survival. They are simultaneously a threat—to the extent to which their society still retains traces of "kinship sociality" (and what society does not?—even if just in the form of cultural fantasies about a golden age, when all behavior was transparent and prosocial and no mindreading acumen was called for)—and a treat for readers who follow their double-dealing tricks with guilty delight.

We find the association between characters' low social status (low, that is, in relative rather than absolute terms: always in comparison with someone else in the story) and their heightened capacity for complex embedment in a broad spectrum of fictional narratives. Some characters embed complex mental states as they mastermind a plot to help their bumbling masters, as do "clever slaves" of ancient Greek and Roman comedies. Some do it as they trick a larger or more violent and dangerous animal in order to save their lives, as do Brer Rabbit of West African folklore and the little mouse of Julia Donaldson's *Gruffalo*. Some seem to lack any agenda and merely display a mastery of innuendo beyond that of their social "betters," as does Algernon's servant Lane in Oscar Wilde's *The Importance of Being Earnest*.[34]

Some have central billing, as does P. G. Wodehouse's Jeeves. Others make only brief appearances in one scene, as Wilde's Lane. Still others, such as the office cleaners from Rachel Cusk's *Saving Agnes* (1993), are episodic characters who lack any identifying features and manage to outclass the main protagonist in the business of mindreading while remaining nameless and faceless:

> Agnes slammed into the house in a state of considerable distemper. She had been forced by the nonchalance with which the editorial department was approaching its deadline to stay late in the office, working alone while the cleaners emptied bins and vacuumed floors around her. Watching them sanitize the unsavory detritus of her day she had been besieged by feelings of shame and guilt, and had attempted to engage them in pleasantries. Not beguiled by her condescension, however, they had roundly rebuffed her overtures and left her feeling that a mysterious exchange of power had taken place, the precise manoeuvres of which she was not able to fathom.[35]

If we map out this "mysterious exchange of power" in terms of its underlying mental states, we can say that Agnes *wants* to make herself *feel better* by engaging in small talk with the cleaners (second-level embedment). The cleaners, however, *know* that she *wants* to use them to make herself *feel better* (third-level embedment) and refuse her that satisfaction. As Agnes apparently expects that her class privilege will automatically translate into superior social acumen (even though she can't see the cleaners as people with faces and names), when their conversation doesn't follow that scripted path, she is left disoriented and angry.

What this example from Cusk's novel shows is that, just as in real life, fictional mindreading hierarchies are situation specific. Our common sense suggests that a protagonist would always be more capable of complex embedment than would be a minor character, if only because what makes them the protagonist is their involvement in the great many social interactions depicted in the story. So if we would merely count the occasions throughout the novel on which Agnes embeds mental states on a high level (which, I hope, we would never do, because that would be incredibly tedious!), there is no doubt that the number of those occasions would trump the number of occasions on which a given episodic character (e.g., an office cleaner, who only appears once) embeds mental states on a high level. But if, instead of thinking in such cumulative terms, we look at specific scenes, we may discover patterns that have less to do with the protagonist's outsize role in the plot and more to do with the novel's engagement with its ideological and generic contexts.

2.6   Race and Embedment in *Invisible Man*

As Ralph Ellison was reflecting, in 1981, on his experience of writing *Invisible Man* (1947), he explained that he had wanted to "create a narrator who could think as well as act." Too many "protagonists of Afro-American fiction" of his day, he felt, "were without intellectual depth, . . . seldom able to articulate the issues which tortured them." Real-life models for individuals who could think were not lacking, but even if they were, "it would be necessary, both in the interest of fictional expressiveness and as examples of human possibility, to invent them."[36]

Other writers, after all, were not shy about inventing deep self-reflexivity for social groups of their choice. Henry James, Ellison observed, had done just that. He had taught his readers "much with his superconscious, 'super subtle fry,' characters who embodied in their own cultured, upper-class way the American virtue of conscience and consciousness."[37] Ellison saw his task as "revealing the human universals hidden within the plight of one who was both black and American," and he considered a crucial step toward that revelation endowing his protagonist with a capacity for "conscious perception" of forces acting on him within and without. As he put it, "[To] defeat [the] national tendency to deny the common humanity shared by my character and those who might happen to read of his experience, I would have to provide him with something of a worldview, give him a consciousness in which serious philosophical questions could be raised."[38]

I find it significant that Ellison was thinking of Henry James as he contemplated ways to give his protagonist a complex and expansive consciousness, particularly in the light of what I am about to show you regarding Invisible Man's capacity for embedment. I have to confess, however, that, so far, I have avoided any references to James, because quoting him feels like cheating. James is one author about whom it can be said that he embeds third-and fourth-level mental states in every single sentence, and I believe that what I have to say is more convincing if I shun such easy targets. I do not want my readers to think, "Well, yes, James, of course, but he is exceptional." When a culture has arrived at the point when its literature cannot function anymore without constantly embedding mental states on at least the third level, prose like James's represents this general tendency, albeit taken, perhaps, to one of its endpoints. It is thus paradigmatic rather than exceptional.

But, guard my argument as I did, James still came in, riding as it were on the coattails of Ellison. So let us establish one thing about both James's and Ellison's representations of fictional consciousness. While the unceasing complex embedment of mental states may not be a sufficient condition for creating James's "superconscious" characters and implied reader, it is a necessary condition. And similarly, while making Invisible Man conspicuously capable of embedding complex mental states may not be a sufficient condition for endowing him with intellectual depth, it may be a necessary condition.

Lest we wonder how Invisible Man's capacity for complex embedment squares with his naiveté, recall that much of mindreading is mind-misreading. Mindreading is a process of *attributing* mental states rather than of telepathic discernment. In fact, as far as mindreading goes, telepathy is its opposite because this fantastic concept presupposes that mental states are *actually there* in people's minds, available for perusal both by the owners of those minds and by those who happen to have the special powers.

Embedding mental states on a high level thus does not make a character particularly penetrating. (It *can*, but it doesn't have to. Just think of how spectacularly misguided James's characters often are.) Instead, this is one way in which literature, as we know it today, signals complex consciousness to its readers, indeed, how it asserts "the common humanity shared by [the] character and those who might happen to read of his experience."

Hence, Ellison's protagonist has a compelling consciousness not when he knows what people around him are thinking—he mostly does not!—but when he allows them to have intentions that are mystifying to him and to themselves. To the extent to which he wonders about their mental states, he *sees* those people.[39] And, conversely, to the extent to which they refuse to wonder about his mental states, they do not see him. As he puts it, Jack, Norton, and Emerson each attempted "to force his picture of reality upon me and neither giving a hoot in hell for how things looked to me."[40] This is to say that they remain willfully blind to the unpredictable complexity of his feelings, which translates, in practice (for, again, we are talking here about practical ways in which literature can represent complex consciousness!) into their inability to embed mental states on a comparably high level.

Here is a scene, at the end of the book, in which Invisible Man becomes aware of the confused perspective of people who have tried, at different times, to control him without actually seeing him. As the leader of the riot

in Harlem, "Ras the Destroyer," commands his men to seize and hang Invisible Man—to punish him for what they think of as his treacherous collabo-
ration with the white people against the Black—the protagonist meditates
on the levels of unknowing that drive the events of this night:

> I looked at Ras on his horse and at their handful of guns and recognized the
> absurdity of the whole night and of the simple yet confoundingly complex
> arrangement of hope and desire, fear and hate, that had brought me here still
> running, and knowing now who I was and where I was and knowing too that I
> had no longer to run for or from the Jacks and the Emersons and the Bledsoes
> and Nortons, but only from their confusion, impatience, and refusal to recognize
> the beautiful absurdity of their American identity and mine. . . . And that I a
> little black man with an assumed name should die because of a big black man
> in his hatred and confusion over the nature of a reality that seemed controlled
> solely by white men whom I knew to be as blind as he, was just too much, too
> outrageously absurd.[41]

This is a very complex passage, and there are several ways to map out its
implied embedments. Here are some of them. The protagonist is keenly
aware that Ras ("a big black man") *doesn't realize* that his reality is being
controlled by white men who themselves are *confused* about the significance of their actions. The protagonist *realizes* that he will no longer be
afraid of or controlled by the people who are *confused and impatient*. By
calling the situation "absurd," he is *aware* that someone capable of a large-
scale perspective (God? History?) would not be *able to see* any meaning in
his death were he to die because of other people's *confusion and impatience*.

Moreover, the protagonist's self-description as "a little black man with an
assumed name" brings to mind not just the "little Fanny" of *Mansfield Park*
but also various picaros and "pipsqueaks" who change their names to survive in a hostile world.[42] Like them, Invisible Man has social circumstances
lined up against him: he is young, poor, and Black in the Jim Crow United
States. And, also like them, he makes his way in the world by actively trying to understand other people's perspectives. He often fails,[43] but he never
stops trying because, unlike people in superior social positions, he can't
afford to remain willfully blind to the subjectivity of others.

Here, for instance, is Invisible Man entering the lobby of the Mens
House, wearing his overalls—which indicate his descent to working class—
and *imagining* people *thinking* that he has *lost his pride* as an upward-bound
college student and, moreover, has betrayed their *expectations* of him: "I
could feel their eyes, saw them all and saw too the time when they would

know that my prospects were ended and saw already the contempt they'd feel for me, a college man who had lost his prospects and pride. I could see it all and I knew that even the officials and the older men would despise me as though, somehow, in losing my place in Bledsoe's world I had betrayed them. I saw it as they looked at my overalls."[44]

Or consider the conversation between Jack and Invisible Man during which Jack reproves him for having organized a mass funeral for Tod Clifton, who was murdered by a policeman. Although Jack repeatedly assures Invisible Man that "he knows what [he] feels,"[45] we end the scene convinced that Jack does not really see him—that is, cannot or will not conceive of him as someone with complex subjectivity. For instance, when Jack hopes that Invisible Man would never find himself in circumstances in which he would have to sacrifice his eye and get an artificial one (as Jack had, in service of the Party), Invisible Man responds with a complex embedment that Jack does not seem to understand. Here is their exchange:

> "Good," he said. "I sincerely hope it [i.e., losing an eye] never happens to you. Sincerely."
>
> "If it should, maybe you'll recommend me to your oculist," I said, "then I may not-see myself as others see-me-not."
>
> He looked at me oddly then laughed. "See, Brothers, he's joking. He feels brotherly again. But just the same, I hope you'll never need one of those."[46]

Jack treats Invisible Man's remark as a joke instead of recognizing it as a biting comment on the selective blindness that enables him not to see, or acknowledge, the complex subjectivity of his "brother." Of course, to recognize it as such a comment, he would have to unpack its soaring levels of embedment. This, after a moment of consideration, he decides not to do. (The narrator indicates that moment by saying that Jack looked at him "oddly").

Jack is hardly a stupid man, so his decision may be a strategic one. His standing as a high-ranking member of the Communist Party is not set in stone. It has to be maintained and defended—for instance, when a charismatic and intelligent Black "brother" appears on the scene. Racism is a powerful factor that would keep Invisible Man in his place, yet alone it may not be enough, especially given the Party's ostensibly egalitarian outlook. So mindreading obtuseness comes in handy, for whoever pretends to know less will (to return once more to Sedgwick) "define the terms of exchange." Jack, by showing that he doesn't need to bother to understand Invisible Man's meaning, seeks to reassert his superiority over Invisible Man.[47]

The Communist Party may thus claim to be color-blind, but it ends up mind-blind, which serves a rather different purpose: that of keeping some of its "brothers" down.

2.7   The Second Model: Inverting the Real-Life Dynamic

Here is what we have done so far. We have looked at fictional case studies in which relative capacity for complex embedment tracks the real-life correlation between weaker social position and more active mindreading. Again, please remember that this correlation gets reimagined in literature in a very particular way. Instead of writers making their downtrodden characters into straightforwardly "better" mindreaders—that is, more perceptive and accurate in their attribution of mental states—they make them into high-level embedders. "Better" mindreading may occasionally happen too, but it's not guaranteed; overthinking others' intentions may just as well lead to one's undoing.

And so I have shown that the young women from Cao's *Dream of the Red Chamber*, the Russian picaro Frol Skobeev, Austen's Fanny Price, and Ellison's Invisible Man all consistently embed mental states on a higher level than do other characters around them who have more power and social clout. Indeed, some of those characters, such as Tom Bertram or Jack, may reaffirm their clout by refusing to navigate complex embedments offered up to them by people in weaker social positions, such as Edmund Bertram or Invisible Man.

I now turn to literary texts that do the opposite. That is, they invert the real-life correlation between lower social standing and active mindreading and portray socially disadvantaged characters as *not* being able to embed complex mental states on the high level of their "betters." I further suggest that, more often than not, such an inversion indicates a particular ideological agenda on the part of the author and that those agendas may range from tacit personal anxiety about one's social status to a fear for one's life, when one happens to be a writer living under a totalitarian regime.

2.8   The Second Model: Bakhtin and the English Comedy of Manners

My first example of the "inverted" model comes from Frances Burney's *Evelina* (1778). *Evelina* is an epistolary novel that, over the past two decades, has become a staple for college courses on eighteenth-century British literature. Written when the author was in her middle twenties, it portrays a beautiful young woman brought up in rural seclusion and thrust onto London's bustling social scene. The story has some dark streaks (those will become more prominent in Burney's later work), but it is largely a comedy of manners. As such, it tends to go over well with undergraduates who enjoy following the romantic adventures of a satirically inclined naïf in a big city.

As befits a romantic heroine, Evelina is a princess in disguise. She is a daughter of a baronet, who abandoned her mother shortly after their marriage and burned the marriage certificate. This means that, though by birth and education she belongs to the aristocracy, her social status is ambiguous, at least until her father publicly acknowledges her as his legitimate heiress. Until that happens, she is subject to amorous advances by men from an unusually wide social spectrum, from tradesmen to aristocrats, each with his own way of speaking and pressing his suit.

In the scene that we are about to look at, one of those men, Mr. Smith, an offspring of shopkeepers who yet wishes to come across as a gentleman, is courting Evelina in a particularly obnoxious fashion. Earlier in the novel, he had invited her to a public ball at the Hampstead Assembly. Although she told him that she didn't want to go, he simply ignored her words and purchased tickets for both of them.

Presented with the tickets, Evelina doesn't just repeat her earlier refusal. Instead, she couches her response in such terms as to show her incompatibility with Mr. Smith.[48] He understands only part of what she says and can't respond properly. This proves her point, because men from the social class to which she anxiously defends her right to belong would have understood and responded in kind (even those of them whose courtship styles are offensive in their own ways).

Here is their conversation. Evelina has just reminded Mr. Smith that she had already told him that she wouldn't go to the Assembly.

> "Lord, Ma'am," cried he, "how should I suppose you was in earnest? come, come, don't be cross; here's your Grandmama ready to take care of you, so you can have no fair objection, for she'll see that I don't run away with you. Besides, Ma'am, I got the tickets on purpose."
>
> "If you were determined, Sir," said I, "in making me this offer, to allow me no choice of refusal or acceptance, I must think myself less obliged to your intention than I was willing to do."
>
> "Dear Ma'am," cried he, "you're so smart, there is no speaking to you;—indeed you are monstrous smart, Ma'am! but come, your Grandmama shall ask you, and then I know you'll not be so cruel."[49]

Evelina and Mr. Smith may as well be speaking two different languages, so loud is the clash of their sensitivities and the social incommensurability that it implies. Yet how is this impression created? That is, what tools do we have at our disposal to explain the rhetorical effect of their amusing exchange?

Eventually, as you can easily guess, I will ask you to look at the difference between Evelina's and Mr. Smith's patterns of embedment. But before we do that, let us consider another, more established and influential interpretive framework: Mikhail Bakhtin's notion of heteroglossia. For, I believe that, in this particular case, the two approaches work better in tandem than my "cognitive" approach would on its own.

Unlike other eighteenth-century authors, such as Fielding, Sterne, and Smollett, Burney was not on Bakhtin's radar when he wrote about the "heteroglot, multi-voiced, [and] multi-styled" language of the novel.[50] Still, her writing seems to exemplify what he called a comic style "of the English sort": one based on "the stratification of the common language" through the "stylistically individualized speech of characters."[51] Burney's first novel, in particular, uses heteroglossia in service of a particular ideology: the way Mr. Smith and Evelina talk underscores their immutable class positions.

Thus, in response to Evelina's polished sentences, the "low-born" Mr. Smith uses short, clipped clauses ("don't be cross") and vulgar expressions that brand him as a shopkeeper aspiring to sound genteel, such as "monstrous smart." His grammar is bad ("you was"). He betrays his crassness, by reminding her that he paid for the tickets ("I got the tickets on purpose"). It's all there, ready for the reader primed to look for sociolectal markers.

But in addition to those obvious markers, we also have here something less obvious, something we would not see without our "cognitive" perspective. Mr. Smith's embedments, both implied and explicit, stay around the second level, whereas Evelina spouts third- to fourth-level embedments one after another. Let us take another look at their exchange, now mapping its embedded mental states:

> "Lord, Ma'am," cried he, "how should I suppose you was in earnest? come, come don't be cross; here's your Grandmama ready to take care of you, so you can have no fair objection, for she'll see that I don't run away with you. Besides, Ma'am, got the tickets on purpose." (Who would *think* that you *meant* what you said? [*two embedded mental states*]. I *know* that you *worry* that there will be no chaperone [*two embedded mental states*].)
>
> "If you were determined, Sir," said I, "in making me this offer, to allow me no choice of refusal or acceptance, I must think myself less obliged to your intention

than I was willing to do." (I might have *felt bad* turning you down *had I thought* that you were *aware* of my *feelings* enough to care to give me a choice [*at least three, perhaps four embedded mental states*]. But because now I *know* that you wouldn't even *care* that I *don't want* to go, I *intend* not to *feel bad* about *disappointing* you [*two parallel sets of three embedded mental states*].)

"Dear Ma'am," cried he, "you're so smart, there is no speaking to you;—indeed you are monstrous smart, Ma'am! but come, your Grandmama shall ask you, and then I know you'll not be so cruel" (I *know* that you are *too smart* for me [*two embedded mental states*]. I *hope* you'll *listen* to your Grandmama [*two embedded mental states*]. I *know* that you will *agree* eventually [*two embedded mental states*].)

When I teach *Evelina*, I ask my students to compare Mr. Smith's pattern of embedment to that exhibited by two other characters, Sir Clement Willoughby and Lord Orville, who belong to the aristocracy, that is, the social class within which Evelina, a daughter of a baronet, will eventually be ensconced. Here are two typical examples of their speech. (I quote them out of context, but it is similar in both cases: each man wants to influence Evelina by disposing her more favorably toward himself.)

Sir Clement Willoughby: "You cannot even judge of the cruelty of my fate; for the ease and serenity of your mind incapacitates you from feeling for the agitation of mine!"[52] We may map this as, I *appreciate* that your *state of mind* makes it impossible for you *understand* how *unhappy* I am (at least three, possibly four embedded mental states).

Lord Orville: "I greatly fear that I have been so unfortunate as to offend you; yet so repugnant to my very soul is the idea, that I know not how to suppose it possible I can unwittingly have done the thing in the world that, designedly, I would wish to avoid."[53] We may map this as, You *must believe* that I am *distressed* to *realize* that I have made you *feel* precisely the way I would never *want* to make you *feel* (at least four embedded mental states).

Mr. Smith's limited capacity for embedding mental states is thus *dialogic*, another key concept from Bakhtin.[54] That is, we may experience it as limited only in contrast with the embedments of other characters, such as Evelina, Sir Clement Willoughby, and Lord Orville. Once we become aware of this contrast, we realize that it is used throughout the novel in two related but not identical ways.

First, it marks bona fide, as opposed to in-name-only, gentility. That is, "real" gentlemen and gentlewomen, such as Lady Howard, Mr. Villars, Mrs. Selwyn, and Mr. Macartney, who also happen to treat Evelina with kindness and respect, consistently embed mental states at and above the third level, while the nominally genteel characters who insult, ignore, and exploit her, such as Lord Merton, Lady Louisa Larpent, Mr. Lovel, and Captain Mirvan, stay around a lower (i.e., second) level.[55]

Besides marking "true" gentility, the differential capacity for embedding is also used to naturalize characters' social status. Shopkeepers and parvenus with shopkeeper mentality don't rise above the second level in their attribution of mental states. Thus, Evelina's low-born cousin, Tom Branghton: "There is nothing but quarreling with the women; it's *my belief* they *like* it better than victuals and drink."[56] Or her ex-barmaid grandmother, Mme. Duval: "*I've no doubt* but we shall be all murdered!"[57] Or Biddy Branghton: "I *wonder* when Mr. Smith's room will be ready."[58] If you consider the dismal treatment that these characters receive throughout the novel, it seems that the lack of capacity for embedding mental states on a high level marks pretty much everyone belonging to this class as not worthy of compassion or sympathy.

The capacity for embedment thus functions as a form of heteroglossia. It can be combined with other sociolectal markers, but only for those characters who are not capable of sophisticated layering of social consciousness. Thus, Tom Branghton's low-level embedments go hand in hand with contractions, clipped sentences, and colloquialisms: "Didn't you [hear of it], Miss? . . . Why then you've a deal of fun to come, I'll promise you; and, I tell you what, I'll treat you there some Sunday noon"[59]; Mme. Duval's, with contractions, double negatives, and bad French: "Pardie, no—you may take care of yourself, if you please, but as to me, I promise you I sha'n't trust myself with no such person."[60] Lord Merton, a newly titled nobleman who lacks true gentility, punctuates his first-level embedments with curses: "I don't know what the devil a woman lives for after thirty."[61] Captain Mirvan, another character whose behavior belies his nominal status of gentleman, sprinkles his second-level embedments[62] with sailor's lingo: "I am now upon a hazardous expedition, having undertaken to convoy a crazy vessel to the shore of Mortification."[63]

In comparison, the speech of unambiguously genteel characters is largely devoid of such markers. The only feature that is reliably present—and thus should be considered a marker in its own right—is the ability to embed mental states on a high level. "Can there, my good Sir, be any thing more painful to a friendly mind, than a necessity of communicating disagreeable intelligence?"; "I am grieved, Madam, to appear obstinate, and I blush to incur the imputation of selfishness"; "The benevolence with which you have interested yourself in my concerns, induces me to suppose you would wish to be acquainted with the cause of that desperation from which you snatched me"; "I am extremely sorry . . . that you think me too presumptuous"; "To what, my Lord, must I, then, impute your desire of knowing [my intentions]?"[64]

Lady Howard, Mr. Villars, Mr. Macartney, Lord Orville, and even Sir Clement Willoughby (except when he's trying to overwhelm Evelina with his dramatic professions of devotion and overblown terms of endearment) sound nearly interchangeable in their complex embedments. It is almost as if the relentlessly demanding pattern of such embedments were too metabolically costly for the text, leaving little energy for further verbal idiosyncrasies to be associated with these characters.

I said before that characters who function on the first and second level of embedment do not, as a rule, elicit much of readers' compassion. As one of my students put it, referring to the cruel prank that Captain Mirvan plays on Mme Duval, "I didn't care about Mme. Duval's suffering. It's one bad character playing a trick on another bad character."[65] It also works the other way around. The characters who are portrayed as being able to afford the cognitive luxury of consistently embedding mental states on this high level come across as more aware of their own[66] and, frequently, other people's feelings[67] and hence more deserving of readers' interest and sympathy.[68]

When my students read *Evelina*, they find it challenging to imagine that real-life eighteenth-century shopkeepers, when it came to their mindreading skills, were *not* inferior to ladies and gentlemen of leisure and that, if anything, their subservient position would have made them more active and perceptive mindreaders than those above them. I believe that I know at least one reason why this is so difficult for them. Burney's novel equates capacity for complex embedment with linguistic capacity. To come across as a sophisticated mindreader, her character must *sound* like one, but to sound like one, the character has got to have had a particular kind of education: no education, no eloquence, no mindreading complexity.

One can imagine an alternative scenario in which readers would *infer* a "low-born" character's complex intentionality, based on their behavior rather than on their words, but Burney does not let it happen. The closest that her novel comes to this is when a tradesman manages to get a free ride out of a gentleman. But then the gentleman demonstrates such tact in responding to this unappealing ploy that the tradesman's capacity for complex embedment is left, once more, in the dust.[69]

To put a sharper point on what Burney is doing here, let us revisit studies that establish the association between lower social standing and more active and perceptive mindreading. To quote from a recent review of those studies,

> A growing body of behavioural and self-report evidence suggests that people who are lower in social standing may be more socially attuned than those of higher social class. Lower social class is associated with greater activation in brain areas involved in understanding the mental states of other people. Working class people may devote more cognitive resources to processing social information and they may encode such information more deeply. Lower social class among college students was correlated with greater activation of the mirror neuron system. . . . Taken together these studies provide strong support for the notion that working class people are more socially attuned and that such attunement may be fairly automatic and visceral.[70]

Moreover, these effects have been observed across cultures, that is, in Russia and China, as well as in the United States.[71] This strongly implies that the situation was not that different in eighteenth-century England and that eighteenth-century English tradesmen did not, in fact, lag as hopelessly behind in their mindreading capacities as Burney is at such pains to demonstrate.[72]

How does one explain Burney's drastic reversal of this real-life mindreading dynamic? We may speculate that it reflected the Burney family's nervousness about their own social standing, for, unlike many other members of their social circle, they had to work for living. Granted, the work was intellectual and not manual, but, still, their survival depended on it.

We may also chalk it up to the young writer's willingness to rely on the conventional association between landed property and "social personality."[73] At least in this particular regard, Burney was perhaps not yet the Burney of her subsequent novels, who, as Margaret Doody puts it, would "examine" and "attack" rather than merely reflect "her society in its structure, functions, and beliefs," especially those pertaining to "social class."[74] Instead, *Evelina* soothed the nervousness of Burney's genteel readers about the incipient porousness of eighteenth-century class boundaries by inventing real-life mindreading dynamics and portraying tradesmen as stunted in their capacity for complex motivation and thus harmlessly amusing to their betters.

2.9   Shakespeare's "Problem Play"

How far back does the association of mindreading acumen with superior social standing go into English literary history? For, inaccurate as this association may be when it comes to real-life communication, it nevertheless took hold in the eighteenth-century popular imagination, informing certain genres of polite literature, such as sentimental plays and novels.

To reconstruct the genealogy of this association, one may turn to Restoration comedy, in which aristocratic wits, such as Dorimant from George Etherege's *The Man of Mode* (1676), embed mental states on the fourth and even fifth level, while their mistresses and hangers-on can barely keep up with them.[75] Granted, for many a Horner—the upper-class plotter from William Wycherley's *A Country Wife* (1675)—there is a Lucy: the clever servant, who steps in at a critical juncture to save her "betters" from catastrophe. Still, after the 1670s, aristocratic high embedders became a recognizable literary type, paving the way for the letter-writing sophisticates of Richardson and Burney. Restoration plays obviously came with their own political agendas—one of which was to please a series of royal patrons and their friends (who would consider themselves the greatest wits of them all)—which demonstrates yet another way in which ideology can drive the inverse-correlation model in fiction.

Going yet further back, one finds a ruler high on the sociocognitive spectrum in Shakespeare's *Measure for Measure* (1604). Shakespeare's men in power are not generally known for mindreading perspicacity, yet Duke Vincentio seems to derive a peculiar personal satisfaction from reading and scripting the complex emotions of his subjects. Thus, he wants Isabella to think that Angelo beheaded her brother, Claudio—even though Claudio is alive—so that, later, when she least expects it, he can reveal to her the true state of affairs and turn her despair into "heavenly comfort":

Isabella [Within]. Peace ho, be here!

Duke. The tongue of Isabel. She's come to know
    If yet her brother's pardon be come hither:
    But I will keep her ignorant of her good,
    To make her heavenly comforts of despair,
    When it is least expected.[76]

The Duke knows that Isabella will be devastated when she hears of her brother's execution. He also *knows* that she will be happy beyond measure when she learns that he is alive—*happier*, presumably, than she *would have been* had she not first *believed* that he is dead (fourth-level embedment). This is to say that the Duke is angling to put himself in a god-like position in which he will have complete access to Isabella's feelings both now and later (i.e., when the truth is revealed). His mindreading hunger is tinged with sadism, even as he wishes to bring Isabella's happiness to the highest pitch (a literary mindreading dynamic that I dub, elsewhere, "sadistic benefaction").[77]

Measure for Measure is considered one of Shakespeare's "problem plays." As Steve Vineberg puts it, "the long final scene can strike an audience as sadistic. . . . And when the Duke proposes marriage to Isabella, after all he's put her through, you may wonder what Shakespeare could have been thinking of." Directors deal with this problem differently. Some play up the Duke's emotional cruelty, showing that Isabella can't catch a break in the patriarchal society of Shakespeare's Vienna; others explain the Duke's behavior by his desire to see if Isabella is capable of generosity—of "moving beyond her own injuries to act on another's behalf"[78]—as when she kneels before the Duke to ask for Angelo's life while still believing that Angelo has killed her brother. However charitable toward the Duke, this reading still can't explain away his stated intention to plunge Isabella to the lowest depths of despair in order to render her subsequent joy more intense. He may claim that he does it for her own good, but he gets out of it an intoxicating fantasy of complete access to her feelings.

What I find striking about the ethical problem that the Duke's behavior presents is that it seems to be mainly *our* problem, rooted in our own particularly historically situated sensibility. Shakespeare himself may not have viewed the Duke's actions as objectionable. The reason I say this is that I can't discern even a hint of punishment meted out to this "sadistic benefactor." The Duke remains beloved by his subjects, and as the play ends, he is on the brink of being rewarded with a marriage to a much younger, beautiful, and virtuous woman. To paraphrase Hamlet, this is hire and salary, not acknowledgment of a problem.

So let us put aside our "ethical" response for a moment. Let's remember instead that real-life rulers stink at mindreading and that Shakespeare didn't need the research of contemporary cognitive psychology to know this, and neither did his audience.[79] This means that, for them, equating mindreading prowess with higher social standing may have had a different political meaning altogether. The space of the play allowed Shakespeare and his contemporaries to fantasize about their social betters who would care about their underlings' feelings so deeply that they would spend their time figuring out ways of getting inside their heads and scripting their emotions. For, as sadistic as this endeavor strikes us today, an early-modern subject might have actually been flattered by the thought of it and wonder if they might not have deserved more political attention from their rulers than they had been getting.

Is this the only possible reading of the Duke's unexpected sociocognitive complexity? Of course not. I don't aim to supply such a reading. Instead I want to stress that this complexity *is* unexpected—and must have been so for early seventeenth-century audiences—and that, more often than not, the association between the capacity for high-level embedment and high social status has specific political underpinnings.

Observe how using insights from contemporary cognitive science (such as the association of better mindreading skills with lower social status) can help us historicize our emotional response to a fictional character, a response that would otherwise seem obvious (as in, "The Duke is sadistic! Poor Isabella! What could Shakespeare have been thinking of?") and thus be ahistorical. A cognitive approach to literature, in other words, comes into its own when it combines insights from cognitive science with sensitivity to specific historical contexts (a paradigm known as "cognitive historicism").[80] My next set of examples comes from the time during which history trod with a particularly heavy step and when the punishment for not aligning the story's sociocognitive complexity just so could lead to the author's death.

2.10 In the Gulag's Vestibule

When, under oppressive political regimes, literary (and cinematic) production becomes explicitly regulated, mindreading sophistication acquires new ideological meaning. Thus, in fiction published in the Soviet Union under the aegis of socialist realism, characters of lower social status would sometimes be portrayed as *less* sociocognitively complex than characters of higher social status. That is, they do *not* engage in high-level mindreading when confronted with the machinations of high-status characters.

This may seem like an unambiguous example of the second model, but it is not. Although technically speaking, these low-embedding characters, such as unskilled factory workers, indigent peasants, and orphaned vagrant children, occupy the lowest rung of a socioeconomic ladder, they are not at all the "pipsqueaks" of yesteryear. Instead they are the new aristocracy— aristocracy of the spirit, as it were—even if they are never referred to this way. The future belongs to them. Due to their currently disenfranchised status, they are ultimately guaranteed privileged access to educational, political, economic, and reproductive resources. In contrast, various "old specialists" ("spetsy" in the half-respectful/half-contemptuous jargon of the 1920s–1930s), who have managed to parlay their education under the tsarist regime into lucrative high-status jobs under the Soviets, are doomed to irrelevancy and extinction. It is those well-heeled characters, as well as their repulsive young protégés, who cheat our low-status protagonists of their rightful share of socialist paradise, but not for long, never for too long.

For instance, Sania Grigoriev, the protagonist of a widely beloved novel by Veniamin Kaverin, *Two Captains* (1938–1945), is shown to be almost completely without guile, and so are his friends and his girlfriend/wife. It is his arch-adversary, a stockbroker under the old regime and school principal / distinguished scholar under the new, N. A. Tatarinov, and Tatarinov's favorite disciple, Romashov, who engage in complex mindreading aimed at destroying the hero. When, at the end of the book, Sania, a former-vagrant-child-turned-arctic-pilot, gains the upper hand, it is because of his determination, courage, and good luck and not because he has more cunning than his enemies do. In 1948–1956, Kaverin re-created this mindreading dynamic in another popular (and also repeatedly televised) novel, *The Open Book*, whose upright protagonist, a poor-scullery-maid-turned-famous-microbiologist, ultimately triumphs over her plotting adversaries. Their old-school Machiavellianism is no match for her talent and "open-book" personality.

Call it the first model with a twist. What we have here is our familiar correlation between lower social standing and high-level mindreading skills, except that low-status characters (i.e., the doomed bourgeois elements) may *initially* come across as high-status characters, while the downtrodden workers, peasants, and vagrants may take some time to reveal themselves as the new aristocracy. And this proletarian aristocracy presumably does not need to excel in mindreading, since the Revolution of 1917 has already stacked the socioeconomic odds in their favor.

Besides, the enemies of this proletarian aristocracy may not be that great at mindreading either. In Haiyan Lee's study of the fate of detective fiction in the People's Republic of China, she provides an important insight into a particular historically specific form that the literary association between high social status and low mindreading skills can take under the watchful eye of the Communist Party. As she explains,

> After the founding of the People's Republic in 1949, [the hitherto thriving] detective fiction was labeled a bourgeois conceit and suppressed. The new society was to be organized as a political communitas in which all were brothers and sisters under the benevolent paternal care of the Communist Party. Everyone had a designated place in society and everyone was a known quantity. Who would have any need for mindreading in such a seen-through society? . . . The only genre fiction permitted to flourish in the socialist period was the spy thriller. Crucially, the mind-game that sustained this genre was directed against "the class enemy," both internal and external. Still, enemy agents were not permitted to truly shine sociocognitively. Rather, they schemed and plotted at a low cognitive level, making laughably naïve assumptions and rudimentary blunders. And it took minimum twists and turns to ensnare them in the vast net of the people's justice.[81]

So while the proletariat had no need to "shine sociocognitively," their enemies were "not permitted" to do that. Did that result in decades of official literary production, in the Soviet Union and the People's Republic of China, with generally lowered levels of mindreading complexity, while works featuring truly sociocognitively complex characters had to find outlets elsewhere: abroad or in the underground/samizdat?[82] And did that mean that the sociocognitive complexity of narrators, implied readers, and implied authors had to be dialed down as well?

One factor that seems to bear out this conjecture is the suppression, in Soviet fiction, of the style of writing that we now describe as unreliable narration. Ilya Ehrenburg's *Julio Jurenito* (1922), Yuri Olesha's *Envy* (1927), and Konstantin Vaginov's *Works and Days of Svistonov* (1929) still featured unreliable narrators,[83] but once socialist realism became the dominant paradigm in the early 1930s, such stylistic experimentation was put paid to.[84] Thus, Vsevolod Ivanov's *U* (1932) was not published in the Soviet Union until 1988, while Leonid Dobychin's brilliant *The Town of N* (1935) was singled out for castigation during the 1936 campaign "against formalism and naturalism," driving its author to suicide. With the latter novel's move away from character-based embedment to embedment emerging almost exclusively from a give-and-take between the implied author and implied reader, it engaged in an experimentation with literary subjectivity that must have come across as politically subversive. Indeed, as one critic observes, it is "something of a mystery how the book was published at all at the height of Stalinism, when dogmatic conservatism, to say nothing of philistinism, ruled the art establishment."[85]

2.11   Socialist Realism: Turning Back the Clock on Complex Embedment in Literature

It is easy for us today to dismiss bona fide socialist realist literature as crude propaganda and a psychological "wasteland."[86] Yet, if we adapt the cognitivist perspective—that is, if we consider socialist realism as a culture-wide attempt to regulate people's mindreading practices—it emerges as a fascinating phenomenon, both politically and literary-historically.

What does it mean, for instance, that the Soviet literary scene could not abide the forms of complex embedment associated with unreliable narration? On the one hand, this seems to exemplify the regime's intolerance for experimentation associated with the modernist aesthetics. (After all, unreliable narration in literature is thought to be a mark of modernist sensibility.) Indeed, the socialist realist condemnation of "decadence" in poetry and art paralleled the crusade against "degenerate art" in Nazi Germany,[87] which implies that both communist and fascist ideologues experienced modernist experimentation with subjectivity as politically threatening.

On the other hand, the relationship between experimental aesthetics and political subversiveness is far from straightforward.[88] There are enough instances of brilliant avant-garde writing and filmmaking (think Mayakovski and Eisenstein) serving ideological agendas of totalitarianism and thus increasing the affective appeal of such agendas. Indeed, as the cultural theorist Sabina Hake has shown, film directors at DEFA (i.e., the main state film studio of the German Democratic Republic, formed in 1946 under the auspices of Stalinism and dissolved in 1992, after the reunification of Germany) used innovative techniques of modernism to maintain the attractiveness of various foundational myths of the GDR, such as the equation of socialism with antifascism. As Hake explains,

> Just as Georg Lukács's pronouncements on the nineteenth-century realist novel as the model for critical realism was used in the 1950s formalism debates to dismiss all modernist experimentation as decadent, the canonization of modernism in the West as inherently resistant has distracted from the affirmative functions of formal innovation. Concretely, in the case of DEFA cinema this means that

an uncritical reliance on the realism-modernism opposition has allowed us to equate filmic experimentation with political dissent. Just as the ideological effects produced by the antifascist classics in the socialist realist mode were never as uncontested as their detractors claimed, the turn to art-cinema traditions never implied automatic opposition to the ideological and institutional structures that relied on antifascism as its founding myth. On the contrary, modernist strategies and techniques often helped to liberate the affective core of antifascism from the ossifications of cinematic illusionism and to redeem the utopia of socialism in aesthetic terms.[89]

This means that if we want to understand why works of literature that foregrounded embedded intentions of implied readers and implied authors (e.g., those featuring unreliable narrators) did not fare well with socialist realist censors, we cannot simply say that such experiments with fictional subjectivity nurtured critical thinking and thus implied political dissent. While this may be true to a significant extent[90] (more about this in the next section), we may also look for a more immediate explanation, one rooted in the principles of socialist realism. What we find there, surprisingly or not, is a certain contempt for cognitive processes of the "proletarian" audiences, signaled by the effective return to what Hans Günther calls "the preliterate tradition" and constituting an intriguing experiment with patterns of embedment in modern literature.[91]

Socialist realist writers were expected to "educate" their readers and indoctrinate them in the ideological precepts of the Party. These goals, however, "could be realized only under the conditions of accessibility (comprehensibility) of literature and art for the popular readers and viewers, under the conditions of conformance to their taste."[92] This led, in practice, to the reclamation of the sensibilities of epic, with larger-than-life heroes engaged in monumental labor: harnessing the power of the machine to transform both the unyielding natural world and the unruly collective. Officially, socialist realist tradition was supposed to be following in the footsteps of the nineteenth-century greats, such as Pushkin and Tolstoy. But, in reality, as Günther points out, "[Insofar as] the nineteenth century distinguished itself by the predominance of a critical and analytical beginning, now images were needed that reflected the optimism of the official Stalinist culture, and these images were primarily sought in the preliterate tradition—myth, folklore, heroic epics, and the like. Paradoxically, a society with an officially declared orientation toward the future, in which the art of the avant-garde left indelible marks and that widely used modern means of communication in propaganda, directs its gaze toward the remote past the result of which was a quaint folklorization of modernity."[93]

Myth, as Günther observes elsewhere, thus emerged as "the soul of proletarian art."[94] What is important here, for our present purposes, is that myth, folklore, and heroic epics do not depend on continuous complex embedment of mental states to the same striking degree to which, say, a novel by Pushkin or Tolstoy does. While we certainly find third-level embedments of mental states in myths and fairy tales, as well as in epics, such as *The Epic of Gilgamesh* and *The Odyssey*, they are relatively rare there—in fact, incomparably so, if we juxtapose these texts with (for instance) eleventh-century Japanese, eighteenth-century Chinese, or nineteenth-century Russian novels.

This is why fiction produced under the aegis of socialist realism is so fascinating from the cognitive literary perspective. Many of its early flagship works, from Fedor Gladkov's *Cement* (1925) and Nicholai Ostrovski's *How the Steel Was Tempered* (1936) in Russia to Eduard Claudius's *People at Our Side* (1951) in East Germany, feature third-level embedments of mental states rarely, staying mainly on the first and second level. Yet, even with these novels' epic (so to speak) unconcern about embedded subjectivity, they can still be affectively engaging. What their presence on the literary scene demonstrates is that fiction did not have to go the route of the hypertrophied embedment and that it was not inevitable that the novel would become as dependent on continuous complex embedment of mental states as it is at the present point in our literary history.

Socialist realist novels, especially in 1920s–1930s Russia and in 1950s GDR,[95] thus represent an important and useful exception to my "rule" that literature, as we know it today, cannot function on a lower-than-third level of embedment. True, their popularity had been enabled by the powerful state apparatus (and, when that apparatus disappeared, they have been forgotten), but, then, many a canonical work of literature depends for its survival on a system of institutional supports. In any case, they were liked well enough by several generations of readers (to which I can attest, having encountered some of them as an adolescent), even when consumed alongside nineteenth-century novels featuring vastly more sophisticated embedment of mental states.

At the same time, it is politically significant that the project of "educating" the proletariat was thus realized by texts with drastically lowered levels of embedment. Think again of Ralph Ellison's decision to match

Henry James's "superconscious, 'super subtle fry,' characters who embodied in their own cultured, upper-class way the American virtue of conscience and consciousness."[96] To endow his Invisible Man with "intellectual depth" and "a consciousness in which serious philosophical questions could be raised," Ellison depicted him continuously embedding mental states on a high level.[97] In contrast, when adepts of socialist realism "conformed" to their readers' taste, they revealed themselves as thinking about those readers as not amenable to contemplating complex subjectivity.

One is reminded here of Burney's construction of working- and middling-class characters, in *Evelina* (1778), as lagging hopelessly behind their social "betters" in the business of mindreading, a construction driven by ideology rather than by real-life mindreading dynamics. Ironically, the socialist realist aesthetics went further and transcended the boundaries of fiction: in a state aspiring to be "a total work of art,"[98] neither characters *nor their readers* were "permitted to shine sociocognitively."[99]

Let us take a closer look at the so-called production novel (i.e., a novel set in an industrial collective) exemplifying these aesthetics. Fedor Gladkov's *Cement* (*Цемент*, 1925) tells a story of a Red Army soldier, Gleb Chumalov, returning home after the Civil War of 1918–1921 and struggling to restart the production of cement at an abandoned factory. Written in the early 1920s, *Cement* went through numerous revisions, which resulted in the drastic paring down of its characters' emotional range. (Indeed, the currently available English translation, published by Northwestern University Press in 1960, seems to be based on one of the earlier drafts and thus may not give the reader an accurate impression of what the novel had become in its last draft, the one most familiar to its Russian audiences.)

Here is an excerpt from the final version of *Cement*, coming from the chapter featuring one of the novel's most intense conversations about the characters' feelings. Gleb wants to hear about the trials that his wife, Dasha, went through at the hands of their class enemies while he was away, the trials that have made her love the Revolution more than she loves her husband (all ellipses are in the original):

> Gleb lay his head on Dasha's knees and saw, above himself, her face, her cheeks, covered with soft down colored by the fiery sky, and her eyes: intent, large, worried, and loving.
>
> "Here, under this sky, one feels a different person, my little Dasha. Here I am, laying in your knees . . . When has it been like that? It seems that I have never

experienced anything like it. I only know that your love was larger and bigger than mine, and I am not worthy of you. I haven't lived through even one hundredth of what you have lived. So tell me yourself about your trials . . . Perhaps then I will get to know myself better.

The air was suddenly lit up by the lightning: big and small stars of light swarmed everywhere. Gleb was swept up by the wave of rapture; excited, he propped himself up on his elbow.

"Dasha, my little dove, look . . . It's so good to struggle and build one's destiny! For, all this—is ours . . . Us! . . . Our power and labor . . . I feel like I am inhaling . . . the way one inhales before the first strike . . . when one wants o swing from high . . .

Dasha again put her hands on his chest. She, too, was excited, and Gleb could hear the heavy muffled pounding of her heart.

"Yes, darling, it is good to struggle for your destiny. Let the sufferings come, let the death come . . . It is scary . . . and not everyone can bear it . . . I had borne it, because my love for you is stronger than fear . . . And then I understood something else, and loved something else . . . perhaps even more than I love you . .

"Speak up, my little Dasha . . . whatever it is—speak . . . I have now learned not just to listen but also to struggle with myself . . ."[100]

The characters' emotions may be larger than life, reaching, as it were, to the stars. The frequent ellipses, too, are meant to signal the grandeur of their feelings, for they seem to experience so much more than they are capable of expressing verbally.[101] (No smooth talkers they—none of that long-winded aristocratic palaver one encounters in old novels!) Yet one struggles to find embedments rising above the second level—an experience extremely unusual when it comes to critical interaction with a novel. Gleb *knows* that Dasha *loves* him. Gleb *feels unworthy* of Dasha. Gleb *knows* that his *happiness* is bound up with the industrial collective. Gleb *feels that* Dasha is *excited* too. He *hopes* that he can *control his feelings* as he listens to her story. Dasha *wants* him to *know* that she *loves* the Revolution more than she loves him.

The last two are perhaps the most unambiguous examples of third-level embedment in this high-wrought chapter, entitled, fittingly, "Inner Interlayers."[102] The depth of emotions explored here (however smacking of agitprop) is rather unique for *Cement*, which tends to report a character's response to a specific challenge and then, immediately, to move on. For instance, captured by the enemies, a woman experiences the arm of the man who is dragging her to her execution as "monstrous"; when she is spared the execution and left alone, she feels "blind terror"; when she subsequently runs into her comrades, she "laughs and cries."[103] This action-reaction rhythm of narrative is unrelenting, which means that there are almost no complex embedments on the level of chapters and very few on the level of paragraphs.

Note the difference between this novel and Evgeny Zamyatin's *We*, discussed in chapter 1 of this book. *We* was written at exactly the same time as *Cement* but first published abroad: in New York, in 1924. Standing pointedly outside the ideological project of "educating" its audiences by talking to them on their (presumably, benighted) level, *We* constantly embeds mental states of characters and implied readers, even while it makes a point of not mentioning emotions. In contrast, *Cement*, true to its peculiar educational mission, refers to emotions frequently yet eschews complex embedment, and it certainly does not engage mental states of the implied reader/author.

Here is another "production novel," Eduard Claudius's *People at Our Side* (*Menschen an unserer Seite*, 1951), "an exemplary work of early socialist realism" from the GDR. (There, writers, too, "were expected to write in a way that was popular (*volkstümlich*) [and] accessible.").[104] The novel's protagonist, a bricklayer named Hans Aehre, is "a forceful person who must persuade his brigade of doubters that they are capable of working collectively on the [ring] furnace without shutting it down, which would cost the factory six months' lost production."[105] In between his bouts of heroic labor, Hans must come to terms with his wife's desire to be seen as "a whole human being" and not just "a woman and a wife."[106] This leads to conversations similar to those Gleb was having with his Dasha, in which Hans learns to listen to his wife and rein in his conservative masculinity.

For most of the novel, however, just as in *Cement*, complex embedments are few and far between. Here, for instance, is Hans's passionate enunciation of his new role at the factory: "Yes, we are workers, Comrade Backhens. We are workers. Even if an engineer or a foreman or the contractor is present—it's we who build the furnace, we! And if there is no foreman who wants to help us or no engineer, well then, we'll build it anyway . . . of course we will . . . it must be possible for us to do it by ourselves."[107]

Take a good look at this remarkable speech. One would think, given the history of complex embedment in literature, that it would be impossible for a modern author to escape the gravitational pull of complex intentionality. This is to say that plenty of experimenting authors, from Alain Robbe-Grillet in *Jealousy* and Muriel Spark in *The Ballad of Peckham Rye* to

Zamyatin in *We* and Cormac McCarthy in *Blood Meridian*, make a point of *not* mentioning mental states explicitly, but their narratives still depend on their readers constantly supplying *implied* mental states to make sense of what is going on. In contrast to those authors, Claudius comes very close to constructing an actual mental-state-free paragraph and thus modeling a new golden age in which word and deed are one and (to quote Hayian Lee again) nobody has "any need for mindreading."[108]

It is worth noting that the "production novel," in and of itself, is by no means antithetical to complex embedment. Already in 1963, the East German writer Christa Wolf published *They Divided the Sky* [*Der Geteilte Himmel*], which takes place, in part, at a train-making factory. Its main protagonist, Rita, witnesses the dramatic endeavor of a "famous brigade" to build "twelve windows per shift," even though, a relatively short time ago, the idea of building a train carriage with ten windows in one shift would strike the workers as "crazy."[109] This industrial backdrop notwithstanding, the novel's complex embedments are off the charts, in typical Wolfian fashion, as we follow Rita's continuous, sometimes oblique, self-introspection [110]

Perhaps not surprisingly, in spite of the novel's clear political allegiances (for Rita, unlike her boyfriend, Manfred, is wholeheartedly committed to the cause of socialism), *They Divided the Sky* was condemned by East German reviewers as politically subversive. Still, it became an immediate best-seller and was soon made into an equally controversial, and popular, movie. Today it is typically featured on such lists as "100 German Must-Reads,"[111] along with novels of such heavy hitters of complex embedment as Musil, Mann, and Zweig.

This is why the socialist realist novel of the early, "exemplary" cut (e.g., Gladkov's *Cement*, Claudius's *People at Our Side*) can be viewed as a fascinating experiment with mindreading. Though relatively short-lived and now largely forgotten, it did turn back the clock on complex embedment and demonstrated the viability of neo-epic subjectivity in literature. What had made this kind of experiment possible was a unique combination of factors: the strong political agenda supported by the punitive state; that state's apparent contempt for the cognitive processes of working-class readers; the cultivation of a regressive dream about a golden age in which minds are transparent; and ubiquitous exposure to well-established literary traditions (exemplified by the novels of Tolstoy and Theodor Fontane etc.), which offered a very different vision of fictional subjectivity yet could, nevertheless, be claimed as precursors to the present one.

Doing full justice to the interplay of these factors is beyond the scope of my argument. It remains an open question, for instance, if the ready availability of the nineteenth-century classics had made the socialist realist experimentation with shallow intentionality more or less compelling or if the intensified censorship trained at least some readers to look for hidden meaning and thus added an unexpected level of implied mindreading to those texts. What I want to emphasize, with this case study, as well as with the preceding ones (i.e., those from English and Chinese literary traditions), is that cognition and ideology are bound with each other in a variety of historically specific forms, most of which have never been acknowledged or explored by cultural historians.

In this chapter, I have focused on one particular way of bringing together cognition and history, namely, on the possibility that patterns of complex embedment in a work of literature may be correlated with the relative social status of its characters and readers. In recent years, literary scholars have advanced other, different, models of cognitive historicism.[112] Yet, on the whole, we have barely scratched the surface. The field of cognitive approaches to literature remains wide open to researchers willing to explore the proposition that a cognitive literary inquiry is, fundamentally, a historical inquiry.

3 "Deep" History: Evolutionary and Neurocognitive Foundations of Complex Embedment

3.1 Thinking on Three Historical Levels at Once

What does it mean to think of complex embedment of mental states as an essential feature of literature, as we know it today?[1] It means thinking on three historical levels at the same time: being aware of the "deep" history of our species, of the more immediate cultural history, and of literary history.

The "deep" history concerns the evolutionary and neurocognitive foundations of complex embedment. Somewhat paradoxically, this perspective may be the least rooted of all, because much of it depends on ongoing research in the cognitive neuroscience of mindreading. To look back at that history means, in effect, to look forward and to be ready to modify one's thinking when more and/or different information becomes available.[2]

The midlevel perspective is also not quite what literary scholars are used to when they think of cultural history. While it does not ignore such familiar factors as the role of the means of textual reproduction and changes in the size and type of the reading public, its main foci are mindreading histories of specific communities, or the local "ideologies of mind." To become aware of those ideologies, I draw on research of anthropologists and ethnographers who study similarities and differences between the cultural practice of "thinking about others' internal states and/or talking about them."[3] As the anthropologist Webb Keane observes, while "theory of mind and intention-seeking are common to all humans," they are "elaborated in some communities [and] suppressed in others."[4] The history of the representation of mental states in literature is profoundly implicated with cultural institutions that "elaborate" or "suppress" mindreading.

Finally, literary history is concerned with the evolution of patterns of complex embedment in literature, as well as with the migration of such patterns across different genres, national literary traditions, and individual texts.

Any given complex embedment of mental states in literature thus relies *simultaneously* on the workings of our evolved cognitive architecture, on a culture-specific "ideology of mind" (which implicitly regulates the public and print discussion of other people's mental states), and on the immediate literary ecology of a particular text.

I can easily imagine how one may focus on one or two of those factors at the expense of other(s). In fact, the structure of this book may encourage this kind of thinking, because, for the purpose of my argument, I consider them in turn—first, the deep-historical (i.e., cognitive), then, the historical, and, finally, the literary—separately. So it is important that we keep in mind that none of those factors can be reduced to others or considered sufficient on their own. For instance, our evolved capacity for complex embedment, alone, *does not determine* the appearance of fictional texts that would ratchet up the frequency of such embedments; neither does a cultural milieu that encourages public speculation about one's own and others' inner states; and neither does the presence of a long-standing literary tradition steeped in complex embedment.[5] The "secret life" of literature is sustained by the interplay of all three.

3.2   Makeshift Metaphors Revisited

"Theory of mind," "mindreading," and "embedment" are useful metaphors for evoking mental functioning involved in daily social interactions. Their utility, however, becomes overshadowed by their clumsiness and inadequacy when we try to understand how the underlying cognitive processes actually work. Take, for instance, the implications conveyed by these terms, that we experience mental states in a propositional, disembodied format (as in, "I know that she doesn't know that I know") and that we are, mostly, aware of our mentalizing.

In reality, this is hardly the case. Mindreading may be said to exist on several different levels. While there is, indeed, "the level of conscious reflection about the mind, what we might call *explicit theories* of theory of mind," the majority of our daily mindreading "happens largely outside conscious reflection and probably conscious control," and it is both grounded in the body and highly context sensitive.[6]

Cognitive scientists are quite aware of how incomplete and misleading the "propositional" view of theory of mind is. For instance, when the social anthropologist Rita Astuti introduces the concept of mindreading, she illustrates it with the following set of images: "[When] you see someone running, you don't just see a physical body in acceleration—you see the intention or the desire to catch the bus or win a medal; when you see a hand reaching for an object, you don't just see a trajectory through space—you see the goal of getting that object."[7]

This is mindreading at its most prevalent: rooted in the body and happening outside of conscious reflection. Similarly, the linguistic anthropologist Alessandro Duranti reminds us that mindreading "does not proceed as a series of self-conscious propositions, as in, I believe that X intends to do Y." Instead, it depends on the embodied, intuitive, prerational understanding of another's actions.[8]

The cognitive anthropologist Pascal Boyer, too, stresses the automaticity, speed, and multimodality of mindreading, as well as the fact that we can't help attributing mental states, whether correctly or not, when he defines it as "a whole suite of specialized systems [that] automatically picks up social information—other people's behaviors, gestures, utterances, but also their facial expressions, choice of words, and so forth—to construct, without any conscious effort, a representation of their beliefs, intentions, and emotional states, all things that cannot be observed and must be inferred."[9]

Slowing down this process, bringing it to conscious awareness, and putting it in words inevitably transforms this experience. The transformation may be enriching, for instance, by forcing us to develop new ways of representing inner states in our discourse.[10] Yet it may also be radically impoverishing, for instance, by stripping mindreading of its contexts and sensory nuances and by misrepresenting it as a linear process. Writers themselves know that linearity distorts our experience of the social environment. As Christa Wolf puts it, "[The] age-old fact that things occur and are felt and are thought simultaneously but that all those things cannot be put down simultaneously on paper in linear writing suddenly rattles me so much that doubt in the realism of my writing grows into a total inability to write."[11]

I can thus understand the position of some of my colleagues in literary studies who are turned off by "theory of mind" and "mindreading" because they object—and rightly so!—to the implications of the conscious, accurate, disembodied, linear, and context-free processes that these terms

convey. Sadly, however, their quests for alternative accounts of social cognition may thus be driven by putting too much stock in those flawed
metaphors instead of paying attention to the actual phenomenon of mindreading studied by psychologists, anthropologists, and ethnographers.

"Embedment" is another one of those problematic metaphors (though
perhaps not as fraught as theory of mind or mindreading). You will notice, as
I proceed with this chapter, that, in different fields of cognitive science, the
capacity for embedding mental states goes by different names. Those include
"recursive embedment," "perspective embedment," "recursive intentions-
reading," "nesting," "level-two perspective taking," "second-order theory of
mind," "second-order false-belief understanding," "levels of intentionality,"
"multiple-order intentionality," and so forth. The images of layers, levels,
hierarchies, and thought bubbles recursively nested within each other, which
such descriptors evoke, are not likely to reflect any actual patterns in the
mind/brain. What they are more likely to reflect, instead, is a long cultural
history of visualization of abstract concepts,[12] now pressed into the service
of rendering instantly intelligible and familiar complicated processes that we
are only beginning to understand.

Is there a way to talk about mentalizing without relying on the language
of representationality? Here is a perspective from developmental psychology, offered by Mark Sabbagh and Dare Baldwin:

> What does a nonrepresentational understanding of intention "look like"? Per-
> haps the most vivid demonstration of an early appreciation of intention comes
> from children's understanding of goal-oriented action. Gergley and colleagues
> (1995) found that even 12-month-old infants were willing to attribute goals to
> shapes that showed signs of being animate (i.e. capable of self-propelled motion).
> Along these same lines, Woodward (1998) has demonstrated that young infants
> construe the actions of humans as goal-directed, though they do not apply the
> same construal to the motions of inanimate objects. Still more convincingly,
> Meltzoff (1995) found that 18-month-olds reenacted events that correspond with
> an actor's likely goals and intentions, even when those actions are not explicitly
> modeled. Across these studies and others (e.g. Carpenter, Akhtar, and Tomasello,
> 1998), very young infants demonstrate an impressive level of sensitivity to the
> fact that others' actions are motivated by internal mental states. However, these
> young infants would fail even the simplest tasks designed to tap a representa-
> tional understanding of mental states, such as the false belief task (Wellman et al,
> 2001). Thus, 18-month-olds clearly understand behavior in a distinctly mentalis-
> tic manner, but it is probably a mistake to ascribe a concomitant representational
> appreciation to these same children at such an early age.[13]

Thus developmental psychologists. Philosophers of mind, too, have grappled with the issue of moving beyond representation. For instance, Mark Johnson speaks of "nonrepresentational theory of mind, where having or entertaining a concept is merely running a neural simulation in which sensory, motor, and affective areas of the brain are activated, not as representations mediating between an inner and outer world, but rather as *the very understanding of the concept*. In other words, the neural activations involved in the simulations within a specific context *just are* what is to grasp the meaning of the concept in question."[14]

One can even speculate (by way of taking this insight to its logical extreme) that propositionally expressed representations of mental states play no role at all in reading literature. Perhaps, when we read, we grasp the density and depth of intersubjective situations that define the experience of literary fiction,[15] and it is only when we try to slow down and describe that experience that we resort to the familiar cultural metaphors of tiers, orders, levels, and embedded mental states.

This view is in broad agreement with that of the cognitive psychologists Hugo Mercier and Dan Sperber, who point out that the procedures involved in unconscious inference are not identical or even similar to those involved in conscious reasoning, and they do not operate on "statements or statement-like representations." The concept of representation, in this view, is understood strictly in terms of its function:

> Representations . . . are material things, such as activation of groups of neurons in a brain, magnetic patterns in an electronic storage medium, or ink patterns on a piece of paper. They can be inside an organism or in its environment. What makes such a material thing a representation is not its location, its shape, or its structure; it is its function. A representation has the function of providing an organism . . . with information about some state of affairs. The information provided may be about actual or about desirable states of affairs, that is, about facts or about goals.[16]

Is it possible or even desirable to avoid the language of representationality, specifically, when talking about complex fictional subjectivity?[17] I am not sure that our gains would outweigh our losses if we were to make a concerted effort to do so, in the name of hypothetical cognitive purism. For keep in mind that writers, too, are faced with this challenge when they try to convey in words deep intersubjective experiences of their characters. They, too, rely on outright descriptions of embedded mental states—albeit some of them to a greater and some to a lesser degree—to bring to life their intuitive visions of intricate social consciousness. To insist that critics must find a way to describe this phenomenon without talking about embedded intentionality—because such representations distort what may be really going on in the brain!—would be similar to insisting that writers must find a way not to talk about it either, for the same reason.

Some writers indeed rely more on implied mental states than they do on explicitly spelled-out mental states, but that alone does not make their writing better or more literary. For instance, as I argued in chapter 1, the difference between literary and popular fiction does not map onto the difference between implied and explicitly spelled-out mental states. Instead, literary fiction is often characterized by the two following features (to a much greater extent, that is, than is popular fiction): first, it embeds mental states of narrators, implied authors, and readers *in addition to* mental states of characters; and, second, it tends to imply mental states *in addition to* and sometimes in place of explicitly spelling them out.

Thus, perhaps a more realistic way to approach the problem of representationality in cognitive criticism is to say that we should strive to retain the view of the multidimensional, multisensory, embodied, and not-yet-well-understood phenomenon of "embedded mindreading" even if we must continue to rely on the terms, currently in wide use, that streamline, decontextualize, and disembody it.[18] So as I go on with my argument, try keeping in mind the perennial gap between the actual cognitive processes (all moving targets, as far as researchers are concerned) and the makeshift, far-from-perfect metaphors that make it possible to talk about those processes.

3.3 Perspectives from Social, Developmental, Clinical, and Evolutionary Psychology and Cognitive Neuroscience

One good starting point for a conversation about embedded mental states is research on the "default network," which looks at the nexus of interacting brain regions involved, among other things, in "inward contemplation and self-assessment"[19] and "conceiving the perspectives of others."[20] What we learn from this research is that, when we engage in a complex social interaction with other people, attributing mental states to them (however unselfconsciously) necessitates attributing mental states to ourselves: "understanding complex social interactions among people who are presumed to be social, interactive, and emotive always involves the processing of self-reflective thoughts and judgments."[21]

Handling communicative intentions is thus "a more complex process than simply thinking about intentions, since we have to recognize that the communicator is also thinking about our mental state. This involves a second-order representation of mental state. We have to represent the communicator's representation of our mental state."[22] Depending on the context of the situation, this may translate into a second level of embedment (as in, "she obviously doesn't know that I know!") or even a higher one, third or fourth (as in, "I wish I had known before that she didn't know that I knew!"). Again, here I am putting it in a propositional format—because I have no way of conveying it to you otherwise!—but during actual social interactions, we do not think propositionally (for one, we *don't have time for it*, except in some special cases).

How involved can such social imagining get? As it turns out, there may be limits to the levels of embedment. The cognitive evolutionary psychologist Robin Dunbar and his colleagues have demonstrated that "fifth-order intentionality" (fifth-level embedment of mental states) represents "a real upper limit for most people," that is, the level after which their understanding of the situation drops drastically.[23] At the same time, there also "is considerable inter-individual variation in the highest achievable levels of intentionality." Those individual differences correlate with such factors as "the ability to correctly attribute blame" and the number of contacts in one's support clique, that is, the number of individuals on whose advice and/or help one would depend at times of great social or financial trouble."[24]

For a quick illustration of how the "upper limit" on levels of embedment manifests itself, when next time, at a party, you find yourself chatting with four other guests, see how long it takes before your group separates into two relatively independent conversational units consisting of two and three people. Presumably it won't be too long, because keeping track of five mental states including your own—which may reach a maximum level of embedment pretty fast—is a cognitive burden. Left to our own devices (as opposed, for instance, to being committed to a more rigid social situation such as a five-person discussion panel at a conference), we *intuitively* try to lessen than burden by modifying the social context that created it.[25]

I emphasize the word "intuitively" because we would not be aware of either carrying any burden or trying to lighten it. It may just so happen that I "spontaneously" discover that what a person next to me is saying is really quite engrossing—or that she is the only one whose voice I can hear over the party's din—so I end up focusing all my attention on a conversation with her, while the three others continue on their own. Negotiating a mindreading overload—what mindreading overload? As the clinical psychologist Philipp Kanske and his colleagues put it, the "ease with which we accomplish [the mindreading] task every day, readily makes us forget the complex computations and processes it entails."[26]

How early does it start? That is, at what age do we begin to attribute embedded mental states to ourselves and others? Until relatively recently, developmental psychologists thought that children begin to appreciate people's false beliefs—that is, realize that people may believe something that is not in fact the case—around the age of four.[27] Then, between five and seven, children become attuned to "doubly embedded" representations; that is, they become aware "not just that people have beliefs (and false beliefs) about the world but that they also have beliefs about the content of others' minds (i.e., about others' beliefs), and similarly, these too may be different or wrong." This awareness is "fundamental to children's . . . understanding of the epistemic concepts of evidence, inference, and truth,"[28] although there are important cultural differences in whether children are encouraged or discouraged to talk openly about their own and other people's mental states.[29]

In experiments involving kindergarteners and first graders writing letters to hypothetical friends who have never experienced some of the things familiar to the authors of the letters (such as, for instance, snow, mentioned in a letter to someone who has never seen snow), the "recursive understanding of embedded mental states" was shown to be implicated with children's growing awareness of a reader's knowledge as distinct from that of the writer's. Around seven years of age, children realize that "an effective writer represents how their reader will interpret their textual meaning (authorial intention) in light of that reader's experience."[30]

The traditional view that before the age of four children are not ready to attribute false beliefs to others was challenged in 2005 in a study by Kristine Onishi and Renée Baillargeon, who showed that fifteen-month-old infants may already understand false beliefs. Since then, numerous other experiments have pushed the age for such understanding even lower.[31] While different theories have been proposed to account for this "puzzle of theory of mind" in infants,[32] for the purposes of my argument, I go with the view that "the infant mindreading system develops gradually, transforming into the adult one through incremental learning and piecemeal conceptual change."[33] The changes that take place around the ages of four, five, and seven may still represent important milestones in theory-of-mind development (especially, as we have seen, in the nuances of perspective taking), but they can now be viewed as steps in a continuous integrated process rather than dramatic breakthroughs.

Embedded mindreading assumes new prominence as children enter adolescence. In 1970 (even before the term "theory of mind" entered the lexicon of cognitive scientists),[34] Patricia H. Miller and her colleagues concluded their essay "Thinking about People Thinking about People Thinking about . . . : A Study of Social Cognitive Development" with the following rueful observation: "often to their pain, adolescents are much more gifted" at "wondering what he thinks of me" and "what he thinks I think of him" than "first graders are."[35] The drama and intensity of alliance building and sexual maturation are inseparable from the reading and, inevitably, misreading of one's own and others' embedded intentions.

When it comes to the cognitive neuroscience of embedment, in 2003, Rebecca Saxe and Nancy Kanwisher published an article, "People Thinking about Thinking People," which showed, for the first time, that there is a particular region of the temporo-parietal junction of the brain that is "involved specifically in reasoning about the contents of another person's mind."[36] There was an increased response in that region when subjects read stories that involved figuring out people's thoughts and feelings, as opposed to stories with no social reasoning. Since then, other studies have addressed questions ranging from whether the same brain region supports thinking about people's "appearance, social background, or personality traits" (it seems that it doesn't)[37] to what neural populations may underwrite the "representations underlying human emotion inference."[38]

To give you an idea about the setup of Saxe and Kanwisher's study, here are two of the stories that the subjects of their experiments were exposed to. The first depicts a woman wanting to get to her office and encountering a construction zone: "Jane is walking to work this morning through a very industrial area. In one place the crane is taking up the whole sidewalk. To get to her building, she has to take a detour." There is some intentionality, yes, but no rich social content and no increased activation in the brain regions under study, as opposed to the subjects' response to the other story: "A boy is making a paper mache project for his art class. He spends hours ripping newspaper into even strips. Then he goes out to buy flour. His mother comes home and throws all the newspaper strips away."[39]

If I put in propositional format my own response to the latter vignette, I can say that the mother *didn't know* that the boy *intended* to do something with the pile of torn newspapers and that the boy may *realize* that the mother *didn't know* that he *intended* to do something with that pile. I can also think what may happen next, imagining that the mother *would want* the boy to *know* that she *hadn't known* that he *intended* to do that and, moreover, that she *would want* him to *know* how truly *sorry* she is.

In other words, both in the process of making sense of this situation and thinking about it further, I recursively embed thoughts and feelings within each other. I also explicitly verbalize it all for you, whereas were I to encounter this vignette in the laboratory, with no time to think it through, I would just process these embedments automatically, without being aware of doing so.

It is not impossible that my personal response to this particular vignette is impacted by my empathizing with its protagonists, for it is easy for me to start thinking about my own school-age child, who would be very upset were I to throw away his project-in-progress. This brings us to empathy, an issue that I haven't addressed at all so far and plan to continue not addressing, after this short interjection.

Whereas I am aware of the variety of fascinating studies of empathy in conjunction with the reading of fiction, I do not work with empathy myself and believe that, for the purposes of studying complex embedment in literature, theory of mind and empathy should be considered separately. Conflating the two would ignore research that points toward important differences between them, and, given how little we still know about cognitive correlates of either, such conflation is not likely to be helpful.

Those of my readers who would like to learn more about these issues may start with a series of recent studies by clinical psychologists and social neuroscientists who looked at the behavioral and neurological markers of theory of mind and empathy in subjects exposed to emotional videos. What they found, in brief, is that "enhanced activation of the ToM [theory of mind] related network was linked to better ToM performance, but not to behavioral measures of empathy. This pattern was replicated when using composite scores of empathy and ToM performance derived from multiple tasks, which corroborates and generalizes the specificity of the brain–behavior relations of the two social capacities."[40]

So while I admire the work of my colleagues in cognitive literary studies, such as Fritz Breithaupt, Suzanne Keen, and Ralph James Savarese,[41] who investigate empathy, I neither engage in such investigation myself nor presume to make any claims about the role of empathetic engagement in our processing of complex embedments.

3.4   Distal and Proximate Causes of Complex Embedment in Literature

To sum up the different strains of research from cognitive psychology and neuroscience, the capacity for embedment of complex mental states is integral to human mindreading. This capacity matures in development, may present enough of a cognitive burden to have something resembling an upper limit set to it, and is supported by specific brain regions.

The deep (that is, the cognitive) history of embedment highlights the *social* aspect of our engagement with literature. While theory of mind evolved, back in the Pleistocene, to track mental states involved in real-life social interactions, on some level, our mindreading adaptations do not distinguish between mental states of real people and of imaginary entities whom we "meet" on the page, on-screen, or on the canvas: as soon as we are faced with behavior, we start attributing intentions to the behaving agents.[42] That literature, in particular (as we know it now), seems to demand that readers *continuously* process complex embedments of mental states leads one to wonder what kind of real-life social challenges, persisting throughout our evolutionary history, this demand on the readers' cognition may be mimicking and exaggerating. Why should it feel good to follow the intricacies of what one person (who doesn't even exist) thinks about what another person (who doesn't exist either) knows about the first person's intentions?

If you are interested in the broader version of this question—which is, why we may actually *enjoy* various cognitive burdens that literature places on our mindreading adaptations—I refer you to my earlier book *Why We Read Fiction: Theory of Mind and the Novel* (2006). To focus more narrowly on the appeal of complex embedments, here are some relevant reflections by cognitive scientists, which provide useful ways of thinking about this issue (yet should not be mistaken for conclusive explanations).

The cognitive psychologist Daniel Nettle offers the following observation about the social rewards of situations in our evolutionary past in which third-level embedments naturally occurred:

> [The] natural situation in which we have three-way mind-reading going on is one that might be rewarding for several reasons. First, if we know what person A is thinking about person B but person B does not know this, then we are in a position of privilege and power. Either person A had taken us into their confidence, which would mean we were a valued coalition partner, or we are very clever, and/or we now have some leverage over person B because we know something important that they do not. If we feel well-disposed to B we may want to warn them, and gain their gratitude and reciprocity; if we are ill-disposed to B we may wish to use it against them or withhold it spitefully. In any event, this is a very significant situation in which we, although a spectator, are now part of a social triangle. This would not be so true if we knew what person A thought about B and B also knew this.[43]

A related explanation comes from the work of the cognitive anthropologist Pascal Boyer, who suggests that in-group cooperation, which was absolutely crucial for the survival of our species, may have favored interest in complex mindreading. Developing "social relations and cooperation among many individuals [allowed] for more efficient cooperation," but that also meant that it was important to discriminate between contexts in which information about others' intentions could be freely disseminated and contexts in which it had to be concealed. As Boyer explains, to maintain "small-scale friendly networks, one needs access to individuals as such and one needs a measure of discretion. Every item of information need not and in many cases should not be broadcast too widely."[44] So keeping track of who has access to whose mental state would be just as important as keeping track of who doesn't and shouldn't have that access.

Moreover, negotiating complex social situations depended on combining explicit discussions of one's own and other people's mental states with implicit attributions of thoughts and feelings to oneself and others. Behavioral neuroscientists are finding today that "implicit and explicit mentalizing processes may be closely related in a healthy population,"[45] which means that when works of literature combine explicitly spelled-out and implied mental states, they mimic and intensify patterns of mindreading that recurred throughout our evolutionary history.

Finally, consider the positive biofeedback associated with the feeling that the awareness of other people's mental states is *our own*. As William

James puts it, "We think; and as we think we feel our bodily selves as the seat of the thinking. If the thinking be *our* thinking, it must be suffused through all its parts with that peculiar warmth and intimacy that make it come as ours."[46] Joining James's insight with those of the neuroscientist Antonio Damasio, the cognitive literary critic Nancy Easterlin suggests that consciousness and self-consciousness—"not only the awareness that we know but also the added awareness that the knowing is specifically one's own—[feel] good. And when knowledge feels good, [we] are apt to seek it actively, to want as much of the thought and feeling of mastery as possible."[47] To the extent to which the processing of embedded mental states of others involves awareness of one's own mentalizing, the pleasure of social inclusion is thus further augmented by this feeling of epistemic mastery.

When we turn to proximate causes, it seems that imaginary representations of third-level embedments model certain thorny types of social challenges that we face in our daily lives today. As such, they may feel particularly attention worthy. It also doesn't hurt that some fictional narratives present us with cleaned-up versions of real-life mindreading problems. That is, in many a work of fiction (though by no means in all!), I actually get to know what a character X thinks about character Z, whereas in real life I have to settle for my imperfect constructions of my own and other people's mental states.[48] Add to this a pleasure that I may feel as I discover new depths of social perception in myself when I think I discern the (implied) author's intention regarding my access to a character's feelings,[49] a discernment that builds on my previous experience with this genre, this author, or this specific work.

Add, too, my happy awareness of myself as a member of a particular community in which such discernment is valued. For instance, I may enjoy realizing that Jane Austen *doesn't want* us to *know* that her Emma is *in love* with Mr. Knightley for as long as possible, perhaps even for as long as Emma herself remains unaware of it. Yet had I been brought up in an environment in which familiarity with Austen's novels were considered a pointless affectation and then happened upon a copy of *Emma*, I may not have brought to it the kind of attention that would allow me to intuitively appreciate the intentions of its (implied) author. It wouldn't have mattered to me that *Emma* taps, in so many intricate ways, into the cognitive adaptations for complex mindreading that may have formed back in the Pleistocene. I would have skipped and skimmed and missed most of its embedments and not considered myself worse off for doing so.

This is to say that when we deal with complex cultural artifacts such as literature, distal causes tend to be conjoined with proximate causes in ways that make it impossible to disentangle the two or to treat one as more important than the other. Specifically, when it comes to patterns of embedment in fiction, we can't just trace them back to the social pressures of the Pleistocene and ignore the immediate circumstances in which writing/performing and reading/watching take place today.[50] We can't focus on the "deep" history at the expense of cultural and literary histories.

4 Cultural History: Ideologies of Mind

The similarities and differences between these two practices—thinking about others' internal states and/or talking about them—are often at the heart of culture.
—Bambi B. Schieffelin

4.1 The "Opacity of Mind" Model

"There is no doubt that humans in all known cultures learn to infer intention . . . from the behavior of other humans," writes the psychological anthropologist Tanya Luhrmann, "yet at the same time, ethnographers observe that the inferences they draw are probably shaped not only by developmental capacity but by cultural specificity."[1] Cultural variation of mindreading practices, underwritten by local ideologies of mind, is not something that literary historians tend to think about when they consider circumstances in which genres arise, develop, and change into other genres. Yet, as I hope to demonstrate by the end of this chapter, this factor is crucial to the production and reception of literature, especially literature as awash in explicit and implied embedments of mental states as ours has come to be.

To start thinking of our daily mindreading practices as reflecting a particular model of interiority, we first have to recognize the existence of other models. Consider, for instance, what is known as the "opacity of mind" model, "found in varying forms throughout the South Pacific and Melanesia." Its most striking feature is a consistent and vocal "refusal to infer what other people are thinking unless they verbalize their intentions."[2] This refusal underwrites a variety of daily practices, ranging from a taboo against eye contact and a tendency to repeat verbatim others' statements about their mental states, without questioning or elaborating them, to the caretakers' avoidance of verbally guessing at the "unclear meanings" of very young children.[3]

Even just these three examples alert us to the possibility that some of our familiar cultural rituals (such as looking people in the eye, publicly second-guessing others' stated intentions, and interpreting toddlers' babbling for them) are locally specific ways to perform mindreading, indicative of a particular ideology of mind. Let us consider this possibility in some detail, by first taking a closer look at the "opacity" model and then thinking through its implications for our cultural and literary analysis.

4.2   Refusing to Talk about Others' Mental States

When anthropologists and ethnographers had initially confronted what they would come to call the opacity model, they wondered if it meant that, in some Melanesian and Micronesian societies, for instance, among the Bosavi (aka Kaluli), Korowai, Ku Waru, and Yap, "it is impossible or at least extremely difficult to know what other people think or feel."[4] This prompted a conversation about methods used for studying mindreading in concrete cultural contexts. The distinction, formulated by Rita Astuti (see chapter 3) between the "conscious reflection about the mind" and the mindreading that happens "outside conscious reflection and probably conscious control" is directly relevant here. For, as she points out, "ethnographic methods are of course well suited to record the former, while experimental methods are best suited to tap into the latter."[5] This means that ethnographers, used to ways in which some forms of "conscious reflection about the mind" are performed in their own communities, should be careful not to substitute their informants' assertions "of how the world should be" for a description of what they may actually "find in the world."[6] Thus, if we focus on the mindreading that happens outside conscious reflection and control, we discover that, while Bosavi, Korowai, and others may avoid public references to other people's minds, they may actually be "more attentive to [their] intentions as a result."[7]

For instance, struck by such a recurrent feature of the opacity model as the taboo against direct eye contact, an ethnographer may assume that Bosavi do not pay attention to each other's facial expressions. She will thus miss the fact that Bosavi keep their foreheads clear of hair or head dresses, letting others "read" their emotions off their foreheads.[8] Similarly, as the ethnomusicologist Steven Feld explains, while Bosavi may not explicitly impute thoughts to others, "there is an impeccable and ubiquitous attendance to what others feel, and that is coded at every linguistic level, but particularly marked by lexical items, emphatics, prosody, and a range of gestural, stance, facial expressive, and other paralinguistic markers coordinated with everyday speech."[9]

Keep in mind, too, a variety of forms that a particular feature associated with the opacity model may take in different communities. For instance, the same injunction against looking "directly into another's eyes" (because that may lead to "inferring privately held intention")[10] manifests itself differently in Yap (Micronesia) than it does in Bosavi. Here is how the ethnographer C. Jason Throop describes it:

> [One] of the first notes I took when arriving in Yap concerned what I held to be a striking lack of eye contact when individuals spoke to one another and a marked tendency for speakers to turn their bodies and heads away from their interlocutors. In fact, it was not uncommon to observe individuals carry on complete conversations with their backs to each other, gazing off in opposite directions. Likewise, during community meetings individuals often sat with their backs against the beams supporting the community house facing out away from the meeting, gazing at the horizon, the dance ground, or other parts of the village center. One conversation I noted early on during my second stay in Yap in the summer of 2001, well before I had acquired the communicative competence necessary to follow along with an ongoing, multi-party conversation, included six individuals speaking for over an hour, none of which were facing one another.[11]

As in Bosavi, so in Yap there is a telling tension between publicly performed and private mindreading. On the one hand, Yap practice what Throop describes as a series of "communicative strategies used to conceal one's thoughts from others." On the other hand, the very fact of using these strategies implies that they intuitively expect that others will attempt to read their words and body language as indicative of underlying thoughts and feelings. Private preoccupation with what people know or don't know about each other's mental states drives the public attempt to prevent mindreading:

> By talking in opposites, being elusive, facing meta-pragmatic restrictions on turn-taking and questioning, only providing the absolutely minimal amount of information necessary, being sarcastic, playing jokes, teasing, avoiding eye contact, or situating one's body such that one's voice is muted and one's facial expressions are concealed from the view of others, individuals are thus able to insure that

their interlocutors are never able to garner a clear idea as to what they are really thinking or feeling. A significant benefit to engendering such communicative opacity, one elder noted to me, is that by putting one's interlocutors off guard and off balance, and by making them uncertain as to one's true feelings and motives, an individual is granted an advantage inasmuch as the speaker is the only one who truly knows what his or her plans are, which could perhaps be importantly used to his or her advantage at some later date.[12]

The gap between publicly following the rules of etiquette associated with the opacity model and the private preoccupation with others' mental states can assume a more obvious form, that is, that of the difference between a public and a private conversation. For instance, on the island of Vanatinai (Papua New Guinea), "Islanders publicly, rhetorically deny the possibility of empathy, of imaginative understanding of and identification with the thoughts/feelings of another being. In private, within a current, constantly fluctuating group of trusted confidants—spouses, lovers, siblings, matrilineal kin—they conjecture at length, in exacting detail, based upon a range of external cues, about what others are thinking and feeling . . . and how this may affect their interactions with others in the recent past, present, or future."[13] This avid conjecturing about others' thinking—conducted with trusted confidants but not in public—reminds us that the opacity model is not an all-or-nothing phenomenon.[14] A community as a whole may adhere to principles of opacity, but those principles don't have to govern every single aspect of social interaction.

4.3   Not Interpreting Infants' Babbling

Another important feature of the opacity model is the refusal, on the part of caregivers, to interpret infants' babbling as expressive of intentions. To put this point into sharper perspective, compare some parenting practices in North America and in Kaluli. North American parents may model their children's articulation of mental states by doing it *for them* early in development, as in, "Aren't you hungry!" or "It must feel very frustrating not to be able to reach that ball!" According to the linguistic anthropologists Elinor Ochs and Bambi Schieffelin, particularly, "In the white middle class developmental story, [assisting the child to clarify and express ideas] is associated with good mothering. The mother responds to her child's incompetence by making greater efforts than normal to clarify his or her intentions."[15] In contrast, when "talking to young children, Kaluli caregivers do not propose possible internal states of their addressees." Thus, when "a child whines or acts inappropriately, caregivers ask, 'Ge oba?!' 'what's with you?!' If a child doesn't eat, they pose a rhetorical question, 'Ge mo:nano?!' 'you don't eat?!' rather than, 'are you hungry?'"

Along the same lines, in Kaluli, when young siblings "do put a referential gloss to the babbles of an infant, [older] caregivers repeat the sounds" but do not use a verb form "which would imply that something meaningful was produced in such vocalizations. Thus through this type of modeling and verb choice, small children are gently socialized to use culturally appropriate ways to verbally report what they hear without attributing meaning, including what constitutes reportable speech and what does not."[16]

This does not mean that Kaluli children are not "encouraged to verbalize their own desires and intentions." They are. It does mean, however, that "they are explicitly socialized," first, not to "verbally guess at or express others' unvoiced intentions and unclear meanings" and, second, not to feel compelled to explain their own motivations when they don't want to.[17] This prepares them for functioning in a society in which, while "almost everything else could be known about a person, people [resist] being coerced into giving moral accounts or making explicit what they were thinking about."[18]

Furthermore (again, in contrast to some North American parents), Kaluli parents use

> no baby-talk lexicon as such, and claim that children must hear . . . 'hard [i.e., real] language' if they are to learn to speak correctly. . . . Kaluli recognize babbling but say that this vocal activity is not communicative and has no relationship to the language that will eventually emerge. Adults . . . will occasionally repeat vocalizations back to toddlers (aged 12–16 months), reshaping them into the names of people in the household or into kin terms of people nearby, but they do not claim that the toddler is saying these names or wait for the child to repeat these vocalizations in an altered form.

The absence of baby-talk lexicon does not result in an impoverished verbal environment. As Schieffelin reports, "although there is relatively little speech directed to preverbal children, the verbal environment of these children is rich and varied, and from the beginning infants are surrounded by adults and older children who spend a great deal of time talking to each other . . . [and hearing] their actions . . . referred to, described, and commented upon by members of the household, especially older children speaking to one another."[19]

Moreover, Kaluli "mothers and infants do not gaze into each other's eyes, an interactional pattern that is consistent with adult patterns of no gazing when talking to others." Instead,

> Within a week or so after a child is born, Kaluli mothers act in ways that seem intended to involve infants . . . in dialogues and conversations with others. Rathe than facing their babies and engaging in dialogues with them in ways many English-speaking mothers would, Kaluli mothers tend to face their babies outward so that they can be seen by and see others who are part of the social group. Older children greet and address infants, and in response to this mothers hold their infants face outward and, while moving them, speak in a special high-pitched nasalized register (similar to that Kaluli use when speaking to dogs). These infants look as if they are talking to someone while their mothers speak for them.[20]

By speaking "for" their infants, mothers socialize them into a community in which expressing their own feelings as well as reporting others' feelings verbatim are acceptable practices while interpreting others' feelings is not As Schieffelin puts it,

> [While] Kaluli obviously interpret and assess one another's observable behavior and internal states, these interpretations are not culturally acceptable as topics of talk. Individuals talk about their own feelings ("I'm afraid"; "I feel sorry"), but there is a cultural dispreference for talking about or making claims about what another might think, and what another might feel, or another is about to do especially if there is no obvious behavioral evidence. Kaluli, however, use extensive direct reported speech, and children use this linguistic resource by 24 months of age. . . . [These] culturally constructed behaviors have several important consequences for the ways in which Kaluli verbally interact with children, and are related to other pervasive patterns of language and social interaction.[21]

4.4   Opacity on a Continuum

Compelling as the concept of opacity seems to be, it is important to remember that we have on our hands yet another case of far-from-perfect terminology. The term "opacity of mind" may seem to imply a sharp break between cultures that are wholly governed by that model and cultures that are not In reality, both types of cultures function on "a not-very-rigid continuum" of opacity.[22] This means, among other things, that features strongly associated with the opacity model in one culture may be present in another, and yet that other culture may gravitate, as a whole, toward the transparency end of the spectrum.[23]

Sometimes such features would be indicative of specific challenges faced by members of the community under particular circumstances. Consider, for instance, that since Bosavi live in close physical proximity to each other, the inside of a person's head is often the only private space available to them. The pragmatics of protecting that space from others are expressed through specific features of verbal etiquette and contribute to the maintenance of psychological well-being.[24] In Bosavi, these features of verbal etiquette are integrated with the ethos of opacity. Yet as the psycholinguist Catherine Caldwell-Harris and her colleagues observe, we find similar features geared toward "allowing people their psychological privacy" in other communities whose members live "in close quarters" but that are not necessarily viewed as conforming to the opacity model.[25]

Or consider the long history of US racism engendering a behavior on the part of oppressed minorities that has features of the opacity model, *within* what we may broadly characterize as the overall North American model, which tends toward transparency. For instance, take the African American practices of "signifyin." As Aaron Ngozi Oforlea explains, signifyin can function as a mindreading strategy aimed at protecting the self by misleading, misdirecting, and outwitting "well meaning acquaintances and powerful adversaries." Thus, he writes, "Zora Neal Hurston describes her way of disguising her mental state to protect her 'business' or personal experiences from white researchers who often visit her to collect folklore. Aware of the interloper's intentions, Hurston strategically decides which information to share or withhold. [She] writes: ['The] white man is always trying to know into somebody else's business. All right, I'll set something outside the door of my mind for him to play with and handle. I'll put this play toy in his hand, and he will seize it and go away. Then I'll say my say and sing my song.'"[26]

When occurring specifically in the context of interaction between the oppressor and the oppressed, signifyin may also reflect the status dynamic discussed in chapter 2, which is that people in stronger social positions don't read minds as actively and perceptively as do people in weaker social positions. This may make the former good targets for signifyin, because they are easily satisfied with the "toys" (i.e., token insights into the minds of the oppressed), set outside for them "to play with and handle." As a protective and defensive measure, opacity can thus be a marker of inequality, social stress, and communal adversity. This is quite different from the role it plays in (for instance) highly egalitarian Bosavi, where it is supposed to contribute to social cohesion.

Here is another example of a particular feature of the opacity model present in a society that does not subscribe to that model. We have already seen that one aspect of Bosavi socialization is the refusal, on the part of caretakers, to expand on their young children's utterances (which would necessitate imputing mental states to them). We encounter the same refusal in Samoa (also already discussed in chapter 2), but there it is driven not by the opacity model but by rigid social stratification. In Samoa, people of higher social standing are not supposed to be guessing what people below them in the social hierarchy mean when they express themselves less than clearly. If the meaning is unclear, "the burden of clarification" is always on the low-status person. As very young children are the lowest-ranking members of their household, people around them do not attempt to read their minds (which would mean lowering themselves to their level).

Here is how Ochs and Schieffelin, working, respectively, with Samoan and Bosavi populations, describe this dynamic: "[Neither] the Kaluli [aka Bosavi nor] the Samoan caregivers . . . appear to rely on expansions, but the reasons expansions are dispreferred differ. The Samoans do not do so in part because of their dispreference for guessing and in part because of their expectation that the burden of intelligibility rests with the child (the lower status party) rather than with more mature members of the society. Kaluli do not use expansions to resay or guess what a child may be expressing because they say that 'one cannot know what someone else thinks,' regardless of age or social status."[27]

"One cannot know what someone else thinks" is a key tenet of the opacity doctrine. "It is not one's business to figure out an underling's meaning" can be, as it were, a key tenet of a rigidly stratified society. That two very different ideologies of mind can lead to pretty much exactly the same observable behavior is something to be aware of as we (i.e., students of literature) begin to test the usefulness of the "opacity of mind" concept for our cultural and literary analysis.

4.5   Opacity and Ethics

We have focused, so far, on the psychology and epistemology of the opacity model, but another productive way of approaching it is to think of its ethics. For, as the anthropologist Webb Keane explains, the taboo on attributing intentions to others often reflects the local notion of personal integrity and inviolability, according to which the loss of ability to keep one's feelings hidden is considered shameful:

> It is not that inner thoughts are inherently unknowable but that they ought to be unspeakable, or at least, it matters greatly who gets to speak these thoughts. . . . [Thus it] is not the case that [the Melanesians] have no capacity to read minds or invent fictions: rather, these capacities serve ethical thought, leading to emphatic denial of something that they are in fact doing. . . . To reiterate, if Theory of Mind and intention-seeking are common to all humans, how these get played down or emphasized can contribute to quite divergent ethical worlds. Elaborated in some communities, suppressed in others, these cognitive capacities appear as both sources of difficulties in their own right and affordances for ethical work.[28]

What happens if we apply these insights—prompted, originally, by the studies of cultures subscribing to the opacity model—to more familiar cultural settings? The reason that I find this idea appealing is that it offers a new perspective on a whole array of social practices that we take for granted. For we do not, usually, go around thinking about how this or that cultural institution (including literature!) "elaborates" or "suppresses" mindreading. Yet, once you adapt this perspective, you realize that mindreading does not take place in a vacuum. Instead, it is shaped by culture-specific ideologies of mind that have both epistemological and ethical dimensions. This is to say that it is shaped, first, by people's beliefs about whether their own and others' "inner thoughts" are knowable and, second, by their assumptions about "who gets to speak those thoughts." To quote Schieffelin again, the "similarities and differences between these two practices—thinking about others' internal states and/or talking about them—are often at the heart of culture."[29] What do we learn about our culture by inquiring into our own practices of talking and thinking about others' mental states?

4.6 Direct Eye Contact and Ideologies of Mind

> She saw something in that image that she hadn't noticed in person. In the photo, his eyes shifted away from the camera. Eventually London had understood that look as one she could not trust.
>
> —Johka Al Harthi, *Celestial Bodies*

Let us revisit the injunction against direct eye contact among Yap and Bosavi. First of all, the psychological intuition that underlies this injunction is the same intuition that, in many Western cultures, makes eye contact a socially desirable behavior. For, just as we believe, to quote Cicero, that eyes are "the mirror of the soul," so do, it seems, the Yap. To quote Throop again, "It is interesting to note the extent to which the face, and particularly the eyes (laen mit, laen awochean) are held, in local configurations of subjectivity and social action, to represent that part of the person that is most susceptible to directly evidencing inner feeling states and thoughts."[30]

In cultures subscribing to the opacity model, infants and young children have to be taught not to look people in the eyes. Bosavi "mothers do not engage in sustained gazing at, or elicit and maintain direct eye contact with, their infants as such behavior is dispreferred and associated with witchcraft."[31] As to Yap, Throop suggests that the etymology of the Yap word "child" (tiir) is tied to the expression "is eyes," and as such, it reflects the fact that it doesn't come naturally to children not to betray their feelings by their gaze: "Children simply look at what they desire; they show no concern for hiding their intentions, emotions, needs, and cravings from others. They have thus yet to cultivate self-governance and have yet to learn to manage their emotions in such a way that there is less of a direct link between their inner feeling states and their expressivity."[32]

Now think about the tendency of Western caretakers to look into their infants' eyes and to encourage reciprocal gazing. We may experience this as a default child-rearing behavior (indeed, associate it with good parenting!) and do not think of it as indicative of some special "transparency of mind" model. Yet, put in the comparativist perspective, this behavior does indicate a certain ideology of mind, one that gravitates, on the whole, toward the transparency side of the spectrum. Were we to explicitly formulate this ideology—which may come out sounding awkward and artificial precisely because we have internalized it—we might say that one *can* know what someone else thinks and that making one's inner thoughts available to others and attempting to penetrate their inner thoughts are generally experienced as prosocial behaviors.

(Thinking of transparency as prosocial may seem to contradict our tendency to value "privacy"—unless one recognizes that the concept of privacy may cover a spectrum of practices. For instance, in contrast to Bosavi, for whom unavoidable physical proximity makes them eager to protect the privacy of their minds, Western cultures may put more emphasis on physical—which may be relatively easy to achieve, as when one is alone in a room—than on mental privacy.)

To see how the unspoken ideology of *knowable* minds undergirds our daily social interactions, think of how we respond to politicians, doctors, salespeople, or even next-door neighbors when they seem to avoid eye contact. Instead of experiencing them as virtuously protecting their own and our personal integrity and inviolability—as we might, were we to operate under the auspices of the opacity of mind model—we perceive them as "shifty-eyed" and thus untrustworthy (or, perhaps more charitably, as painfully shy).

In fact, there seems to be a gap between the broad range of our reactions to direct eye contact—which are not always positive!—and the cultural ideology that codes such contact as mostly good. This is to say that even in societies associated with the transparency model, direct eye-gazing can provoke mixed emotional responses. While it can be experienced positively—as signaling motivation to approach or romantic interest—it can also be taken as indicating "hostility and impending peril."[33] Still, this variety of actual reactions notwithstanding, the dominant expectation seems to be, and has been for some time, that people who look at us directly are "more caring, trustworthy, harmonic, inclusive and respectable" than are those who avert their gaze.[34]

Accordingly, consider Western parents' discomfort when their children refuse to make eye contact. A popular website that offers "11 Reasons a Child Cannot Look You in the Eyes" may acknowledge that the dispreference for eye contact may be the result of "cultural differences,"[35] yet the majority of the listed reasons still reflect the belief that all is not well when a child cannot meet your gaze. The child may be suffering from "social anxiety" or "low self-esteem" or may be "lying about something."[36] (Ironically, in some cultures of opacity, it is direct eye contact that indicates an intention to deceive.)[37]

Generally, it does not take very long for the ideology of knowable minds to turn ugly. What is felt as the right to read other minds "can run in tandem with a need for mastery over others that has been the cause of great suffering over the . . . long course of our history."[38] As the cognitive narratologist Porter Abbott reminds us, what in the context of the European colonial project was presented as "the heroic quest to penetrate the unknown can be hard to separate from the desire to appropriate and tame—in effect to

spread knowability." But this "illusion of knowability" is built "on preexist-
ing terms," that is, those legible to the colonizer. As Abbott puts it, "when
one is sent into a land where one not only does not know the language of
the people but [also] cannot read their faces, the effect goes deep." The shat-
tered "illusion of knowability" augments "the fusion of fear and fury that
grip a soldier fighting in a strange land."[39]

On a different note, recall the notorious practice of diagnosing autistics
as "mind-blind" (that is, "lacking" in theory of mind) because they fail
to perform such a culturally sanctioned way of mindreading as focusing
on their interlocutors' eyes.[40] In a critique of this practice, the cognitive
neuroscientist Gregory Hickok reminds us that "behavior does not *automatically* reveal its cause and can be misleading."[41] For instance, the autistic
individual's Fusiform Face Area could be hyperactive—as opposed to the
conventional view, implied by the "mind-blindness" hypothesis, according
to which it is hypoactive (i.e., inhibited): "Hyperresponse to social stimuli
can be explained in terms of the emotional intensity of the signal, which
triggers anxiety and avoidance responses. [This means that the person's]
active avoidance of eye contact provides just as much evidence for [the
person's increased] sensitivity for the information contained [in the eyes]
as does active engagement of eye contact."[42]

Or, to quote Lucy Blackman, a nonspeaking autistic writer, "It may be
that the social deficits which are the cornerstone of an autism spectrum
diagnosis tell us far more about the person who made them markers for
such a diagnosis than about the child whom she observes. . . . That is, the
whole testing procedure is somehow actually constructed on whether the
tester observed the person to socialize in a way that the tester understood
to be socialization. . . . We often use the term 'communication' when really
we mean that we have observed in another human being a behavior from
which we derive meaning."[43]

Because a Western culture may assign a very particular meaning to direct
eye contact, it takes a comparativist perspective to be reminded that it is a
culturally constructed behavior, associated with what we may call an "ideology of transparency," or the idea that other minds are knowable and that,
under most circumstances, we have a right to know them. When we are
denied the valuable social knowledge that, we believe, can be obtained that
way—or, as it were, denied the right to that knowledge—we may feel a
range of negative emotions toward the person who seems to deny it to us.

Keeping in mind this cultural construction of direct eye contact as a sign of prosocial behavior, imagine how different our art, movies, poetry, and novels would be if reading assiduously "the language of the eyes" were considered inappropriate, antisocial, and dangerous: associated, for instance, with the intent to harm by witchcraft (especially in societies in which suspected witches used to be killed, as in Bosavi and Korowai).[44]

Thus, when the protagonist of Samuel Richardson's novel *Clarissa* (1747) observes that she and her would-be seducer, Robert Lovelace, "are both great watchers of each other's eyes," we make sense of her comment within the context of a "transparency of mind" model largely governing Western representations.[45] We know that Clarissa and Lovelace don't trust each other and hope to catch a glimpse of another's true intentions during unguarded moments; and we are also aware of erotic overtones of their behavior. But such interpretations are a product of a particular ideology of mind. Were we to read the same body language in the context of the "opacity of mind" model, Lovelace's and Clarissa's deliberate eye watching might acquire different overtones, ranging from the socially uncouth to the physically dangerous.

Here, then, is one preliminary observation about fictional narratives. When works of literature foreground the language of the eyes in their representation of characters' mental states, they build on a particular aspect of the mindreading adaptation that can be considered universal. For, both in cultures of opacity and in cultures of transparency, "the face, and particularly the eyes" are considered a direct conduit to the person's "inner feeling states and thoughts."[46] But it is reasonable to expect that, in cultures of opacity, fictional situations featuring direct eye contact would often be bundled up with contexts and expectations that are less indicative of prosocial behavior than they would be in cultures of transparency.[47]

4.7   From the "Monastic Theory of Mind" to the Academic One

If communities indeed elaborate some mindreading practices and suppress others, we can view a variety of cultural institutions as implicated in this project. For instance, a recent study, Paul Dilley's *Monasteries and the Care of Souls in the Late Antique Christianity: Cognition and Discipline* (2017), builds on Luhrmann's view of culture-specific models of mindreading to suggest that "the training of thoughts practiced by early Christian monks led to the gradual acquisition of a new and particularly monastic theory of mind."

Some of the key precepts of this monastic theory of mind were that the mind was both permeable and accessible. This is to say that monks had to learn that their cogitations arose "not only from the interior self, but also through divine guidance or demonic temptation" and that "God was aware of their private thoughts, which were also known to certain inspired saints."[48]

Learning these precepts demanded introspection, physical exercise, and communal activities. Also, interestingly, monks were encouraged to read hagiographies, which, Dilley argues, constituted a particularly instructive and pleasurable training in mindreading.[49] Hagiographies facilitated the acquisition of "the monastic theory of mind, by offering a privileged perspective on the saints' internal deliberations, including the use of clairvoyance and other revelations in their disciplinary decisions."[50]

Where would the "monastic theory of mind" fall on the continuum of opacity? It seems that, on the whole, it gravitated toward the transparency end of the spectrum. Other people's minds were considered inherently knowable, and a particular virtue was attached to being able to figure out the source of one's own and other people's thoughts (i.e., divine or demonic). Moreover, one's "secret thoughts" were not really secret, for God was aware of them and so were the saints.[51] This means that "thinking about others' internal states and/or talking about them" (Bambi Schieffelin) was both a useful and an ethical thing to do.

If we remember that monasteries were "the centers of learning before the rise of the universities"[52] and that Sorbonne, Oxford, and Cambridge continued to be theological schools "until the middle of the fourteenth century,"[53] it makes sense to think about the "academic theory of mind" as influenced by the monastic one. We can consider, for instance, the role of the dual belief that other people's thoughts are knowable and that there is a particular virtue associated with tracing the provenance of those thoughts in the development of some academic disciplines; and we can also talk about the gradual suppression of explicit mindreading as a prerequisite for the emergence of others. I can't hope to do full justice to this topic here (it would require a separate book), but let us take a quick preliminary look at some forms of mindreading associated with academic learning, in a culture that edges toward the transparency end of the opacity spectrum.

4.8   Patterns of Mindreading in Conversations about Literature

One academic subject that is unthinkable today without mindreading is literature. Talking about mental states of fictional characters is something that secondary-school students begin to do quite early. By the time they reach college, they are, at least in principle, primed for the kind of sophisticated mindreading that will be expected from them in literature courses.

To see how some of them rise to such expectations, consider works of fiction that intuitively experiment with theory of mind by suppressing all mentalizing references, explicit or implied. Take, for instance, Alain Robbe-Grillet's *Jealousy* (1957), which is notorious for its depiction of actions drained of mental states. Here is a characteristic excerpt from the chapter describing the banana plantation where the action takes place:

Prolonging this patch toward the bottom, with the same arrangement of rows, another patch occupies the space included between the first patch and the little stream that flows through the valley bottom. This second patch is twenty-three trees deep, and only its more advanced vegetation distinguishes it from the preceding patch: the greater height of the trunks, the tangle of fronds, and the number of well-formed stems. Besides, some stems have already been cut. But the empty place where the bole has been cut is then as easily discernible as the tree itself would be with its tuft of wide, pale-green leaves, out of which comes the thick curving stem bearing the fruit.

Furthermore, instead of being rectangular like the one above it, this patch is trapezoidal; for the stream bank that constitutes its lower edge is not perpendicular to its two sides—running up the slope—which are parallel to each other. The row on the right side has no more than thirteen banana trees instead of twenty-three.

And finally, the lower edge of this patch is not straight, since the little stream is not: a slight bulge narrows the patch toward the middle of its width. The central row, which should have eighteen trees if it were to be a true trapezoid, has, in fact, only sixteen.

In the second row, starting from the far left, there would be twenty-two trees (because of the alternate arrangement) in the case of a rectangular patch. There would also be twenty-two for a patch that was precisely trapezoidal, the reduction being scarcely noticeable at such a short distance from its base. And, in fact, there are twenty-two trees there.

But the third row too has only twenty-two trees, instead of twenty-three which the alternately-arranged rectangle would have. No additional difference is introduced, at this level, by the bulge in the lower edge. The same is true for the fourth row, which includes twenty-one boles, that is, one less than an even row of the imaginary rectangle.[54]

How does one respond to a work of literature that makes it this difficult to read intentionality into it? It turns out that some readers may actually redouble their efforts to discern complex mental states in such texts. Thus, according to David Richter, who teaches literature at CUNY, when he assigns *Jealousy* to his undergraduates, "they read the repeated narrative about the centipede that horrifies A and is killed by Franck as coming from a jealously obsessive narrator noticing and recalling over and over Franck's responsiveness to A. They even read the chapter in which we are told about how many banana trees are in each row in each segment of the plantation as coming from a mind that was forcing itself to pay attention to objective facts about his banana plantation in an attempt to stop himself from obsessively thinking about his wife A and her possible relation to Franck."[55]

Think about what these students are doing. Broadly speaking, they are "naturalizing" a difficult text, making it easier to comprehend.[56] Yet the particular way in which they are achieving it—that is, by constructing complex embedments of mental states—is a product of a specific culture. This culture has institutional settings that reward people for speaking and writing about intentionality. This means that they learn to approach texts marked as fiction with the expectation of mindreading, and of a particularly elaborate kind at that, if they happen to encounter those texts in a literature course.

As a corollary to Richter's experience, consider the history of critical readings of Herman Melville's short story "Bartleby the Scrivener" (1853), whose protagonist behaves in such a way that neither other characters in the story nor its readers can attribute any mental states to him. But, as Porter Abbott puts it, the "experience of unreadable fictional minds, meant as such, is very hard to maintain."[57] So one strategy for responding to an "unreadable character" is to interpret him as a "generic stereotype," as in, Bartleby is insane, and that explains his incomprehensible behavior. Another strategy is to shift "the mode of reading" altogether and cease regarding Bartleby as a human being (or a representation of a human being), whose mental states can be inferred. Instead, he becomes a "catalyst" for understanding *other* characters or an idea, a symbol, as in, "Bartleby is the ghost of social conscience haunting the precincts of the ruling class."[58]

Note that a symbolic reading also involves mindreading. For, when we say that "Bartleby is the ghost of social conscience haunting the precincts of the ruling class," we *still* attribute a mental state—such as a vague feeling of guilt—only now not to a specific person but to a more abstract entity such as the "ruling class."

As Abbott observes, the shift to the symbolic "allows meaning to rush in" (read: opens up a whole new cluster of mental states), and this is "what has happened almost invariably in the critical response to Bartleby." Taken as a human being, "and not as a symbol, Bartleby remains unreadable." But this state of affairs is "unendurable" for Melville's audiences, so, "one way or another, [they] will generally find some strategy to make it go away."[59]

Yet another course of action for making the unendurable go away is to use the difficult-to-read characters as catalysts for generating readers' *own* complex mental states. Consider the experiment run by the cognitive literary critic Emily Troscianko, who studied readers' response to a short story by Franz Kafka, "Jackals and Arabs" (1917). "Kafka's fictions," Troscianko explains, "never really give us privileged access to the workings of his protagonists' minds." Instead they confront us with characters "whose capacities for introspection . . . or capacities for insight into other's minds . . . are limited." Troscianko found that her subjects were "fascinated" by this "scarcity of insight" and that they compensated for it by constructing embedments that involved their own embedded insights. As one of them put it (emphasis added): "I *find it intriguing, fascinating*, to be guided through the story without ever fully *understanding* what the narrator *feels*."[60]

Thus, while Richter's students made sense of *Jealousy* by force-reading into it thoughts of its characters, and Abbott's readers reached out to the minds radically outside the story (e.g., the mind of the "ruling class"), Troscianko's subjects responded to Kafka by imagining their own mental states. Elsewhere, I have discussed a similar dynamic structuring our response to paintings that actively prevent us from attributing mental states to anybody/anything within them.[61] Finding ourselves in situations such as college courses or critical conversations, in which we are expected to talk about such paintings, we begin to attribute mental states to their creators (by trying to figure out what the artist meant), or to ourselves (by explaining how these paintings make us feel), or to some external entities (by treating the work in question as a cultural symbol).

This is to say that while we may be "designed by nature," as Abbott puts it, to read mental states into behavior, we still need to be "trained by culture" in the locally appropriate ways to perform such readings.[62] Thus, we respond to cultural incentives to engage in mindreading—but also remain sensitive to the disincentives—as we learn that intense mindreading is a prerequisite of success in some academic disciplines but not in others.

It is not surprising that the technique of close reading—or, as I argued earlier, close mindreading—is closely related to the history of religious exegesis and, most immediately, to the history of biblical textual criticism. Still, we can't quite say that talking about the minds of fictional characters, their authors, other critics, and our own in college literature courses is the exclusive legacy of the monastic theory of mind. Traditions of monastic mindreading may have shaped formal practices of Western literary interpretation, but the tendency to talk about mental states when discussing literature is not limited to communities influenced by Christian monasticism.

To take a quick look at the forms that such conversations may take in the absence of monastic influences, we turn to literary traditions of the Bosavi and Ku Waru. We will use as our starting point Webb Keane's observation that, while cultures of opacity may suppress explicit intention-seeking in their discourse, "it is not the case that [their members] have no capacity to read minds or invent fictions," and we will see what kind of mindreading is encouraged by their "fictions."

We start with the Bosavi. On the one hand, "prior to missionization," which began in the 1970s, there "were no equivalencies in . . . metalinguistic and metapragmatic repertoire for reporting the private thoughts or internal states of others,"[63] unless one repeated verbatim what the other person had said about their feelings and used a source tag—an "evidential marker"—to clearly indicate the original speaker.[64] On the other hand, there *was* one important exception: a linguistic context that allowed reporting others' hidden thoughts. That exception was the "traditional story genres that recounted Bosavi origins, or the bawdy adventures or social dilemmas of fictitious cultural heroes, schlemiels, and animals."[65] Such narratives appeared to "mobilize different linguistic resources as part of the register of the genre." For instance, a "morpheme–mosoba ['I wonder'], relatively rare in spontaneous speech, was found more in stories" (as in, "o:no gasa a:no: eno: ko:lo: go:mosoba?"; "that dog I wonder if it was his?"). In addition, storytellers disclaimed "responsibility for the information" about the characters' mental states, by reminding listeners that this was all "in the story."[66]

Or consider the Ku Waru, who live to the east of Bosavi:

[While] in-principle assertions of the opacity doctrine are common [among the Ku Waru], they are contradicted by other things that people do, including the stories that they tell. For example, in a genre of sung tales of courtship that are composed and performed in the region, at the point in the story when the lovers first meet, there is often a passage such as this: "Right then he wanted to marry her. / That's what the man was thinking. / And she thought the same about him. / The minds of both, you see / Were working completely as one." In other words, given the lovers' strong mutual attraction, it is possible for each of them to know what is in the other's mind because it is the same as what is in his or her own.[67]

Another important example of mindreading involved in literary production—especially if we understand "literature" broadly and include performative genres as well[68]—has to do with performers attributing mental states to their audiences and adjusting their behavior as they go along to reflect their perception of those mental states. Consider Gisalo, a song and dance ceremony practiced by the Kaluli, that is, the people of Bosavi. (Note that, although I talk about it in the present tense, my discussion of it refers to the period of the 1960s–1980s, for it is not clear if Gisalos still take place today.) Gisalos are designed to evoke strong feelings of nostalgia, sorrow, and loneliness in their audiences by integrating into their sung narratives references to specific locations that have profound personal meaning for the listeners. A Gisalo is considered successful if listeners weep and try to hurt (i.e., burn) performers in a ritualistic way, to make them pay, as it were, for having thus gotten under their skin.[69] As Edward Schieffelin explains, "The listeners' feelings and reactions are not merely a response to the performance; they are integral to its structure and significance. The dancing and singing by the performers and the weeping and burning by the audience stimulate and aggravate one another. If the [listeners] fail to respond to the songs, even enthusiastic performers soon lose interest, and the ceremony falls apart before the night is over."[70]

Once the ceremony is over, the mindreading continues, albeit now in a more explicit form. Here, recall again that the Kaluli subscribe to the opacity model; that is, they consider it inappropriate to talk about other people's mental states. Yet they do talk about those mental states—with a vengeance!—when discussing recent Gisalo songs. Those remain the subject of conversation for many days after a performance, as appreciative members of the audience keep uncovering "subtlety and complexity in the [singers'] interweaving of geography and personal allusion."[71] In situations when a tape recording of a Gisalo made by an ethnographer is available, hearing this tape may prompt a "discussion session," which would last "for hours" and in which "several older Kaluli men" would listen "repeatedly to the same song, . . . recalling the history of its performance, who had wep and why, and how the song [reached its emotional climax]."[72]

It seems, in other words, that to talk about cultural representations tha build on our mindreading adaptations—prose fiction, certainly, but also performance genres whose success is judged by their capacity to evoke emo tional responses in their audience—we *have to* talk about mental states be they those of fictional characters or those of performers and audience members. Societies closer to the transparency end of the mindreading spectrum, such as ours, may have codified formal venues for doing so (including college courses in film and literature), but societies closer to the opacity end may engage in such conversations even in the absence of historically entrenched institutional structures designed to elicit and facilitate them.

Ironically, public exercises in communal mindreading that occur in a literature classroom may be accompanied by disavowals of interest in intentionality that would not be out of place in a community subscribing to the opacity model. It is not inconceivable that, were an ethnographer to approach a literature professor and ask her how knowable she considers various minds under consideration in her course, the professor would deny any special access to those minds. She might say, for instance, that we have no way of knowing what the author was thinking, that characters don't exist, so they can't really have thoughts and feelings, and so on.

We may think of this response as underwritten by healthy epistemological skepticism, by the ethics of personal integrity and inviolability, or, more broadly, by what the linguistic anthropologist Alessandro Duranti characterizes as a "defense strategy against the accountability that comes with making claims about what others think or want."[73] But however we choose to account for it, the larger point remains. Even if some of us (i.e., teachers of literature) sincerely believe that we are *not* in the business of mindreading, our classroom conversations revolve around mindreading, focusing on our own and other people's (including fictional characters') mental states.

And so do our scholarly conversations. Consider this brief sampler of quotes from prominent literary critics (with attributions of mental states italicized). What it shows is that the thoughts and feelings of characters, authors, and audiences have been their prime subject since Aristotle and that to talk about those thoughts and feelings, critics have always had to construct complex embedments of their own. The "monastic theory of mind" must have both tapped this tendency (what with the monks following avidly mental states of saints, in hagiographies) and given it a more defined institutional expression.

- Aristotle mentions disapprovingly those who "make an unreasonable prior *assumption* and, having themselves made their decree, . . . draw their conclusions, and then criticize the poet as if he had said whatever they *think* he has said if it is opposed to their *thoughts*" (*Poetics*).[74]
- Wayne Booth observes in his analysis of Jane Austen's *Persuasion* that upon meeting Captain Wentworth "after their years of separation that follow her refusal to marry him," Anne Elliot "is *convinced* that he is *indifferent*," while the reader "is likely to *believe* that Wentworth is still *interested*" ("Control of Distance in Jane Austen's *Emma*").[75]
- Gayatri Chakravorty Spivak discusses Victor Frankenstein's "ambiguous and miscued *understanding* of the *real motive* for the monster's vengefulness" ("The Women's Texts and a Critique of Imperialism").[76]
- Susan Sontag wonders if "perhaps Tennessee Williams *thinks* [*A Streetcar Named Desire*] is about what [Elia] Kazan *thinks* it to be about" ("Against Interpretation").[77]

You may notice that these critics range widely in their choice of people whose minds they read: Aristotle talks about embedded mental states of readers; Booth, about those of characters and implied readers; Spivak, about those of characters; Sontag, about those of the author. It so happens that the last three scholars discuss works of literature that seem to offer plenty of room for moving between different types of minds. But in some cases, the decision to read a text in terms of mental states of its implied readers signals more than just an immediate interpretive choice of a particular scholar. It may indicate a change in the wider cultural perception of the text, such as a redefinition of its genre or a renegotiation of its place in the literary canon. To put it differently, a cultural repositioning of the text is usually accomplished through switching mindreading targets associated with that text.

For instance, the late sixteenth-century anonymous Chinese novel *The Plum in the Golden Vase* (*Chin P'ing Mei*) has long occupied an ambiguous place in Chinese literary history. Lay readers consider it pornography, while scholars treat it as a literary masterpiece. It is reasonable to assume that readers who turn to this novel for its explicit sex scenes register mainly mental states of its characters, thus missing the complex mutual awareness between the implied reader and the implied author. In contrast, students of classic Chinese literature pay a great deal of attention to mental states of those nebulous entities, speculating about their intentions vis-à-vis each other.[78]

Thus, Andrew Plaks cites the critical responses to *The Plum*, provided by medieval Chinese commentators, which contain such observations as, "the author definitely has his own intentions" and "there is an object to [the text's] ironic stabs." Plaks himself discusses at length "the possibility of hidden intentions" implied by the author's use of "borrowed material," such as songs and poems, as well as the role that "frequent interpolations of authorial asides" play "to periodically remind the reader of the presence of the narrator somewhere between himself and the story."[79] What this focus on the mental states of *The Plum*'s narrator and implied reader indicates is that the novel deserves to be taken seriously as part of the Chinese literary canon.

To return to European literary history, consider Eliza Haywood's novella *Fantomina* (1725), an amatory romp following sexual stratagems of a young aristocratic woman in early eighteenth-century London. *Fantomina* had remained outside the canon until the 1980s, when feminist literary critics adjusted drastically the mindreading lens associated with it. Instead of continuing to read it focusing on the mental states of the inventive Fantomina and her clueless lover, Beauplaisir, they began using those as jumping-off points for a conversation about the cultural work accomplished by this piece of genre fiction—this is to say, about the mental states of the novella's implied author and its original readers. For instance, the feminist literary critic Ros Ballaster writes about the novella's capacity to change the self-perception of women in a world in which they did not have much power. As she puts it, "by dehistoricizing and mythologizing the public sphere, the romantic fiction writer provided the female reader with a sense of feminine power and agency in a world usually closed to her participation."[80]

Observe what happens here. Making *Fantomina* a subject of scholarly conversation and, consequently, putting it on our course syllabi depend on opening up a new vein of mindreading associated with it. We talk about the (hypothetical) mental states of the author, her readers, and the broader English public (which we imagine here as *not willing* to grant women much power or agency). In other words, as with *The Plum in the Golden Vase*, the admission of a text into the canon involves recalibration of the mindreading effort associated with it.

Moreover, in an environment as fundamentally dependent on elaborate attributions of mental states as are departments of literature, such recalibrations may be par for the course. Casting out for new minds to read, or else for new ways to read the minds already associated with a particular text, constitutes the bread and butter of literary interpretation.

4.9   Critical Thinking and the "Transparency Model"

To give the screw yet another turn, recall that advocates of the humanities often say that taking courses in literary and film studies develops students' critical thinking and thus contributes to the well-being of the community at large.[81] Yet what is "critical thinking," in the particular context of these disciplines,[82] but the heightened capacity for convincingly questioning and elaborating people's intentions? If in Bosavi, the statement of one's intentions is taken as precluding further public speculation about them (what goes on in private and how much others actually believe those stated intentions are, of course, different matters),[83] in Western culture, such a statement often serves as an invitation for open scrutiny. Clever public contestations of other people's mental states are applauded. An ability to construct a convincing argument about what a politician or a writer *must have really meant*—in direct opposition to what they claimed to have meant or even may have sincerely believed to have meant—is a prized skill. As Elinor Ochs observes,

> In legal and other contexts, if it is established that a negatively valued behavior was consciously intended, then sanctions are usually more severe than if the speaker/actor "didn't mean to do it." . . . [While] establishing intentionality is not always critical to sanctioning . . . , [the] important point is that . . . what a person means or meant to do or say is an important cultural variable. For this social group, what a person means to do is distinguished from what he does. This orientation leads members to take seriously, and to pursue the establishing of, individuals' motivations and psychological states.[84]

But even when taken "seriously," the pursuit of someone else's "motivations and psychological states" is a deeply fraught process, both in legal contexts and beyond them. The rise of today's therapeutic culture, for instance, seems to reaffirm the value of opacity, for the notion of "sharing" one's emotions emphasizes the deliberateness of the personal choice of how and when to render oneself transparent to others. And, in general, if you think of the ethos of transparency as an unalloyed social good, just recall situations in which someone else (a family member, a colleague, a reviewer of your book) made assertions about your motivations—this is to say, *interpreted* your mental states for you—instead of merely reporting something that you said. As far as I see, such assertions do not necessarily lead to greater social cohesion, either in personal communication or on the global stage. Still, plenty of our cultural institutions—indeed, those that we may think of as fundamental to a liberal democracy (e.g., the prized right to "free speech")—are geared toward rendering people's motivations transparent, or temporarily legible, by various eloquent others.

This is to say that the "transparency" model works better in some contexts than in others, just as, presumably, does the "opacity" model. To adapt Webb Keane's formulation, both models are "sources of difficulties and . . affordances" for their respective communities, meeting their needs in some respects and failing in others. That both prove to be, fundamentally, mixed blessings is, perhaps, unavoidable, given the precarious nature of the phenomenon that they attempt to regulate and describe (i.e., people's mental states).

4.10   Mindreading in the Social, Natural, and Physical Sciences

Academic disciplines, in their current cultural configurations, differ widely in their attitudes toward using mental states—or referring to intentions—in their discourses. This means that when a student decides to major in history, mathematics, chemistry, evolutionary biology, or literary and film studies, they effectively commit themselves to a mostly unspoken paradigm of mindreading specific to a particular academic environment. We have already seen how this paradigm plays itself out in literary studies. Let us now take a closer look at several other academic environments.

Departments of history depend on mindreading in their construction of narratives of cause and effect (although not everybody is happy about this state of affairs).[85] Indeed, historians routinely attribute feelings and intentions not just to people but also to geopolitical entities. Here is a random excerpt from Michael Howard's *The First World War* (2002), with emphasis added, in which countries feel "proud" and "anguished," coalitions "wish" they could "ignore" certain political realities, and the world is busy keeping a running total of its great empires:

A liberal-radical coalition [that] came to power [in Britain] in 1906 . . . *could not ignore* the paradoxical predicament in which Britain found herself at the beginning of the century. She was still the wealthiest power in the world and the *proud* owner of the greatest empire that the *world had ever seen*; but she was more vulnerable than ever before in her history. . . . Ideally [successive British governments] *would have wished* to remain aloof from European disputes, but any indication that their neighbors were showing signs, singly or collectively, of threatening their naval dominance had for the previous twenty years been a matter of *anguished* national concern.[86]

In contrast, the physical sciences have worked long and hard to remove references to intentionality, divine or human, from their discourses and have largely succeeded. Still, if you pick up a standard science textbook, you notice that its authors sometimes liven up their material with appeals to their readers' theory of mind. Consider this passage (emphasis added) from Nivaldo J. Tro's *Chemistry: Structure and Properties* (2017): "Table E1 shows the standard SI base units. For now *we focus* on the first four of these units: the meter, the standard unit of length; the kilogram, the standard unit of mass; the second, the standard unit of time; and the kelvin, the standard unit of temperature."[87] The phrase "we focus" conjures up a momentary image of joint attention, a speck of sociality in a sea of data. Because of this brief evocation of mental state, the data may now be easier to process, especially for readers who find this material only moderately exciting.[88]

Medical schools present an interesting case. On the one hand, they seem to actively suppress mindreading, at least in their written discourse, by discouraging students from referring to their own and their patients' mental states. According to the physician and literary scholar Rita Charon, as "students are groomed to speak in medicine's language," the style of their "written language flattens out." She offers the following example of an exercise produced by a third-year student (in which "HPI" stands for "history of present illness"): "HPI: 51 yo man with HIV (diagnosed in 20xx, recently began HAART in February, March 20xx CD4 204 / 27%, VL UD, CD4 nadir 191 in 11/20xx, no OIs, RF: multiple transfusions), hemophilia A, HTN c/b ESRD on HD w/ TLC c/b multiple MSSA infections, HCV (genotype 1b, untreated), with recent prolonged hospitalization 02/4/xx–04/7/xx for MRSE MV endocarditis c/b MCA CVA 2/2 septic emboli who presents with high blood pressure and headache."[89]

On the other hand, there is a growing recognition that draining medicine of language that serves as "a means to access a person's inner sensations and thoughts" denies the humanity of both patients and doctors and is having devastating effects on the profession.[90] Thus, the new field of narrative medicine,[91] spearheaded by Charon, challenges this status quo by reintroducing a conversation about mental states into interactions between the doctor and the patient.[92]

References to mental states may also find their way into other disciplines whose very foundation depended on excising any notion of intentionality from their discourse. For instance, an evolutionary biologist may write an article on the genetic basis of color adaptation—a subject in which intentionality has no place—and yet find a way of encouraging mindreading in her audience. "Thus Hopi Hoekstra (emphasis added): 'Many aspects of modern evolutionary research are *motivated* by the *desire* to *understand* how diversity arises and is maintained in nature. How and why do organisms look and act so differently, and in some cases, so strangely? In fact, these are the same questions that *inspired* Darwin, but thanks to Watson and Crick *we now can look* for the answers in the language of DNA.'"[93]

Hoekstra's writing has long been admired by her students and colleagues, and we can see one reason why. She evokes mental states: those of the implied researcher, her readers, and other scientists. The effect is such that, while not detracting from the rigor of her insights, it makes those insights easier to follow. A bit of sociality, created by references to mental states, makes the account of the genes involved in color adaptation reader-friendly.

What I wanted to show with this set of examples is that, even in a culture that gravitates, on the whole, toward the transparency end of the spectrum, attitudes toward explicit mindreading remain in flux. Even in a narrowly circumscribed institutional setting, such as the university, forms targets, and ethical meanings of various mindreading practices are subject to constant renegotiation.

This is not terribly surprising, given the fundamental ontological instability of the phenomenon in question: after all, mental states are not "really" there—they are something that we cobble together as we move along, to make sense of our social environment. While communities that subscribe to the opacity model respond to this instability by claiming that minds are *not* knowable (even as their private practices may belie the official doctrine), communities that subscribe to the transparency model insist that minds *must be* knowable and scramble to construct those "knowable" minds, with very mixed results, or else declare certain areas of (academic) inquiry mindreading-free. The historical approach to cognition that I advocate in this book thus proposes to take into consideration this spectrum of attitudes toward other people's minds and to view specific cultural developments (e.g., the rise and fall of certain literary genres and practices of interpretation) in relation to these inescapably flawed models of social reality.

Here is, then, how my approach differs from that of more traditionally minded literary historians. They may inquire into ways in which, for instance, the growth or decline of adult literacy rates or the repeal or introduction of censorship laws may affect a cultural career of a particular literary genre. What I would also want to know in such cases is whose minds are rendered as more or less knowable as a result of those changes or, to put it differently, which mindreading practices are newly perceived as more or less publicly acceptable, desirable, and ethical. Community-specific ideologies of mindreading may be all but invisible to members of the community, but they shape both daily social practices and literary reimaginings of these practices.

5   Literary History: The Importance of Being Deceived

5.1   Realism: Nothing but Trouble

To recap, thinking of complex embedment of mental states as an essential feature of literature (as we know it today) calls for operating on three historical levels simultaneously. The first level is the "deep," that is, cognitive, history. The second is the more immediate cultural history, that is, implicit expectations about forms, targets, and ethics of mindreading and the social institutions that support these expectations. In this chapter, we turn to the third level—literary history—that is, the evolution of patterns of complex embedment within and across specific literary traditions.

How does one go about reconstructing this kind of history? On the one hand, even a quick look at ancient epics, novels, and plays, as well as literary texts that defy clear generic classification, shows that third-level embedment of mental states has been around for a long time. It is already there in *Gilgamesh* and the Bible, in Homer, Petronius, Apuleius, Heliodorus, Wang Shifu, and Luo Guanzhong. So, in principle, one should be able to show how literary embedments change over time: how instances of complex embedments become more frequent (for, in *Gilgamesh*, they are relatively rare), how they come to depend more on particular elements of style, and how their evolution is driven by specific social and cultural contexts.

On the other hand, the meager number of surviving texts from ancient literary traditions makes it difficult to construct a responsible argument about the early history of this trend in different genres. Take fifth-century BC Greek drama. One may be tempted to contrast Aeschylus with Sophocles—because the latter seems to embed complex mental states more frequently that the former does, especially of the explicit kind—and to develop a claim about an important milestone in the history of embedment that was reached at that time. Given, however, how few of either Aeschylus's or Sophocles's plays came to us intact and how little of a broader context we would have for such a claim (with only 1 percent of ancient Greek literature having survived), its value would be dubious.

Or consider a seemingly straightforward argument that one can make about the relationship between embedment and the rise of what is commonly called the "psychological" or "psychologically realist" novel. On the one hand, there seems to be little doubt that the sheer scale of complex embedment—its increasing cascading frequency—in such authors as Murasaki Shikibu, Cao Xueqin, Samuel Richardson, Laurence Sterne, Jane Austen, George Eliot, Gustave Flaubert, Fedor Dostoevsky, Lev Tolstoy, Virginia Woolf, Marcel Proust, James Joyce, Thomas Mann, Lu Xun, and Henry James dwarfs all preceding patterns of embedment, making their writing feel drastically different from that of Homer, Apuleius, Heliodorus, Nizami, and Luo Guanzhong.

On the other hand, the association between psychological realism and hypertrophied complex embedment is more complicated than it appears to be. The terminology itself is problematic. If we acknowledge complex embedment of mental states as an important feature of psychologically realist novels,[1] then one is compelled to ask *for whom* this experience is "realist." It may be so for characters themselves, for they can function on the first and second level of embedment, with only occasional third- and fourth-level spikes. But in what sense is it realist for readers—who have to cope with the *ongoing onslaught* of mental states embedded on at least the third level (if they hope to stay on the text's wavelength)—which is, arguably, *not* something that they are called on to do in their "real" life?

As a quick illustration of what I mean by the onslaught, consider an excerpt from Alexander Pushkin's novel in verse *Eugene Onegin* (1833), which is often characterized as a great realist, or pre-realist, work of Russian literature.[2] Here is a description of its title character, who, at eighteen, is already well versed in the "art of soft passion" of love:

> How early he was able to dissemble,
> conceal a hope, show jealousy,
> shake one's belief, make one believe,
> seem gloomy, pine away,
> appear proud and obedient,
> attentive or indifferent!
> How languorously he was silent,

> how flamingly eloquent,
> in letters of heart, how casual!
> With one thing breathing, one thing loving,
> how self-oblivious he could be!
> How quick and tender was his gaze,
> bashful and daring, while at times,
> it shone with an obedient tear![3]

What a tour de force of complex embedments! When the situation calls for it, Onegin dissembles (i.e., he *wants* the pursued woman to *think* that he *feels* something that he doesn't really feel); shows jealousy (*wants* her to *believe* that he is *afraid* that she may *love* someone else); seems gloomy (*wants* her to *think* that he is *miserable*); pines away (*wants* her to *think* that he is *despondent*); appears proud (*wants* her to *think* that he *believes* himself to be above the situation), obedient (*wants* her to *think* that he will do anything she *wants*), attentive (*wants* her to *believe* that he can only *think* of her), or indifferent (*wants* her to *think* that he *doesn't care* whether she *loves* him). He is "languorously silent" (i.e., he *wants* her to start *wondering* what's *on his mind*), "self-oblivious" (he *wants* her to *think* that he *is not in control of his feelings*), "bashful and daring" (he *wants* her to *think* that he is *embarrassed* of his passion yet can't help it), or tearful (he *wants* her to *think* that he is *deeply moved* by the situation).

And on top of that, we have the complex embedments arising from the interaction between the narrator and the reader. For that is what all those frequent "hows" accomplish (as in, "how languorously he was silent"). The narrator invites the reader to share in his amused admiration of the hero's antics: not only does Eugene *want* the woman to *wonder* what's *on his mind*, but the narrator *wants* the reader to be *aware* of Eugene's *wanting* the woman to *wonder* what's *on his mind*!

So it appears that, in *Eugene Onegin*, one single stanza can make us process fifteen or so tightly compressed[4] complex embedments in about ten seconds (which is, roughly, the time that it takes us to silently read it). How often do we do that in the course of our daily life?[5] That is, how often do we find ourselves processing complex embedments with anything resembling this frequency? Ironically, the works of Homer, Heliodorus, and Nizami may be said to be *more* psychologically realist (or, to quote Patricia Miller and her colleagues again, more "ecologically plausible") in this respect because their rate of complex embedment—occasional as opposed to nonstop—may be closer to what we experience in our daily social interactions.

In fact, if the world conjured by Pushkin feels more psychologically realist to us than the world of Heliodorus does, it may be because reading novels has skewed our idea of what "real" or "realist" is. Perhaps we have even been flattered into thinking that this is what our daily mindreading *might* look like, if only we would find ourselves in the right place with the right (i.e., introspective, sophisticated) people. But is that indeed the case?

Think of situations in which we are confronted with numerous complex embedments in a short span of time. Do not consider special professional contexts: some occupations, such as family lawyer, psychologist, poker player, and professor of literature, routinely depend on intense bouts of complex embedment.[6] Instead, recall more mundane occasions. In our everyday life, when we find ourselves in circumstances that call for processing numerous complex embedments (for instance, when we have to remind ourselves, first, not to say something about one person's intentions in front of another person, who, we know, may use that information to thwart the first person's plans, and, then, not to say anything about the second person's intentions in front of the first person, and so forth), we do not perceive that as particularly realistic. In fact, we may complain that there is "too much drama" in our life just then or observe that there is a "soap-opera" quality to our experience.

In other words, our "real" life begins to feel rather special when we find ourselves inexorably processing one complex embedment of mental states after another, even though—and I hope you appreciate the irony of it—one of the key components of literary "realism" seems to be its thick sociality, created by the "ecologically implausible" piling up of complex embedments.

Moreover, literature does not just pile up complex embedments of the soap-opera-ish kind, as in, for instance, "I must remember that she must not know anything about his intentions." Instead it often conceals and masks embedded intentions and prompts us to ascribe them to entities that are not involved in actual social interactions that take place in a story, such as narrators and implied authors/readers.[7] Whereas this is not unusual in real life—indeed, contextual irony can be richly present in some of our daily conversations—what is unusual is the scale on which it happens in literature, where a single paragraph, for instance, from Lu Xun or Henry Fielding, can give us multiple high-level embedments of this kind.

So when my undergraduates, who increasingly (alas) haven't had much previous experience reading novels, throw up their hands and tell me that they don't know what is going on in the text, even though they say that they understand the meaning of individual words, perhaps it is not because their social life is impoverished and they are not used to complex embedments as such. Perhaps it is—at least in part, that is—because the frequency and kind of such embedments in literature place demands on their mindreading skills that may exceed what they are used to in their daily social exchanges, and it takes both time and effort to adapt to those demands.

This is why, from a cognitive literary perspective, it makes particular sense to speak of the novel as experimenting with, rather than reflecting, "realistically," this particular aspect of human psychology. I have argued something along similar lines in chapter 2, in which I showed that writers can intuitively follow the real-life dynamics of associating more vigorous mindreading with lower social standing, but then they also can, just as easily, ignore and subvert this particular feature of real-life mindreading. Realism, it seems, is what realism does, particularly in a genre as tightly bound to it in cultural imagination as is the novel.

"Realism," of course, is a term that is notoriously slippery and subjective.[8] There is something paradoxical, as Troscianko reminds us, about the fact that we require it "to converge with our expectations about cognition, which may themselves be subject to (systematic and interesting) errors."[9] Perhaps we are better off shifting the terms of our discussion and considering the critical obsession with realism as a fascinating cognitive cultural phenomenon in its own right—worthy of studying as such but not something one would want to lean on too heavily in a critical analysis.

5.2 Novels: Still Nothing but Trouble

But let us say that we push aside the pesky issue of realism. Still more trouble awaits us as we consider the relationship between embedded mental states and the novel as such. Especially in the novel's more recent incarnation (i.e., Tolstoy's *Anna Karenina* as compared to Heliodorus's *Aithiopika*), its treatment of consciousness makes it *the* genre most dependent on complex embedment. As Andrew Plaks puts it, "to say that the novel as a genre deals with human consciousness . . . does not set it off from other literary genres, but, as a matter of proportion, the degree to which the novel does so is indeed rather unique."[10] Just so, while no work of literature can construct human consciousness without embedding at least some complex mental states, the degree to which the novel embeds them is indeed rather unique.

Recall, for instance, that one of the "defining criteria of the genre" is irony and that an author's "ironic reflection on the product of his own creation" calls almost incessantly for the reader's processing of high levels of intentionality.[11] Or consider works that do not cultivate irony but are still characterized by "radical reflexivity," for instance, autobiographical novels about autobiographies, such as Christa Wolf's *Patterns of Childhood* (1976).[12] While containing "autobiographical traces," *Patterns of Childhood* focuses "more on autobiographical writing as a theme, elaborating and challenging the genre from within," a challenge that directly depends on the reader's awareness of the author's embedded intentionality.[13]

Yet to claim that the novel as a genre is most obviously associated with complex embedment is to ask for trouble. The reason for this is that the critical discourse of the "rise of the novel" comes with its own controversies, and if I hitch my cognitivist wagon to that discourse, I inherit those controversies. Specifically, by saying that massive-scale embedment of mental states is an essential feature of the psychological novel, I can be seen as courting the charge of determinism, which has been haunting historians of the novel. Let us take a closer look at that charge.

As Plaks explains, determinism used to be associated with scholars of the epic—who "observed the appearance of that form in widely separate cultures and therefore assumed it to be an *inevitable* phenomenon of human creativity"—but it has now migrated to the novel. Determinism rears its ugly head when one notices the "striking correspondence between . . . essential qualities of the novel" in the European and Chinese traditions and "the fact that these comparable developments occur" around the sixteenth to eighteenth centuries, that is, "at a time of *limited* mutual influence."[14]

In addition, one may look at socioeconomic factors that correlate with the rise of the novel in some cultures and notice that, in other cultures, the novel arose in their absence. For instance, "the relation demonstrated by many Western scholars between the rise of the novel and the social and economic development of the pre-modern period also describes quite well the context of the emergence of full-length prose fiction in China,"[15] but this relation doesn't obtain for the history of the early Japanese novel.[16]

Consequently, one may be tempted "to conclude that the emergence of such a genre of . . . prose fiction may represent an inevitable function of human culture, bound to appear in any literary civilization regardless of its particular course of historical development."[17] And, with that, the torch of determinism appears to have been successfully passed from scholars of the epic to scholars of the novel.

At first blush, the cognitive approach only makes things worse. A cognitive literary theorist, such as myself, who sees the massive embedment of mental states as constituting an "essential quality" of the eighteenth-century European and Chinese novel *as well as* the eleventh-century Japanese novel, may be tempted to see the advent of such an embedment as a predetermined "outcome of human creativity." The temptation may be particularly strong because it is so easy for us to focus on the universalist aspect of the cognitivist discourse—which is, to quote Webb Keane again, that "theory of mind and intention-seeking are common to all humans"—while losing sight of the crucial qualification of that universalist stance, which is that those cognitive adaptations "are elaborated in some communities and suppressed in others."[18] Both elaboration and suppression can take myriad forms and be integrated with such factors as socioeconomic conditions, political agendas, and intellectual history.

But if the massive complex embedment of mental states that we associate with the novel happened to arise in societies that encourage particular forms of mindreading, then there is nothing predetermined about it. Societies that regulate their mindreading energies differently end up with different clusters of mind-modeling artifacts. I mentioned already the Gisalo songs of Bosavi. These are deemed successful if the performers manage to get under the listeners' skins, while the listeners *both want to be affected by a song and resist it.* This give-and-take between performers and listeners assumes particular poignancy because it takes place in a culture of opacity, in which people are not supposed to be attributing mental states to each other.[19]

Once you learn of such complex forms of literary production, an argument about the "inevitable" rise of the novel as the pinnacle of sociocognitive complexity becomes even less compelling. Because human cultures' engagement with theory of mind is dynamic and open-ended, so are the forms that mindreading takes in a given community. Hence, when we talk about the complex embedment of mental states in plays, novels, and narrative poetry, we must remember that this is literature as it happened to be here now and not the expression of some platonic ideal of what it should be.

To conclude, reconstructing the history of complex embedment in literature is a tough balancing act. One is hampered by the scarcity of surviving texts. And even when there are enough texts to go on, one has to resist the grand narrative of the *inevitable* rise of a particular genre that would feature large-scale continuous embedment of complex mental states. One focuses instead on the *probability* of the emergence of self-reflective literary narratives in communities that encourage particular forms of mindreading. Keeping these limiting factors in mind—*"not inevitable but probable under certain circumstances"*—one may come up with a series of preliminary hypotheses These can then be tested and corroborated by others—or refuted!—if the evidence from a particular literary tradition weighs in against them.[20]

5.3 "Men Were Deceivers Ever"

> Utnapishtim said to his wife, "All men are deceivers, even you he will attempt to deceive."
> —*The Epic of Gilgamesh*

> When my love swears that she is made of truth, I do believe her, though I know she lies.
> —Shakespeare, Sonnet 138

> How early he was able to dissemble . . .
> —Pushkin, *Eugene Onegin*

> Lying, in essence, is theory of mind in action.
> —Victoria Talwar, Heidi Gordon, and Kang Lee, "Lying in the Elementary School Years: Verbal Deception and Its Relation to Second-Order Belief Understanding"

Here, then, is one such working hypothesis. It appears that the further back one goes in time, the likelier it is that third-level embedments in literature are created by portraying characters who intentionally deceive other characters.[21] This is in contrast to more "modern" literature, in which third-level embedments are created by a much wider variety of social contexts, which include deception but are by no means limited to it.

Such is my hypothesis, and, right away, I foresee more trouble. For instance, I put scare quotes around the word "modern," to stress that modernity, thus understood, is diachronic rather than synchronic. This is to say that if a transition from a primarily deception-driven embedment to a more varied type does take place (for I don't claim this to be a universal phenomenon), different national literary traditions go through it at different time periods. One should thus be wary of seeing some form of cultural influence and hence causality in what is likely to be a coincidence, as, for example, the fact that both the English and Chinese novel seemed to have gone through that kind of transition around the same time period.

The flip side of the danger of explaining too much by cultural influence is explaining too little. Over the past thousand years, very few national literatures existed in isolation from each other. As Haun Saussy puts it, "many of the most influential works in any tradition are translations, not 'native' compositions."[22] And even those that can be considered "native" compositions bear numerous debts to foreign predecessors. Take for instance, Henry Fielding, one of the avowed "fathers" of the English novel, whose 1749 *Tom Jones* echoes *Don Quixote*, the ancient "foundling" romances, and *The Iliad*. One cannot, in good faith, speak about a discrete "English" literature: depending on which genealogical path we choose, we can trace a history of a particular genre—and thus its patterns of complex embedment—to the French, Spanish, ancient Roman and Greek, or biblical literary tradition.[23] As I see it, it is impossible to use English literature to test my hypothesis about deception as the primary engine of complex embedment at some early point in its history. For what would be considered "an early point" for such a hybrid tradition? *Don Quixote*? Plutarch's *Lives*? *Aithiopika*? *The Iliad*?

This is why we should count ourselves very lucky on the rare occasion when we come across a relatively well-preserved national literature that functioned, for a long period of time, in isolation from other literary traditions and whose formative influences during the shift from complex embedment driven exclusively by deception to complex embedment driven by a wider set of representational means are well documented. Such is the case with Russian literature, in which one such shift can be traced to 1760–1830.[24] During that period, Russian writers began imitating French and English models and by doing so drastically changed the pattern of embedment hitherto prevalent in works of fiction. In the next section, I first briefly recount the history of this shift and then look at some patterns of embedment in the works of nineteenth- and twentieth-century Russian writers.

5.4   What Happened in Russia

If we look at Russian medieval texts, explicitly positioned as literature (as opposed, that is, to historical chronicles and hagiographies), such as Fedor Kuritzyn's *The Tale of Dracula* (ca. 1490), Ermolay-Erazm's *The Tale of Peter and Fevroniya* (1547), the anonymous *Tale of Misery-Luckless-Plight* (seventeenth century), and the picaresque *The Tale of Frol Skobeev* (1680–1720), we notice that all of them achieve complex embedment through plots of deception.

For instance, the blood-curdling *The Tale of Drakula* (*Povest' o Drakule*) tells the story of a Romanian prince, Vlad Drakula, who deceives a Turkish king. When the king sends his ambassador to Drakula, demanding tribute, Drakula hosts the ambassador lavishly, dazzles him with his wealth, and asks him to pass the following message to the king: "Not only am I ready to pay the tribute, but I also want to become his vassal, putting my army and my wealth at his beck and call. Only tell him that when I go to him, he must make sure his people don't harm me and my army, and I will follow you very shortly, along with my tribute." Drakula *wants* the Turkish king to *think* that he *intends* to become his vassal. When the gullible king lets Drakula and his army deep into his territory, Dracula attacks the unprotected cities, plunders their wealth, sadistically murders their inhabitants, repatriates the Christians who used to live there, and sends the king a sarcastic message asking if he wants more of Drakula's service. "And the king couldn't do anything with him and was only covered with shame."[25]

In *The Tale of Peter and Fevroniya* (*Povest' o Petre i Fevronii*), an evil dragon assumes the appearance of a local prince and starts visiting that prince's wife, forcing her to have sex with him. When the wife tells her real husband about those visits, he implores her to use her "seductive charms" to learn what keeps the dragon alive and how he can be killed. "Holding the husband's words in her heart," the wife then approaches the dragon with flattering speeches—she *wants* him to *think* that she *admires* him—and asks if his omniscience extends to knowing what would cause his death. The dragon tells the woman that he is destined to be slain by a man named Peter, which is the name of the prince's own brother. Then, one day, as Peter is visiting his brother and his wife, he is confused because, having just seen the prince in his sister-in-law's chamber, he then encounters him immediately afterward in a different room. But when the prince tells him that he has been in this room all along, Peter *realizes* that the dragon *wants* him to be *afraid* of killing his own brother and so appears to him as the prince. ("Those, brother, are the intrigues of the sly dragon: he assumes my appearance, so that I would be afraid to kill him, thinking this is you—my brother.")[26]

The anonymous seventeenth-century narrative poem *The Tale of Misery-Luckless-Plight (Povest' o Gore-Zloschastii)* tells the story of a young man from a well-to-do family who doesn't listen to admonitions of his parents and as a result finds himself alone and destitute, far away from his hometown.[27] He works hard, gains wealth and respect, and is about to marry a young woman of his choice, but then he makes the mistake of boasting at a party about his recent successes. Misery overhears this bragging and decides to show him that nobody can outwit it and escape its hold. After giving some thought about the best way to influence his victim,

> evil Misery devised cunningly
> to appear to the youth in his dream:
> "Young man, renounce your beloved bride,
> for you will be poisoned by your bride;
> you will be strangled by that woman;
> you will be killed for your gold and silver!
> Go, young man, to the tsar's tavern,
> save nothing, but spend all your wealth in drink;
> doff your costly dress, put on tavern sackcloth.
> In the tavern Misery will remain,
> and even Luckless-Plight will stay—
> for Misery will not gallop after a naked one,
> nor will anyone annoy a naked man,
> nor has assault any terrors for a barefooted man."[28]

Misery *wants* the youth to *think* that his fiancée only *wants* his money and that to stay safe from people who are after his wealth, he ought not to have any. When the young man doesn't believe his dream, Misery hatches a more devious plan:

> The young man did not believe his dream,
> but evil Misery again devised a plan,
> appeared as the Archangel Gabriel,
> and stuck once more to the youth for a new plight:
> "Are you not, youth,
> acquainted with poverty and immeasurable nakedness,
> with great paucity and dearth.
> What you buy for yourself is money wasted,

But you, a brave fellow, will still survive!
They do not beat, or torture naked people,
or drive them out of paradise,
or drag them down from the other world;
nor will anyone annoy a naked man,
nor has assault any terrors for a naked man!"

Misery *wants* the youth to *think* that Archangel Gabriel himself *wants* him to give up his wealth. This time the deception works, and the young man falls right into Misery's clutches:

The young man believed that dream:
he went and spent all his wealth in drink.[29]

And we have already seen how a plot of deception plays itself out in the late seventeenth-century *The Tale of Frol Skobeev*. As a "likable and clever delinquent," Frol rises to wealth and nobility through bribery, crossdressing (see figure 5.1), and blackmail, that is, through social situations rich with opportunities for deception—and complex embedments.[30]

Figure 5.1
Frol Skobeev, dressed as a woman, is plotting his seduction of a courtier's daughter. Scene from the production of the Moscow State Historical-Ethnographic Theater. (Copyright © 2013 МГИЭТ; http://etnoteatr.ru/komediya-o-frole-skobeeve.html)

The early 1760s saw a watershed moment in the development of the national literature because, for the first time, works of European fiction entered Russian cultural imagination. A group of writers, associated with the Cadet School, "set about the systematic translation of English and French novels": "Lukin and Elagin translated Antoine Prévost's *Adventures of Marquis G., Or, The Life of a Nobleman Who Abandoned the World* (1756–61), and Semyon Poroshin translated the same author's *English Philosopher* (1761-7). The novels of Henry Fielding, René Lesage, Pierre Marivaux, and Daniel Defoe's *Robinson Crusoe* were also translated. These [translations] provided the Russian public with entertaining reading in addition to acquainting it with those works which had already become part of the culture of every literate person in western Europe."[31]

And then, almost overnight, Russian literature changed. Alongside embedment driven by deception, there appeared embedment driven by the buildup of complex emotions. Fedor Emin (1735–1770) a prolific writer of foreign extraction (his original name may have been Mahomet-Ali Emin), known as the first Russian novelist, started publishing works of fiction imitating French sentimentalism. Here, for instance, is a plea of a young man from Emin's 1766 epistolary novel *Letters of Ernest and Doravra* (*Pisma Ernesta i Doravry*), in which the anguished lover *hopes* that his beloved will *pity* the man who *knows* that he won't be able to stop *thinking* about her even when they part forever:

> Forget my fault and know that the love that's devouring me deserves punishment, not contempt. No one is angry at a person condemned to death; everyone pities him; and if you, heavenly beauty, follow the way of worldly justice, you will pity the miserable, from whom this letter will be the last, who can't cause you more chagrin, and who, going to his eternal confinement, carries with him the fiercest memory of your charms, which will never cease tormenting all his thoughts, his feelings, and his whole nature.[32]

Complex emotions continue to drive embedment in perhaps the most famous late eighteenth-century tale, "Poor Liza" (1792), by Nikolai Karamzin, the writer known as the "Russian Sterne." "Poor Liza" is a story of a love affair between a gentleman and a peasant girl who kills herself after he abandons her. It is told, crucially, by the narrator, who wants his readers to know early on that he "loves the objects that touch [his] heart and make [him] cry the tears of tender sorrow."[33]

Emotional responses of this sentimental narrator color every important scene. Here, for instance, Liza is sitting on the riverbank, imagining what would happen if Erast, the kind and handsome gentleman she met recently were a poor shepherd and, hence, her social equal—"He would look at me affectionately—perhaps take my hand in his. . . . A dream!"—when she hears the splash of oars and sees Erast approaching her in a boat:

> All her little veins trembled, but, of course, not from fear. She rose, wished to go and couldn't. Erast leaped onto the shore, approached Liza and—her dream having come partially true—*he looked at her affectionately and took her hand in his.* . . . Ach! He kissed her, kissed with such fervor that the whole universe appeared to her to be on fire. "Darling Liza!," said Erast, "Darling Liza! I love you." These words resonated in the depth of her soul as a heavenly, ravishing music; she hardly dared to believe her ears and . . . But I throw down the brush. I will only say that that minute Liza's timidity disappeared. Erast learned that he was beloved, beloved passionately by a fresh, pure, open heart.[34]

Words fail the narrator, repeatedly. When Liza's "dream comes true," he is so fused with the speechless protagonist that all he can say is "Ach!" And when Erast confesses his love, the narrator simply "throws down the brush." That is, he *wants* us to *know* that he is as *overwhelmed with emotion* as is his innocent, deeply feeling protagonist.

This pattern of embedment continues throughout the story. The narrator keeps drawing readers' attention to his own feelings as he paints his characters' emotional reactions. Or he claims to be incapable of doing so and hence invites the reader to imagine those reactions. Although the story still contains its share of lies—for instance, Erast will eventually abandon the "poor Liza" in spite of all his promises—complex embedments generated by deception are dwarfed by embedments generated by the give-and-take between the narrator and readers.

5.5 Unreliable Narrators and Eavesdropping Characters

Karamzin's fiction as well as his autobiographical *Letters of a Russian Traveler* (1789–1790), modeled on Sterne's *A Sentimental Journey through France and Italy* (1768), had a profound influence on several generations of Russian writers. But, even more important, those writers continued to be shaped by their contact with European literature, for once those floodgates opened, they never (fully) closed.[35] This meant a constant exposure to the eighteenth-century European writers' experimentation with new ways of representing fictional consciousness and hence to new ways of embedding complex mental states.[36]

We can briefly speculate here about various historical factors—such as the economic and political reforms of Peter the Great, who forced his compatriots to open up to the world beyond their geographical borders—which may have made some communities in the early days of the Russian Empire particularly keen on elaborating their mindreading practices. We can further say that this new interest in their own and other people's intentions may have continued to contribute to the development of literature throughout the respective rules of Elizabeth and Catherine II, what with their ties to Europe and their support for the arts and higher education. Conversely, we can say that, when under socialist realism in the 1930s–1980s, the range of other people's intentions, both within and outside the national borders, was largely constricted to "for us" and "against us," it hampered the ironic self-reflectivity of the novel and narrowed down the range of minds to be read into it. (This argument works as long as we are aware of its limited scope, for, important as sociopolitical history may be to the history of mindreading in literature, it neither defines nor determines it.)

So keeping in mind those distal historical causes, as well as the more immediate literary contexts, both European and national, we can say that one way in which Russian writers of the first half of the nineteenth century expanded their repertoire of embedments was by focusing on the mind of the narrator. For to look for complex embedments in the works of Alexander Pushkin, Mikhail Lermontov, and Nicolai Gogol is to come across, again and again, an idiosyncratic or even unreliable narrator.

Consider, for instance, the opening of "The Shot," the first short story from Pushkin's *The Tales of Belkin* (1831): "We were stationed in the small town of ***. Everyone is familiar with the life of an army officer. In the morning, drill and riding practice, dinner at the regimental commander's or in a Jewish tavern; in the evening, punch and cards."[37]

This is our first sighting of the narrator, who hastens to tell us not just that the life of an army officer is boring but also that "everyone" knows it's boring. At this point, we don't yet know why it is so important for him to get this point across. It becomes clear later on, when we realize that this young officer has "a romantic imagination" and that the tedium of army life may have made him particularly susceptible to romanticizing his acquaintances.[38]

To map out this opening in terms of its embedded mental states, the narrator *wants* us to *think* that anyone would be *bored* with this routine. Moreover, the implied author *wants* us to *notice* the narrator's *eagerness* to establish the dullness of army life as an incontrovertible fact.

Take another opening sentence, that of Nicolai Gogol's "The Overcoat" (1842):

> In the department of . . . but it would be better not to say in which department. There is nothing more irascible than all these departments, regiments, offices—in short all this officialdom. Nowadays every private individual considers the whole of society insulted in his person. They say a petition came quite recently from some police chief, I don't remember of what town, in which he states clearly that the government decrees are perishing and his own sacred name is decidedly being taken in vain. And as proof he attached to his petition a most enormous tome of some novelistic work in which a police chief appears on every tenth page, in some places even in a totally drunken state. And so, to avoid any unpleasantness, it would be better to call the department in question a certain department. And so, *in a certain department there served a certain clerk*.[39]

What is going on here? In the words of another devotee of unreliable narration, Vladimir Nabokov, "The Overcoat" can be summed up thus: "mumble, mumble, lyrical wave, mumble, lyrical wave, mumble, lyrical wave, mumble, fantastic climax, mumble, mumble, and back into the chaos from which they all had derived."[40] The narrator starts off briskly enough—"In the department of"—but immediately changes his mind: "it would be better not to say in which department." He then hastens to justify this mumbling with more mumbling: you know how those officials are; they get offended easily; just look at that police chief of I-don't-remember-which town. By the time we get back to the actual story of the clerk, we are, to quote Nabokov again, deep in "a grotesque and grim nightmare making black holes in the dim pattern of life."[41]

But let us leave off those lovely metaphors and see what kind of "thinking about thinking people" this paragraph may expect from its readers. The narrator *doesn't want* to name the department because he is *afraid* of being persecuted by people who *don't understand* the difference between a novel and a denunciation. The implied author, meanwhile, is doing something even more interesting. He wants his reader to be that narrator. That is, he *wants* his reader to *imagine* what it feels like to be a person who is *compelled* to tell a story yet is *anxious* about the social implications of the whole business of storytelling.

Thus Pushkin and Gogol. More odd characters itching to tell their tales are waiting for us on the pages of Lermontov's *A Hero of Our Time* (1840). This novel is divided into five parts, which are narrated by three different people—in the words of James Wood, not a single "reliable storyteller among them."[42] Since other scholars have explored this aspect of Lermontov's writing in depth, I will focus here on something else. Observe that, even as he experiments with such sophisticated strategies for embedding complex mental states as the narrator's unreliability, Lermontov doesn't shun other, older and (arguably) cruder ones. Thus, in addition to lying, his novel often relies on its junior cousin, eavesdropping. For, what is eavesdropping but a shortcut for a very particular mindreading dynamic? If lying is *wanting* others to *think* that your *thoughts* are something other than what they really are, then eavesdropping is *not wanting* others to *know* that you *know* something important about their real *thoughts*.

Lermontov is no worse an offender here than Cao, Austen, or Emily Brontë. If Dai-yu can eavesdrop on Bao-yu and Xiang-yun; Anne Elliot on Captain Wentworth and Luisa Musgrove; and Heathcliff on Catherine and Nelly,[43] then, surely, Lermontov's protagonist is entitled to one or two—or, as it happens, eight—instances of fateful overhearing of other people's conversations in "Princess Mary" alone. ("Princess Mary" is one of the five stories that make up *A Hero of Our Time*.) So frequently does Lermontov arrange putting his narrator in the know through eavesdropping that, according to Nabokov, we soon stop registering it as something out of the ordinary: "the author's use of this device is so consistent throughout the book that it ceases to strike the reader as a marvelous vagary of chance and becomes, as it were, the barely noticeable routine of fate."[44]

As cognitive literary critics, we must recognize eavesdropping as a handy sociocognitive tool available to writers. If used sparingly (or, as in the case of *A Hero of Our Time*, brazenly), it complements both that old workhorse of complex embedment—lying—and the shinier, newer machinery of unreliable narration. It takes all kinds of complex embedments to construct a literary subjectivity, so a writer, even one destined to enter a pantheon of national literature, can ill afford to spurn any of them.

Speaking of not spurning old workhorses, recall the stanza that describes the protagonist's lovemaking in Pushkin's *Eugene Onegin*: "How early he was able to dissemble, / conceal a hope, show jealousy, / shake one's belief, make one believe," and so on. What Onegin is doing here is putting on one false front after another. Yet he is neither a picaro in the mold of Frol Sko-beev, nor what Haiyan Lee would describe as a groveling "pipsqueak,"[45] nor a liar (indeed, he may challenge to a duel a person who would accuse him of lying). Instead, he is a literary heir of Dorimant and other aristocratic wits from English Restoration comedy, who signal their depth and complexity by playing mind games with the willing ladies of their acquaintance. A bet-ter social class of deceivers thus comes into play as Russian Romantics keep mining the mother lode of deception even while discovering new ways to embed complex mental states.

5.6   The Poetics of Shame and Self-Deception

Back in the 1830s, the idiosyncratic narrator was not the only exciting new path to third-level embedment explored by Russian writers.[46] Other paths involved portrayal of manipulative behaviors, such as hypocrisy; of tangled motivations, such as self-deception; and of complex social emotions, such as shame.[47]

We start with shame. No national literary tradition is ever the same after it discovers the sociocognitive potential of shame, especially if it is also compounded with lying. But before we get to the man who made the most of it, Fedor Dostoevsky, let us see what shame did for Pushkin in the early days of modern Russian literature.

Take again "The Shot," from Pushkin's *Tales of Belkin*. Its plot centers on a gentleman named Silvio, encountered by the narrator during his stint in the army. One evening, Silvio, who has a reputation for being a crack shot, is insulted by another officer and, instead of challenging him to a duel, lets it pass. The narrator, who used to think of Silvio as a mysterious and intrepid Romantic hero, now feels awkward around him: "But after that unfortunate evening the thought that his honor was stained and by his own fault had not been washed clean never left me and prevented me from behaving with him as before; I was ashamed to look at him."[48] Being ashamed on another's behalf presupposes a very complex embedment: the narrator *imagines* what it *feels* like to *know* that other people *think* that you are a coward.

Silvio easily intuits his young friend's feelings: "Silvio was too intelligent and too experienced not to notice it and not to guess the reason for it. It seemed to pain him; at any rate I noticed a couple of times that he wished to talk with me; but I avoided such occasions, and Silvio gave it up."[49] The story thus continues to unfold through a series of complex embedments. The narrator *realizes* that Silvio *knows* that the narrator *feels awkward* around him and that Sylvio wants to talk to him, and he makes a point of avoiding such occasions.

How do we make sense of his behavior? We may assume, for instance, that the narrator thinks that Silvio would try to justify his reluctance to fight a duel and cannot conceive of any justification that would make any difference in his perception of the situation. That is, the narrator is afraid of feeling more shame on Silvio's behalf after their conversation and so does everything he can to prevent it.

But if shame is a highly generative social emotion when it comes to embedded mental states, so is self-deception, an offshoot of deception. In "The Shot," it turns out that the reason that Silvio didn't want to cleanse his "stained honor" was that he felt that he couldn't put his life even at minimal risk because of *another* duel that he was hoping to fight one day. A while back, a dashing young aristocrat had incensed Silvio by seeming to be indifferent to danger while standing there waiting for Silvio to pull the trigger during their duel, and Silvio decided to take a rain check on his shot until the Count would have more reasons to value his life.

When that hour does come (the narrator will learn about it later, from a different source), Silvio has the satisfaction of seeing his formerly dauntless adversary tremble while waiting for his shot, because, being newly married to a lovely young woman, he now indeed has strong reasons for not wanting to die. Silvio spares his victim because he hopes that, from now on, the Count will live his life writhing in shame, unable to forget his instance of less-than-manly behavior. The Count, however, is not the type to obsess over the past. As John Mersereau Jr. explains, "Of course, the mental anguish with which Sylvio [*sic*] seeks to poison the Count's life is based on a reading of how he, Sylvio, would react in the Count's place, and the Count behaves otherwise. Ironically, the diabolic revenge to which Sylvio devotes years of preparation proves worthless."[50]

Silvio *assumes* that the Count will *feel as anguished* about his humiliation as Silvio would have *felt*, but he is mistaken. He is deceiving himself—a bright early specimen in the gallery of Russian literary protagonists who find ever new ways to turn their cages into fool's paradises.

5.7   The Original Cringe Factor

Later in the century, shame becomes a wellspring of complex embedment in the novels of Dostoevsky. As Deborah A. Martinsen puts it in her study *Surprised by Shame: Dostoevsky's Liars and Narrative Explorers*, "In mobilizing shame as a narrative strategy, Dostoevsky adds shame's affective and cognitive synergy to the recursive relations among author, reader, and text. The activity of writing exposes characters to readers' views; the activity of reading positions readers as witnesses."[51] In other words, Dostoevsky doesn't merely want his characters to be aware of their own or others' shame but he also wants his readers to be ashamed—and to *know* that they are *ashamed*—on behalf of those ashamed characters.

Martinsen sees Dostoevsky as prefiguring insights of the later-day philosophers of shame, such as Emmanuel Levinas and Jean-Paul Sartre, who wrote about the "reflected assessment of the self" involved in shame.[52] From the cognitive literary perspective, what I find particularly fascinating about the dynamic that Martinsen describes is that Dostoevsky exploited one of the most powerful social emotions known to humans to expand the repertoire of fictional embedments beyond what may be familiar to us from our daily life. His characters wallow in layers of embarrassment and self-exposure until no one around them is able to take it anymore, and *then* they add more to make it yet worse.

Think of this original cringe factor as yet another case of a writer's experimentation with the reader's social brain—experimentation, that is, as opposed to a faithful reproduction of any "real-life" dynamics. Can we process these emotionally gripping complex embedments of mental states? Yes, we can. Are they "realistic"? If your answer is, "Well, not in my personal experience, but I wouldn't put it past those crazy Russians," I suggest checking in with a Russian of your acquaintance.

As far as *this* Russian remembers, the ever-widening and ever-deepening circles of mortifying self-awareness that Dostoevsky cultivates in his novels is not something that I have encountered in reality. But, of course, now, thanks to Dostoevsky, I can imagine surfing those dark waters and suspect that one day a conversation with friends and family may yet veer in that direction. As the literary critic Lidiya Ginzburg puts it, "Dostoevskian sensibility [Достоевщина] as a moral and ideological phenomenon is highly repugnant to me, and not because it is alien, but because, it is, to a degree, inherent in me."[53]

I think I understand the reason why we may treat Dostoevskian sensibility as, "to a degree, inherent" in us. On some level, our mindreading adaptations do not differentiate between attributing mental states to real people and to fictional characters.[54] Having processed those complex embedments in a novel—that is, having experienced ourselves as being capable of such deep, involved, yet coherently articulated mental states—we may now, indeed, believe that a day may yet come when we will find ourselves luxuriating, with a sickening abandon, in the embarrassment caused to others and ourselves by our self-exposure. That the day keeps being indefinitely postponed does not contradict the reality of having had those feelings one fine afternoon while reading *The Idiot* or *The Brothers Karamazov*.

Reenter lying. Here is a passage from *The Idiot* (this one happens to be relatively low on the cringe factor), in which the protagonist, Prince Myshkin, is reflecting on the conversation he has just had with the old General Ivolgin. The General, a drunkard and inveterate liar, has left the house thinking that Myshkin, a naïve young man, believed his tall tale about the General's former tender friendship with the Emperor Napoleon: "He also understood that the old man left the house intoxicated by his success, yet he also had a presentiment that he was one of those liars who, though lying up to the point of voluptuousness and even self-oblivion, at the very peak of their euphoria, still suspect deep inside that others do not believe them and cannot possibly believe. In his present state, the old man could come to his senses, be extremely ashamed, surmise that [Prince Myshkin] was boundlessly compassionate toward him, and become affronted."[55]

I won't bother spelling out the obvious complex embedments of mental states that structure this passage. The reason I quoted it (following Martinsen's lead) is that I wanted to illustrate the new role that lying, once it joins forces with shame, begins to play in the Russian novel.

In medieval Russian literature, lying was instrumental, antagonistic, and private: it helped protagonists to survive or gain an upper hand over their enemies. In contrast, in nineteenth-century literature, shame-driven lying becomes, paradoxically, prosocial, occurring, as Martinsen points out, largely "in the public sphere." Dostoevsky's liars, such as the old Ivolgin, "lie because they are ashamed of themselves. They do not intend to [defraud] others but to create a public persona that will be accepted and admired. They lie to affirm their own self-worth and thus their social worthiness."[56] There is a performative aspect to their lying, which implicates others as (more or less) appreciative spectators.

Other Russian writers, such as Turgenev, Tolstoy, and Chekhov will take shame and self-deception—already brimming with embedded mental states—and add something else to the mix: imperfect introspection. Their characters will not quite trust their own emotional reactions. Their torturous vitality will often come from querying their motives when they feel ashamed of themselves or on behalf of others, from being aware of their double consciousness (i.e., aware of seeing themselves through the eyes of imagined others), and from suspecting that they deceive themselves.

Such, then, is one story one can tell about the early history of Russian literature if one focuses squarely on the role of embedded mental states in the development of literary imagination. We start out, in the sixteenth and seventeenth centuries, with complex embedment driven mainly by antagonistic lying. Then, in the 1760s, the influx of western European novels significantly expands the range of representational strategies for embedment. The expansion continues in the 1790s–1830s with new embedments arising from interactions between various idiosyncratic narrators and their implied readers, as well as from the fictional exploration of hypocrisy, shame, and self-deception. Then Dostoevsky perfects the cringe factor and recasts lying as a public performance, and later yet, Tolstoy and Chekhov experiment with nuances of self-deception and imperfect introspection. To sum it up, while lying as the engine of complex embedment in literature never goes out of fashion, it gets continuously reinvented, now by being layered with the author's ironic self-reflection, now by being integrated with a variety of complex social emotions.

5.8   What Happened in China

Let us now turn to another national literary tradition, one that has developed, until relatively recently, independently from European influences and can, as such, be particularly illuminating as a test case for our working hypothesis about lying and literary history. Can we say that the further back one goes in time, the likelier it is that third-level embedments of mental states in Chinese fiction arise mostly from situations in which characters intentionally deceive other characters? And can we also say that after a certain point in time, more and more complex embedments are created by social contexts other than lying, as well as the ones that reimagine lying, integrating it with a variety of social emotions and with nuances of authorial self-consciousness?

Broadly speaking, yes, it seems that we can make such an argument, but with some qualifications. For instance, as we have just seen, in Russian literature, the breakthrough increase of embedment techniques in the late eighteenth century owed to the introduction of French and English models in the 1760s. Chinese literary history developed along a very different path. One way to trace its patterns of complex embedment is to look at the experimentation with literary forms that took place within the "literati" (i.e., scholar-official) culture in the Tang dynasty (618–907 AD) and to compare its patterns of embedment with those that we find in the fiction of the preceding and following centuries.

One factor that makes this comparison challenging is the lack "of general agreement on criteria by which to identify [the] earliest examples" of Chinese fiction.[57] While some critics believe that the first examples of texts that "ceased to be classed as history" and were instead "considered as fiction" appeared only toward the end of the Tang period,[58] others trace it further back, for instance, to the third century, that is, the early years of the Six Dynasties era,[59] or even "to the list of works labeled *hsiao-shuo* in . . . *The History of the Western Han Dynasty*, completed shortly after A.D. 92."[60]

Another complication arises from the expectation of the linear development that seems to be implied by my working hypothesis. Especially given the variety of genres that fed into literature, the fictional status of which remains contestable, we cannot expect to see a "gradual straight evolution" from embedment arising almost exclusively from lying to embedment arising from a broader variety of contexts.[61] The process is more complicated and allows for returns to the earlier forms of embedment, especially in various hybrid genres, including historical fiction (as we will see shortly).

With these caveats in mind, let us compare patterns of embedment in some of the earliest stories that can be arguably identified as fiction, with those in the later Tang period and beyond, and speculate about circumstances that may have triggered the Tang authors' experimentation with contexts for embedment.

Cao Pi's "Scholar T'an" (談生) is dated to the late second–early third century. It tells a story of an old bachelor suddenly blessed with a beautiful wife, who, however, asks him not to "shine any lights" on her at night for three years. They live together and have a son, but when the child is two years old, T'an's curiosity gets the better of him: "One night, lurking and waiting after his wife had gone to bed, he stealthily shone a light on her. From the waist up she was just like any human being, but from that point downward there was no flesh, only dried-out bones."[62]

T'an will lose his wife but, eventually, gain riches and palace employment, for the woman turns out to have been the late daughter of a local prince. The story is very short—about one-third of a page—and T'an's preparing to disobey his wife's injunction is, it seems, its only instance of third-level embedment. T'an *doesn't want* his wife to *know* that he *intends* to find out who she really is—hence all the "lurking" and "waiting after she had gone to bed" and shining a light "stealthily."

Niu Seng-ju's "Scholar Ts'ui" (崔書生), another very short story, is dated to the early ninth century. Its protagonist falls in love with a beautiful woman and marries her without informing his mother. That leads to a deception that will have fatal consequences. Ts'ui *doesn't want* his mother to *think* that he married without her *knowledge*, so he tells her that he had merely "taken a concubine."[63] The mother eventually breaks up the couple, as neither she nor her son know that the young woman is a daughter of a goddess and that staying married to her for at least a year could bestow immortality on Ts'ui and his family.

Yuan Zhen's "Ying-ying's Story" (鶯鶯傳), also from the early ninth century, is a longer piece, centrally preoccupied with its characters' tangled motivations. It tells about the seduction and subsequent abandonment of a beautiful girl from a good family by a young scholar, although the questions of who seduced whom and whether the abandonment was justified remain open. There is no shortage of lies. For instance, Ying-ying doesn't want her mother to know that she loves student Zhang; Ying-Ying may be deceiving either herself or Zhang when she wants him to think that she only summoned him to their initial rendezvous because she wanted to chide him for his improper advances; Zhang may be deceiving himself and his friends when he claims that he decided to abandon Ying-ying in order to guard his virtue against her "bewitching beauty," and so forth.[64]

Yet interweaved with the complex embedments driven by deception and self-deception are those arising from the interaction between the implied reader and the implied author. As Pauline Yu puts it, "Ying-ying" is "a consummate writerly text, one that seems to be talking self-referentially as much about what it is doing as text as about what it as text contains."[65] By having student Zhang explain to the narrator why he "hardened [his] heart" against his mistress (an explanation that may come across as feeble and self-serving); by including a long manipulative letter from Ying-ying; and by having Zhang, Ying-ying, and the narrator write stylized poems reflecting on the romance, the text draws the readers' attention to the "arbitrariness of what it is doing." There are plenty of "behavioral motivations" to choose from, and none is really adequate.[66] As a result, "Ying-ying" becomes a narrative about *constructing* a narrative—rather than about why the characters did what they did—which presupposes an ongoing mutual awareness between the implied author and the implied reader.

That the protagonist and his friends write poems about the affair firmly situates "Ying-ying's Story" within the literati culture. Yet this may not be the most important sign of the narrative's indebtedness to the mid-Tang poetic tradition. A key feature of that tradition, as Stephen Owens explains, was the poets' insistence on shifting their readers' attention from what is being interpreted to the *act of interpretation*. Thus, Du Fu (712–770), Han Yu (768–824), Jia Dao (779–843), Bai Juyi (772–846), and Xue Neng (ca. 817–880), while contemplating something pointedly insignificant, such as a miniature pond, a porch in need of repair, or a tiny patch of bamboo plants, conjured up observers—now disapproving, now sympathetic—even when claiming that those observers' responses do not matter to them. Hence Du Fu in "Deck by the Water" (764), thinking of his intention to fix a sagging porch:

> I suspect I'll be laughed at by those who know it.
> . . . .
> But people are moved by familiar things,
> and I am overwhelmed by grief."[67]

And hence Han Yu in "Pond in Basin":

> I mean it, this old man
> is acting just like a kid,
> he buried a basin and drew some water
> to make a little pond.
> . . . .
> Don't tell me my pond in a basin
> is not completely done,
> I began planting slips of lotus root
> and now they are growing evenly.[68]

Owen sees this "version of the 'private,' with its constant attention to being observed from the outside, [as] ultimately a form of social display, depending on the amused approval of others who are playfully excluded."[69] What I want you to notice is that this social display depends on complex embedment of mental states. The poet invites us to watch him as he watches other people as they watch him. He wants us to know that he is aware of their perspective, perhaps even encouraging us to side with him against that communally sanctioned perspective.[70]

One may speculate that the interest in conflicting interpretations cultivated by the mid-Tang poets found further expression in ninth-century tales of romance, such as "Ying-ying's Story." As Owen argues, the "rise of romance [was] closely related to the development of individual acts of interpretation or valuation" in poetry. Thus, "Ying-ying's Story" "begs us to pass judgment" on its protagonists; yet, "in the end, the disputants are deadlocked," and so, instead of siding with either, readers are left arguing about the validity of those conflicting perspectives and even the possibility that the author may have been personally invested in the situation.[71] This focus on the process of interpretation is what marks "Ying-ying's Story" as an early example of "a fully developed fictional form."[72]

But "Ying-ying" also seemed to go beyond the interest in the act of interpretation cultivated in mid-Tang poetry. It expanded that interest in a direction that was not available to contemporary poets. According to Owen, working in prose allowed Yuan Zhen to delve into minutiae of motivation that might not be amenable to poetic treatment. Looking at "verse renditions of romantic stories, both in quatrains and long ballads," including a poem that Yuan Zhen's friend Yang Juyuan (755–?) wrote about Ying-ying, Owen observes that "prose narratives often give complicated and nuanced accounts of human behavior, [while poetry] for all its undeniable virtues, . . . flattens these complications into purified roles."[73] Thus, in Yuan Zhen's quatrain, the complications of Ying-ying's manipulative letter—in which "the Ying-ying who wants to show the self-effacing concern of a model wife is in conflict with another Ying-ying who is both desperate and enraged"[74]—"are reduced" to such stock description as "her broken heart."[75]

Were we now to construct a straightforward narrative of a gradual expansion of strategies for complex embedment of mental states in Chinese literature, this narrative might go like this: The exploration of readers' consciousness in mid-Tang poetry led to the opening up of new social contexts for complex embedment in ninth-century tales of romance. One of them, "Ying-ying's Story," depends on a much broader range of complex embedments—particularly those driven by the give-and-take between the implied reader and the implied author—than we have encountered in, for instance, "Scholar T'an" and "Scholar Ts'ui," in which embedment was driven exclusively by lying.

Then there is also a long tradition of Chinese drama, in which complex embedments arise from the embodied presence of actors onstage, for instance, from comic disjunctions between the sentiments conveyed by the characters' words and their body language. To see how these three contexts for complex embedments (that is, lying, the give-and-take between the implied author and the implied reader, and the disjunction between words and body language) come together, think of Wang Shifu's thirteenth-century comedy *The Story of the Western Wing* (西廂記). Based on "Ying-ying's Story," the *Western Wing* replicates some of "Ying-ying's" plot-based lies, and it also continues to cultivate the awareness between the implied author and the implied reader through its steady stream of references to classical texts, including poetry.

We would turn next to *The Plum in the Golden Vase*: more intricate lies and references to poetry and philosophy. And then we would inevitably end up with Cao Xueqin's *Dream of the Red Chamber* (ca. 1750–1760), which features lies and classical references, foregrounds its characters' body language, and also talks obsessively about their thoughts and feelings. Voilà! Behold the steadfast movement toward increasingly diverse ways to embed complex mental states in Chinese literature.

I believe that this narrative has merits as long as we also acknowledge fictional texts that disrupt its seemingly smooth course by *continuing* to rely almost exclusively on lying to generate complex embedment. Consider Luo Guanzhong's *Romance of the Three Kingdoms* (三國演義). Written in the fourteenth century—that is, after "Ying-ying's Story" and *The Story of the Western Wing* had demonstrated the possibilities of the expanded repertoire of contexts for complex embedment—this eight-hundred-thousand-word novel features complex embedments relatively infrequently,[76] and when it does, they are mostly driven by lies. Specifically, they are driven by stratagems and manipulations perpetuated by various warring factions, which necessarily involve lying.[77]

To give you a flavor of those stratagems, here is one of *The Three King* *doms'* frequently retold episodes. Wang Yun, a high-level official in the Han government, gives his adopted daughter, Diaochan, to the evil warlord Dong Zhuo as a concubine, in order to sow discord between Dong Zhuo and his adopted son, Lu Bu. Once Dong Zhuo and Lu Bu are both besotted with Diaochan, she takes turns lying to both of them in order to manipulate them: "One day Lu Bu went to inquire after his father's health. Dong Zhuo was asleep, and Diaochan was sitting at the head of his couch. Leaning forward she gazed at the young man, with her hand pointing first at her heart, then at the sleeping old man, and her tears fell. Lu Bu felt heartbroken."[78]

Diaochan *wants* Lu Bu to *think* that she *loves* him and not Dong Zhuo. Later on, when Dong Zhuo accuses her of consorting with Lu Bu, she pretends to want to commit suicide to prove her devotion to Dong Zhuo. That is, now she *wants* him to *think* that she *loves* him and not Lu Bu. And so it goes on, until, driven by anxiety and jealousy and secretly aided by Wang Yun, Lu Bu kills his foster father.

Why did *The Three Kingdoms* rely on the "old" form of embedment instead of building on the new forms compellingly explored by such works as "Ying-ying's Story" and *The Story of the Western Wing*? One possible explanation is that, in contrast to both of them,[79] *The Three Kingdoms* had stronger roots in historical chronicles and folk literature.[80] In fact, notwithstanding its iconic status as one of China's "Four Classic Novels," when critics talk about *The Three Kingdoms*, they often qualify its status as novel, referring to it now as "historical fiction,"[81] now as "China's first successful historical novel,"[82] or even as (note the extra quotation marks!) a "historical 'novel.'"[83]

From the cognitive literary perspective, such qualifications are fascinating. They may reflect, among other things, our intuitive awareness that texts that we call novels today embed mental states at a higher frequency and by a greater variety of means than *The Three Kingdoms* does. Perhaps one reason that *The Three Kingdoms* is considered a novel, and not, say, a fictionalized warfare chronicle, is that it often enters the cultural imagination through *other* works of fiction and thus through a much more variegated repertoire of contexts for complex embedment. For instance, the story of the beauteous and devious Diaochan has been retold in opera, plays, films, and manga series.[84] (One of such intermedial incarnations, *DiaoChan: The Rise of the Courtesan*, was performed on the London stage in 2016 and described by critics as soaring "to the heights of Shakespearean tragedy . . . and never more so than when each character reveals his inner thoughts through soliloquy")."[85] I will revisit this point in chapter 6, with an even more drastic example, showing how being reimagined through other media may lead to a text being considered a novel in the absence of *any* complex embedments (even those driven by deception).

And, meanwhile, we return to our narrative about the gradual expansion of literary contexts for complex embedment. We do so by revisiting *The Plum in the Golden Vase*. Written in the last decades of the sixteenth century and building on touchstones of literati culture for its numerous classical references, it emerged as "the first Chinese novel that was wholly the creation of one author and had no antecedent in the oral tradition."[86]

5.9   Golden Lotus Drives a Servant to Suicide

As it so happens, this famous or, rather, notorious candidate for the role of "first" Chinese novel has a very special relationship with lying. *The Plum* tells the story of an upwardly mobile merchant, Hsi-men Ch'ing,[87] and his six wives and concubines, whose lives are steeped in "deception, bribery, blackmail, profligacy, flamboyant sex, and even murder."[88] Among those familial pastimes, lying occupies a pride of place. Every couple of chapters, a new intrigue blossoms, often starting with a sexual transgression and then snowballing as the characters keep eavesdropping on and framing each other.

What do scholars of Chinese literature make of those swarms of lies? Some view them as integral to the author's larger project of critiquing the corruption of the contemporary imperial court. For, while the story "is set during the reign of Emperor Huizong of Song (1101–1126 CE)," as a political allegory, it "points clearly to contemporary Ming rulers as well."[89] Others consider the characters' eager intriguing as a warped expression of "competing claims of individual feeling and the constraints of conventional morality."[90] Yet others, such as the seventeenth-century commentator Chang Chu-p'o, appreciate the elaborate architectonics of the three-thousand-page novel, in which every little detail becomes a "structural device" used by the author "to accomplish his aims without leaving a trace."[91] For instance, as Chang explains, the "author needs [one character to be driven to suicide after having been framed] in order to bring out as completely as possible the viciousness of [another character]," for it is this second character's "double-tongued troublemaking" that precipitates that "needless suicide."[92]

All of these are compelling arguments, and my "cognitive" perspective by no means invalidates them. Instead, it complements them. For it makes sense to assume that in a complex artifact (such as a novel), a recurrent feature (such as a plot of deception) would end up serving multiple cultural and structural purposes. Let us, then, take a closer look at "the double-tongued troublemaking" that Chang refers to and see how this subplot allows the anonymous author to continuously embed complex mental states *and* engage in a multilevel critique of the parties involved.

In chapter 25, when Hsi-men Ch'ing's purchasing agent, Lai-wang, comes back from a business trip, he learns that while he was away, Hsi-men Ch'ing started an affair with his wife, Sung Hui-lien. The person who informs Lai-wang of his wife's infidelity is one of Hsi-men Ch'ing's concubines, Sun Hsüeh-o. Lai-wang confronts his wife, but she claims that her enemies "made up this tale."[93] This seems to placate him. It may help that by now he has started his own affair with Sun Hsüeh-o.

Another of Hsi-men Ch'ing's retainers, Kan Lai-hsing, has a grudge against Lai-wang. He has a chance to act on his grudge when he overhears Lai-wang, in his cups, railing angrily against Hsi-men Ch'ing and one of his wives, P'an Chin-lien, who (as Lai-wang has been told by Sun Hsüeh-o) has provided cover for the affair between Sung Hui-lien and Hsi-men Ch'ing. Lai-hsing goes to P'an Chin-lien, tells her (falsely) that Lai-wang tried to pick a fight with him, and gives her an exaggerated account of Lai-wang's threats.

The incensed P'an Chin-lien reports Lai-wang's (presumed) threats to Hsi-men Ch'ing. Hsi-men Ch'ing questions Sung Hui-lien, but she swears that Lai-wang "never said any such thing" and that Lai-hsing has "made up this story out of whole cloth."[94] Hsi-men Ch'ing believes her and promises to send her husband off on another long-term business trip. Sung Hui-lien and Hsi-men Ch'ing then agree on a lie that she will tell when others notice a new gift that Hsi-men Ch'ing is about to give her.

When P'an Chin-lien learns that instead of punishing Lai-wang, Hsi-men Ch'ing plans to trust him with another prestigious errand, she convinces him that Sung Hui-lien lied to him about her husband's intentions and that, sooner or later, Lai-wang will take revenge on his master. Hsi-men Ch'ing decides to drive Lai-wang away. He frames him and has him imprisoned. What follows is a long series of lies aimed at making Sung Hui-lien believe that her husband is doing fine, when, in fact, he is being severely beaten in jail.

Sung Hui-lien eventually learns the truth and kills herself. To avoid an official investigation of her death, Hsi-men Ch'ing bribes the court magistrate and concocts a story of Sung Hui-lien being put in charge of the household's silver utensils and hanging herself in fear of retribution when a cup goes missing.

You can see, based on just one episode, what an important role deception plays in *The Plum*. Now let us take a look at how the characters' shenanigans generate complex embedments. I will keep this part of my argument very brief because, at this point, what I have to say here may already be self-explanatory.

When Lai-wang first confronts Sung Hui-lien about her affair with their employer, she *wants* him to *believe* that her enemies *wanted* him to *think* that she has been unfaithful ("some backbiting . . . person . . . must have put you up to abusing your old lady").[95] Later, Lai-hsing *wants* P'an Chin-lien to *think* that Lai-wang *intends* to kill her, and then P'an Chin-lien, in her turn, *wants* Hsi-men Ch'ing to *think* that Lai-wang is *keen* on revenge. Then, when Lai-wang is in jail, Hsi-men Ch'ing *doesn't want* Sung Hui-lien to *know* that he *intends* to force Lai-wang to run away by making his life unbearable. Finally, after Sung Hui-lien kills herself, Hsi-men Ch'ing *wants* the magistrates to *think* that the young woman was *afraid* of being punished for misplacing a silver cup.

Note that although I speak of Hsi-men Ch'ing's *wanting* to shape the magistrates' *thinking* about the reason a young woman in his household *would want* to kill herself, it falls to the reader to reconstruct those and other mental states in this fashion. The novel itself offers almost no explicit references to characters' thoughts and feelings. Instead, as Tina Lu observes, in *The Plum*, "bodies are depicted from the outside, and there is very little internal monologue." What we have, instead, are implied embedments. That is, characters' interiorities emerge from the "matrix of negotiation, of motivation perceived through the prism of other peoples' motivations."[96]

Earlier I listed only a few such implied embedments. There are many more, both in Sung Hui-lien's story and elsewhere in the text. At every turn of the plot, another one springs to life. To make sense of what is going on, readers have to constantly keep in mind what one character *wants* another character to *think* about their own or someone else's *intentions*.

Once we notice this pattern, we can speak of various ways in which it is put to use in *The Plum*. We can say, along with the other scholars quoted earlier, that it serves to present Hsi-men Ch'ing's household as rotten to the core and thus deserving the awful retribution that awaits them, that it critiques the corruption of contemporary rulers; and that it shows what twisted forms individual initiative can assume when (as in the case of women in the patriarchy) it has no better outlet than selfish intriguing. We would do well, however, even as we commit to any of those interpretations, to acknowledge the role of the "cognitive" factor in structuring our response to the story. For, while lying in fiction does not always call for moral condemnation,[97] it invariably opens the door to complex embedment of mental states and, with it, to a very pointed and energetic engagement with readers' theory of mind. Thus, I do not think that it is a coincidence that the text, which is considered to be an important milestone in Chinese literary history, experiments with this kind of intense engagement. Lying is a serious cognitive business, which is why the relentless massive lying that we encounter in *The Plum* is a serious cognitive literary business.

5.10 Lies and "Face"

We have seen, with Russian literary history, how representation of complex social emotions, such as shame, can transform a cognitive literary landscape. Again, Chinese literature developed along a very different trajectory. Still, it is worth noting how often the concern with one's dignity, that is, "face"—which is structurally similar in its effects to shame—motivates characters in *The Plum* and is implicated with their lying.

As Haiyan Lee has shown, the notion of "face" is in and of itself an effective generator of complex embedment in fiction because it conjures the perspective of a character thinking about how they would be perceived by an imagined observer.[98] In *The Plum*, given Hsi-men Ch'ing's social ambitions, the worry about face is ever present. Consider, for instance, a debacle in chapter 12, when P'an Chin-lien first fools around with a page boy and then claims that it never happened and that her enemies cooked up the whole story. P'an Chin-lien's lie works because her loyal servant, P'ang Ch'un-mei, exploits Hsi-men Ch'ing's fear of losing face with his neighbors if he would punish P'an Chin-lien on (as she claims) false premises.

As Ch'un-mei puts it, "This is all something fabricated by someone who is jealous of [P'an Chin-lien] and me. [Hsi-men Ch'ing], you ought to think what you're doing, or you'll only make an ugly reputation for yourself, which won't sound any too good when it gets abroad."[99] Ch'un-mei *wants* Hsi-men Ch'ing to *imagine* what other people will *think* when they find out about his rash behavior.[100] Her manipulative invocation of those judgmental others bolsters a lie—for, in the same breath, she also *wants* Hsi-men Ch'ing to *believe* that the reason P'an Chin-lien was accused of adultery is that other wives *want* to bring her down.

A lie, thus, can gain in persuasiveness when paired with a reminder of one's social vulnerability (i.e., dependence on other people's opinions). This is what happens in chapter 25, that is, the story of the banishment of Lai-wang and suicide of his wife, which I discussed earlier. When P'an Chin-lien wants Hsi-men Ch'ing to believe that Lai-wang considers him his enemy, she makes Hsi-men Ch'ing worry about what other people will think about him. Thus, she refers to "allegations" that Lai-wang makes "in front of people" and assures him that "such allegations would not redound to [Hsi-men Ch'ing's] credit."[101]

Similarly, when P'an Chin-lien wants Hsi-men Ch'ing to think that Sung Hui-lien conceals from Hsi-men Ch'ing the true extent of the enmity that Lai-wang bears him ("Whatever that woman has had to say for some time now has only been spoken on behalf of that slave of yours"), she, once again, brings in public opinion. If Lai-wang defrauds Hsi-men Ch'ing of his money (something that, P'an Chin-lien implies, he surely intends to do), Hsi-men Ch'ing will be too embarrassed to "accuse him of anything," because everybody will have known that he has stolen Lai-wang's wife.[102]

And, again, when Lai-wang is already in jail, tortured for a crime he didn't commit, and P'an Chin-lien learns that Hsi-men Ch'ing is writing a note to the judge asking for his release, she lobbies for "[polishing] off this slave once and for all" by planting an image of jeering neighbors in Hsi-men Ch'ing's mind. Lai-wang, she claims, shall always hold a grudge against his master, even if Hsi-men Ch'ing will go as far as marrying him to someone else, to make up for having taken Sung Hui-lien from him.

For instance, if Lai-wang comes to "report something" to Hsi-men Ch'ing and sees him together with Sung Hui-lien, wouldn't Lai-wang get "angry"? And would Sung Hui-lien then have "to stand up" to greet her ex-husband?

Wouldn't that be embarrassing for Hsi-men Ch'ing? As P'an Chin-lien puts it, "Just to start out with, this alone wouldn't look right. If it got around, not only would your neighbors and relatives laugh at you, but even the members of your own household, high and low, would not be able to take you seriously."[103]

Finally, P'an Chin-lien uses the appeal to face to finish off the poor Sung Hui-lien. She does it by making Sung Hui-lien imagine that other people will never believe her side of the story. Here is how this is set up by the text: After Lai-wang is driven away, just as P'an Chin-lien hoped he would be, she goes between Sun Hsüeh-o and Sung Hui-lien, reporting lies that can't fail to stir up a "sense of grievance and desire for revenge." First, she *wants* Sun Hsüeh-o to *think* that Sung Hui-lien *knows* that Sun Hsüeh-o told Lai-wang about Sung Hui-lien's affair with Hsi-men Ch'ing (which is not true) and that she blames Sun Hsüeh-o for making Hsi-men Ch'ing angry and for making him want to get rid of Lai-wang. Then she goes to Sung Hui-lien, whom she *wants* to *believe* that people in the compound *think* that she *has never cared* about her husband. So she reports to Sung Hui-lien—falsely—that Sun Hsüeh-o tells everyone that Sung Hui-lien is an "old hand at inveigling" her masters "into adultery" and that the tears that she sheds about her husband "are only crocodile tears."[104] These lies precipitate an ugly standoff between Sung Hui-lien and Sun Hsüeh-o, which pushes Sung Hui-lien over the brink and leads to her second, and this time successful, suicide attempt.

5.11   Beyond Lies and Shame

Some comparisons between the early Chinese and the early Russian novel are worth highlighting here. For instance, both *The Plum* and *Eugene Onegin* (Pushkin's "novel in verse") feature continuous embedment of complex mental states. Both cultivate such embedment by having their characters behave deceitfully ("How early he was able to dissemble") and by motivating them through complex social emotions, such as shame. (Onegin kills his best friend in a meaningless duel because he is afraid that others will consider him a coward if he attempts to seek peace.) One important difference between the two texts is that Pushkin *talks* about mental states—those of his characters, his readers, and his poetic persona—incessantly, while the anonymous author of *The Plum* leaves it to the reader to *infer* thoughts and feelings behind behavior.

Another point of comparison, as far as the construction of complex embedments is concerned, is the role played in both novels by references to other texts. *Eugene Onegin* is deeply entrenched in the European literary tradition.[105] The conversation about French, English, and German prose and poetry that the implied author is having with the reader supplies its own steady stream of complex embedments. For instance, the narrator wants his reader to consider the ironic implications of the fact that the main female protagonist, Tatiana, imagines Onegin as the hero of the last novel by Samuel Richardson, *The History of Sir Charles Grandison*, a man who is *not* motivated by shame and is opposed to dueling in principle. As the narrator assures us coolly, "our hero, whoever he might be, / quite surely was no Grandison."[106]

A similar conversation is taking place in *The Plum*. Its frequent evocations of classic Chinese songs and poems presuppose ongoing mutual awareness between the implied author and the implied reader. For instance, when Sung Hui-lien wants to convince Hsi-men Ch'ing that her husband would never curse and threaten Hsi-men Ch'ing behind his back, she asserts that, were Lai-wang to do such a thing, he would effectively be biting the hand that feeds him and he is not that stupid. As she puts it,

> If he should:
>> Live off King Chou's largesse,
>> And yet call King Chou a villain,
> on whom could he depend to make a living?[107]

Sung Hui-lien's mention of King Chou comes close on the heels of an earlier reference to the ancient *Book of Documents* (*Shu-ching*, 書經). That reference, according to *The Plum*'s translator, David Tod Roy, tacitly likens Hsi-men Ch'ing to King Chou, the "evil last ruler" of the Shang dynasty.[108] So, here, while Sung Hui-lien seems to want to emphasize the implausibility of her husband's bad-mouthing Hsi-men Ch'ing, she accomplishes quite the opposite with her quote: she badmouths him herself. For, as Roy explains, the "unmistakable implication" of what she says "is that Hsi-men Ch'ing himself is an evil last ruler."[109]

That neither Sung Hui-lien nor Hsi-men Ch'ing is aware of this implication makes their mutually pleasing exchange profoundly ironic. The implied author wants the reader to know he considers Hsi-men Ch'ing evil, but he also wants us to know that Hsi-men Ch'ing doesn't realize that the argument that he apparently finds convincing is a classical reference that condemns him. Nor is he aware of the grave innuendo of being likened to a *last* ruler—something that the implied author wants the reader to keep in mind as we follow the household's rejoicing at the birth of Hsi-men Ch'ing's son, Kuan-ko. Finally, we know that Sung Hui-lien doesn't know, when she unwittingly calls her lover a villain, that he is about to behave like one toward her and her husband—another nuance in the ongoing give-and-take between the author and the reader.

Earlier, in chapter 4 of this book, I pointed out the ambiguous position of *The Plum* in Chinese literary canon—which some readers view as a work of pornography and others treat as a literary masterpiece. I also showed that scholars who consider it a masterpiece focus on the novel's intentionality, arguing, for instance, that the text's implied author seems to want to remind "the reader of the presence of the narrator somewhere between himself and the story."[110]

That the "cognitive" perspective would strongly support these kinds of hyperintentionalist interpretations of *The Plum* is not at all surprising. The history of the novel as a genre involves two developments entwined in such a way that it is difficult to say where one ends and the other begins. Perhaps I may be allowed to call it a "coevolution" of readers and writers. On the one hand, the novel may be said to constantly cast about for new compelling ways to embed complex mental states of its characters, implied author, and the reader. On the other hand, at least some of its readers may be said to constantly cast about for new ways to read complex mental states into the text. Those are readers who have had significant exposure to literary fiction and thus tend to find characters' motivations not as clear as do less experienced readers, which is to say that they are more comfortable with ambiguity than are less experienced readers.[111] They are also more eager to look for intentionality cues in their social environment, which may translate into a greater awareness of the conversation that the implied author is having with the reader.

It should be pointed out that the comparison between *The Plum* and *Eugene Onegin* still holds when we think of these different types of readers. For instance, *Eugene Onegin* was a staple of the high-school syllabus in Soviet Russia, but the depth of its engagement with the European literary tradition was not acknowledged. It took Vladimir Nabokov's *Commentary* (1964) to place Pushkin's novel in a sustained conversation with its European predecessors and thus open up a new layer of mindreading involving the author and his audience. That Nabokov was an émigré writer—unconstrained by

Soviet nationalistic censorship[112] *and* professionally trained (so to speak) in the intricacies of literary mindreading—shows how the political and the personal can get intertwined in the quest for new ways of reading complex embedment into a text.

I am also certain that, today, plenty of readers of *Eugene Onegin* are happily unaware of the embedments arising from its implied author's oblique references to the European literary tradition, just as plenty of readers of *The Plum* do not think twice of the significance of Sung Hui-lien's mention of Ling Chou. Although some of the mindreading practiced by literary scholars makes it to the cultural mainstream, plenty of it remains in a category of its own.

So let us say that we belong to the group of readers who see *The Plum* as "a model of the literati novel genre maturing in the sixteenth century" and that we thus acknowledge its unprecedentedly innovative appeal to late-Ming-dynasty readers' theory of mind.[113] We can further ask how different aspects of this novel's mindreading profile—which include deception, psychological manipulation of one character by another, and the implied author's ironic appeals to the implied audience—were amplified in such mid-eighteenth-century classics as Wu Ching-Tzu's *The Scholars* (*Rulin waishi,* 儒林外史) and Cao's *Dream of the Red Chamber.* Let us take a quick look at these two novels, focusing specifically on their potential to keep their readers steadily embedding mental states on a high level.

5.12 "Lust of the Mind"

Lying continues to drive complex embedment both in *The Scholars* (1750) and in *Dream of the Red Chamber* (ca. 1750–1760). Indeed, a separate study can be written on how much these novels depend on lying and on the role of social class and gender in the construction and consumption of lies. Also, just like in *The Plum,* lying often goes hand in hand with a concern about face. Yet both lying and fear of losing face are also treated now in ways that make possible distinctly new forms of complex embedment. For instance, in *The Scholars,* they can be combined with a nearly direct appeal to the reader, while in *Dream,* they are used to highlight important features of its characters' psychology.

Let us start with *The Scholars,* a satirical novel about educated gentlemen vying for plum positions in civil service. At one point, early in the story, a magistrate named Shih wants Wang Mien, a peasant who paints exquisite pictures of flowers, to pay him a visit, so that one Mr. Wei, a distinguished scholar and Shih's superior, can meet this homespun prodigy. Wang Mien turns down the magistrate's invitation (which he correctly recognizes as a thinly veiled order) because he is a man of independent spirit who doesn't want to curry favor with the high and mighty. His refusal, however, creates a problem for several people who now worry about the effect that this insubordination may have on other people's perception of their social status.

One of those people is bailiff Chai, whom the magistrate employed as his messenger. To help Chai save face, Wang's friend and neighbor, Old Chin, suggests that Chai lies to Magistrate Shih and tells him that Wang Mien is ill. That is, Old Chin *doesn't want* Shih to *know* that Wang Mien *doesn't consider* his invitation an honor. Instead he *wants* him to *think* that Wang Mien *would like* to come and only his illness prevents him from doing so.

Magistrate Shih, however, does not believe the bailiff's report:

> When Magistrate Shih heard the bailiff's report, he thought, "How can the fellow be ill? It's all the fault of this rascal Chai. He goes down to the villages like a donkey in a lion's hide, and he must have scared this painter fellow out of his wits. Wang Mien has never seen an official before in his life. He's afraid to come. But my patron charged me personally to get this man, and if I fail to produce him, Mr. Wei will think me incompetent. I had better go to the village myself to call on him. When he sees what an honour I'm doing him, he'll realize nobody wants to make trouble for him and won't be afraid to see me. Then I'll take him to call on my patron, and my patron will appreciate the smart way I've handled it."
>
> Then, however, it occurred to him that his subordinates might laugh at the idea of a county magistrate calling on a mere peasant. Yet Mr. Wei had spoken of Wang Mien with the greatest respect. "If Mr. Wei respects him, I should respect him ten times as much," Magistrate Shih reflected. "And if I stoop in order to show respect to talent, future compilers of the local chronicles will certainly devote a chapter to my praise. Then my name will be remembered for hundreds of years. Why shouldn't I do it?"[114]

This passage is an avalanche of complex embedments. Magistrate Shih *thinks* that Wang Mien only *wants* him to *think* that he is ill because he is, in reality, afraid of government officials. This leads him to *believe* that if he visits the humble rustic in his own august person, Wang Mien *will realize* that nobody *intends* him any ill. Readers, of course, *know* that Shih is *mistaken in his assessment* of Wang Mien's *feelings*—a bit of dramatic irony here.

The real joke of the situation, however, comes with the sly conversation that Wu Ching-Tzu is having with his readers. Shih fondly imagines that "future compilers of local chronicles" will devote a whole chapter to his praise. And, as a matter of fact, Wu does devote a couple of pages to him, and these are the pages that we are reading. Wu *wants* us to be *aware* that Shih *imagines* that future generations will *think* that he *wanted* to "show respect to talent," and he also *wants* us to *suspect* that Shih's *hopes* may have been disappointed. The magistrate is not an unsympathetic character, but because we *know* that he *wanted* us to *admire* him for his *respect* for talent, we are not sure anymore that he is worthy of our admiration. What has started out as a series of complex embedments arising from the lies and the characters' concerns about "face" is gradually turned into an exploration of the mutual awareness between the implied reader and the implied author.

The fear of losing face is also a powerful motivator for many characters in *Dream of the Red Chamber*, which tells the story of two lovesick cousins, a girl named Lin Dai-yu and a boy named Jia Bao-yu, kept apart by their karmic destiny and, more immediately, by their family's ambitions. For Dai-yu, however, the concern about face can take a peculiar form of neurotic overthinking of other people's intentions. Such overthinking is, of course, Dai-yu's trademark psychological trait, something that has been known both to exasperate and attract the novel's readers. Let us consider some examples of Dai-yu's anxious social projections aimed, ostensibly, at saving face; driven, at least partly, by self-deception; and embroiling the reader in guessing and second-guessing of everyone's intentions.

At one point, Dai-yu and Bao-yu go to visit their other cousin Xue Bao-chai, whom Dai-yu considers her rival for Bao-yu's affections, not only because of her beauty and sophistication but also because her mother, old Mrs. Xue, is a rich widow, whose fortune would come in handy were Bao-yu to marry Bao-chai. As they are sitting at Mrs. Xue's house, chatting and drinking tea and wine, Dai-yu's maid, prompted by another maid, brings her a hand warmer. Dai-yu then scolds her for it. Neither Bao-yu nor Bao-chai says anything, though for different reasons. Bao-yu knows "perfectly well" that Dai-yu's carefully phrased rebuke was "really intended for him," but he makes "no reply, beyond laughing good-humoredly," whereas Bao-chai, "long accustomed to Dai-yu's peculiar ways," simply ignores her words.

Mrs. Xue, however, is deaf to such intricacies and takes Dai-yu's complaint at its face value. She points out to Dai-yu that it was "nice" of Dai-yu's maids to think of her, because she often feels chilly. Dai-yu responds thus: "You don't understand, Aunt. . . . It doesn't matter here, with you; but some people might be deeply offended at the sight of one of my maids rushing in with a hand-warmer. It's as though I thought my hosts couldn't supply one themselves if I needed it. Instead of saying how thoughtful the maid was, they would put it down to my arrogance and lack of breeding."[115]

Dai-yu claims to be *imagining* people who'd *think* that she *thinks* that her hosts are not taking good care of her. Though presented as an attempt to save face, her own and Mrs. Xue's, this complex embedment is an expression of the exhausting self-monitoring carried on by the neurotic and powerless Dai-yu. Not surprisingly, instead of appreciating her sentiments, Mrs. Xue can only respond with the head-scratching, "You are altogether too sensitive, thinking of things like that. . . . Such a thought would never have crossed my mind."[116]

Here is another example of a face-saving enterprise devolving into an anxious overattribution of intentions. At Bao-chai's birthday party, while the family is watching a play performed by a group of professional child actors, her aunt, Wang Xi-feng, comments slyly on the resemblance between "someone we know" and a beautifully made-up child who plays the main heroine. Bao-chai and Bao-yu merely nod without responding (once again, they know better), but another young relative, Xiang-yun, is "tactless enough" to blurt out that the actor looks like Dai-yu. Bao-yu shoots "a quick glance in [Xiang-yun's] direction; but [it's] too late," for now the other guests catch on to the resemblance and start laughing.[117]

Shortly after the party breaks up, the offended Xiang-yun orders her maid to start packing. Bao-yu overhears it and attempts to make her change her mind, explaining that the only reason he gave her that look is that he "was worried for [her] sake." He claims to have known that Xiang-yun didn't know how sensitive Dai-yu can be and to have been "afraid that [Dai-yu] would be offended with [Xiang-yun]." Xiang-yun won't have any of it. She knows that Bao-yu is not being emotionally honest with her, though she can't, perhaps, identify the exact meaning of his maneuvers. The way she reads it (or claims to read it) is that Bao-yu's glance implied that everyone thinks that she is "not in the same class" as Dai-yu and hence mustn't make fun of "the young lady of the house."[118]

I condense their conversation here, but you can see even from this condensed version that it consists of a series of complex embedments all involving Xiang-yun's perception of Bao-yu's intentions regarding Dai-yu's feelings and leaving it up to readers to decide which interpretation of those intentions they would find most plausible.

But then it turns out that Dai-yu overheard Bao-yu's conversation with Xiang-yun, so the real fun begins. First, Dai-yu "coldly" explains to Bao-yu that even though he didn't compare her with the child actor and didn't laugh when others did, his secret thoughts, of which she's apparently the best judge, implicate him severely. In the quote that follows, the italics are in the original:

> "You would *like* to have made the comparison; you would *like* to have laughed," said Dai-yu. "To me your way of *not* comparing and not laughing was worse than the others' laughing and comparing!"
> Bao-yu found this unanswerable.
> "However," Dai-yu went on, "that I could forgive. But what about that look you gave Yun? Just what did you mean by that? I think I know what you meant. You meant to warn her that she would cheapen herself by joking with me as an equal. Because she's an Honourable and her uncle's a marquis and I'm only the daughter of a commoner, she mustn't risk joking with me, because it would be so degrading for her if I were to answer back. That's what you meant, isn't it? Oh yes, you had the *kindest intentions*. Only unfortunately she didn't *want* your kind intentions and got angry with you in spite of them. So you tried to make it up with her at my expense, by telling her how touchy I am and how easily I get upset. You were afraid she might offend me, were you? As if it were any business of *yours* whether she offended me or not, or whether or not I got angry with her!"[119]

The reason that Bao-yu and Dai-yu often find themselves pulled into this kind of labyrinthine social reasoning is that their psychological profiles—or, shall we say, their mindreading profiles—are uniquely and tragically suited to each other. While Dai-yu overthinks people's intentions, Bao-yu over-reads them, being afflicted with the condition described in the novel as "lust of the mind" (yiyin, 意淫). This condition has been interpreted by critics in a wide variety of ways, so the interpretation that I give you here reflects specifically my "cognitive" perspective. From this perspective, "lust of the mind" means that Bao-yu feels the need to know and share the emotions of girls, dozens of them, servants, cousins, and young aunts, populating the Jias' sprawling aristocratic households—an empathetic drive hardly compatible with his position as the heir on whom the family's hopes of future prosperity are pinned.[120]

The male protagonist's passionate desire to understand the feelings of women is something that is hard to imagine in the universe of Hsi-men Ch'ing (from *The Plum in the Golden Vase*). Indeed, the scenes of intense mindreading and misreading that we get in *Dream* are something quite unprecedented in the literary history of medieval China.[121] I will conclude this section with another one of such scenes, which starts, once again, as an ostensible endeavor to save face, implies self-deception, and embroils the reader in complex and ambiguous mindreading attributions.

Bao-yu, having spent his early childhood cosseted by his loving grandmother and other relatives, is finally forced to start his formal education. On the first day of school, he decides to visit Dai-yu to say good-bye, for he won't see her now for most of the day. After chatting with her for a while, he is ready to tear himself away, but Dai-yu stops him to ask if he's "going to say good-bye to [his] cousin Bao-chai" too. In response, Bao-yu smiles but says nothing and goes "straight off to school with [his friend] Qin Zhong."[122]

How are readers to make sense of this exchange? While there are several different ways to interpret Bao-yu's smile, it's important to note that all of them seem to involve complex embedments, some reaching even to the fourth and fifth level. For instance, we may say that Bao-yu smiles because he *thinks* that he *knows* that Dai-yu *doesn't really want* him to stop by Bao-chai's room to say good-bye. That is, he *thinks* that he *knows* that Dai-yu (sensitive as she always is to how her behavior may be perceived by others) *doesn't want* anyone to *think* that she *thinks* she has any right to usurp Bao-yu's attention on this particular morning.

Moreover, by telling us that Bao-yu goes straight to school instead of indeed stopping by Bao-chai's room first, Cao *wants* us to be *aware* not just of the clear *preference* that Bao-yu has for Dai-yu but also of the tortuous way in which the admission of this preference was extracted from him. Bao-yu certainly hasn't planned to play favorites this morning—it's not likely that he'd even been thinking about it when he stopped by Dai-yu's room—but Dai-yu's self-conscious remark has made him express his feelings. Ironically, this is what Dai-yu would have wanted—even though she would never admit that to anyone. Bao-yu's smile thus can also be interpreted as his *realization* that Dai-yu has just made him newly *aware* that he *likes* her more than he likes Bao-chai—and that she did it without being implicated in doing so and perhaps not even intending it.

I expect that not every reader will agree with my interpretation of Bao-yu's smile. What's important, however, is that even if you disagree with this interpretation and propose your own, yours is still likely to feature a complex embedment of mental states. That is, to do justice to a nuanced psychological dynamic conjured up by Cao, we have to embed mental states on at least the third level, even if their exact content and configuration differ from one reader to another.

5.13 Conclusion: "Cheater Detection" or "Destruction of the Subject-Matter by the Form"?

As we are nearing the end of our conversation about lying, here is the question that this long chapter has been begging for a while: *Why* is lying so integral to representation of literary consciousness? One answer offers itself immediately. As developmental psychologists put it, "lying, in essence, is theory of mind in action."[123] Given the centrality of mindreading and misreading to human communication, it is not terribly surprising that writers would exaggerate this aspect of human sociality to make their narratives more engaging.

We can stop at this, or we can indulge our critical perversity and dig deeper. *Why*, we may ask, should this particular aspect of sociality be of such interest to readers? After all, literature can (and does) play with mindreading uncertainty in many other ways. Why keep returning to deception?

One way to respond to this question is to roll out a couple of heavy guns. By that, I mean turning to cognitive adaptations—all connected with mindreading—that evolved hundreds of thousands of years ago and still underlie much of our social functioning today. We may start with the concept of cheater detection. According to the founders of cognitive evolutionary psychology, John Tooby and Leda Cosmides, detecting cheaters in situations involving social exchange was an adaptive problem faced by our ancestors, and the solution to this problem was the evolution of a "cheater-detection mechanism." This mechanism "looks for cheaters"; that is, "it looks for people who have *intentionally* taken the benefit, specified in a social exchange rule, without satisfying the requirement [of the cost]." The appraisal of intentions is crucial: the mechanism "is not good at detecting violations caused by innocent mistakes, even if they result in someone being cheated."[124]

Now remember that, on some level, our mindreading adaptations do not distinguish between mental states of real people and those of fictional characters. This means that once we attribute an intention to cheat to a fictional character, this cognitive output feeds into the cheater-detection mechanism. Evolved for detecting cheaters in real life, this mechanism now has no choice but to start detecting them in made-up stories as well. So we can say that one way in which works of fiction compel our attention is that they keep our cheater-detection mechanisms up and running.[125]

Think of one of the earliest known examples of lying in literature. Gilgamesh promises to Utnapishtim to stay awake for six days and seven nights in exchange for the secret of immortality; then he promptly falls asleep; then, upon awakening, seven days later, he denies that he has slept at all. Behold a cheater! Gilgamesh wants the benefit (i.e., immortality) without having satisfied the requirement (i.e., not sleeping).

Of course, in spite of Utnapishtim's grim observation that "all men are deceivers," *The Epic of Gilgamesh* doesn't actually feature many instances of cheating. So we should not overstate the role that our cheater-detection mechanism may play in our interaction with this text. It is merely one of numerous inducements to pay close attention that the story offers—important (no question about that, in the case of the four-thousand-year-old artifact!)—but, still, one of many.

Here comes another heavy gun. Our species also evolved to pay attention to sexual deception. (Ancestors who didn't do that aren't our ancestors, because they didn't leave descendants.)[126] This means that, today, we are attuned to a broad variety of mental states involved in sexual deception, including mental states that we attribute to fictional characters. When we are certain that Othello is wrong in thinking that Desdemona is in love with Cassio, it is because we have been carefully checking Iago's allegations against what we know about Desdemona's feelings, while also pondering Iago's motivations.

It may seem that just these types of cheating—seeking to get a benefit without incurring a cost and sexual deception—would account for a lot of lying that takes place in literature. Thus, we have another possible answer to the question of just *why* lying is so integral to representation of fictional consciousness. We can say that complex embedments of mental states still often arise from plots of deception—even though other, more "sophisticated" contexts of embedment have long been available—because writers intuitively rely on social contexts that are guaranteed, by our evolutionary history, to sustain their readers' attention.

Then there is also the question of genre. Some genres, such as detective and spy stories, suspense thrillers, and romances, derive most of their emotional punch from deception. This is to say that such stories are deemed successful to the extent to which their readers are caught up emotionally in the project of identifying liars, understanding their motivation, and assigning different moral values to different instances of lying. In contrast, other genres (and here we are, once more, on the treacherous critical ground of drawing a distinction between "popular" and "literary" fiction) may still exploit lying for its capacity to generate complex embedments, but the affective charge of those narratives is not tied to their plots of deception.

Let us look at a couple of such texts and see what kind of emotional response they seem to be eliciting from their readers. For instance, in Shakespeare's Sonnet 138, "When My Love Swears That She Is Made of Truth," lying repeatedly serves as a source of complex embedments, yet it also appears that its readers are encouraged *not to care* about the grave sexual and social repercussions of deception that the speaker and his beloved practice on each other:

> When my love swears that she is made of truth,
> I do believe her, though I know she lies,
> That she might think me some untutored youth,
> Unlearnèd in the world's false subtleties.
> Thus vainly thinking that she thinks me young,
> Although she knows my days are past the best,
> Simply I credit her false-speaking tongue:
> On both sides thus is simple truth suppressed.
> But wherefore says she not she is unjust?
> And wherefore say not I that I am old?
> Oh, love's best habit is in seeming trust,
> And age in love loves not to have years told.
> Therefore I lie with her and she with me,
> And in our faults by lies we flattered be.

What is going on in this sonnet? Or, to put it differently, what complex embedments do we process in order to make sense of it? The speaker's beloved wants him to think that she doesn't lie to him. He realizes that he is willing to deceive himself by trusting her because he likes to think that she thinks that he is young enough to believe her. But he also knows that she doesn't really think that he is young, which means that perhaps she knows that he has his own reasons for wanting to believe her even when he knows that she is lying. Moreover, she knows that he wants her to think that he trusts her (for "love's best habit is in seeming trust"), and so forth.

Now let us look at the emotional value of the sonnet. One narrative generated by all those complex embedments is quite sad. The speaker's beloved is cheating on him, while he is meditating on his old age, her youth, and the vagaries of self-deception. (Were we to go for a crude pseudoevolutionary reading, we'd even say that this is a downright tragedy: the guy is a genetic dead end.) Yet the same poem also tells another story: that of a poet enchanted by the pliability of the word "lie" and rounding it all off triumphantly with a double entendre built around that word. Readers, too whatever negative emotions this account of sexual infidelity and powerlessness may be expected to elicit in us, the last two lines invite us to join the fun that is to be had when strong emotions fade into delightful wordplay.

To see what Shakespeare is doing here—and why our reflexive interest in keeping tabs on a cheater does not account for it—we may want to turn to the German poet and philosopher J. C. Friedrich von Schiller and the Russian cognitive psychologist Lev S. Vygotsky.[127] It takes a "true master," wrote Schiller in 1794, to know how to "destroy the subject-matter by the form."[128] Yes, agreed Vygotsky in 1925, and the way this destruction works is that the reader is made to experience two opposing emotions, "developed together and with equal force": one elicited by the subject matter of the poem, another "by the artistic form and the particular arrangement of the material."[129]

A cognitive literary critic may add here that the destruction of subject matter by the form introduces more embedded mental states. For instance, the embedments discussed earlier (e.g., "she knows that he wants her to think that he trusts her") all focus on the thoughts and feelings of the speaker and his beloved. But the concluding double entendre involving the verb "lie" shifts that pattern by orienting us toward mental states of the speaker and the reader. The reason for this shift is that puns come with their own built-in intentionality: they signal the punner's desire to draw the reader's attention to the form of the word.

Let us now bring it all together: evolution, Schiller, Vygotsky, lies, and embedment. Our evolutionarily conditioned interest in deception may very well be integral to our interaction with Shakespeare's sonnet, but it contributes little to the sonnet's artistic value. That value is generated, at least in part, by the clash of the two contradictory affects: one driven by the content of the poem, another, by its form. The melancholy affect arising from the content is entangled with the complex embedments associated with the speaker, who reflects on various mutual deceptions that make the relationship possible.

But the joy arising from the form is *also* entangled with complex embedments, and it starts developing even before we arrive to the final lines that contain the double entendre. For, is not our awareness of the complexity of the speaker's emotions in and of itself a source of positive affect, as it reminds us that we are all interesting beings here, endowed with rich inner lives, attuned to the intricacies of our social environment, apparently with cognitive resources to spare? The concluding pun adds a nice nuance, but do not overestimate its role! Most of the poem's heavy lifting—that is, of "destroying" the depressing subject matter by the delightful form—has already been done by the time we read the pun.[130]

Have we seen this dynamic before? Yes, we have. Recall the stanza from Pushkin's *Eugene Onegin* that lists lies practiced by the main protagonist as he seduces various ladies of his acquaintance. There, too, the "heavy" subject matter of lying is undercut by the poetic form. For the stanza that starts with "How early he was able to dissemble" contains not just multiple instances of deception but also the narrator's amused reflection on Eugene's amorous machinations—not to mention the sheer delight induced by its pattern of sounds and rhymes in those who read this "novel in verse" in its original language, for that, too, goes a long way toward destroying the negative affect that the protagonist's treacherous, antisocial behavior could, in principle, induce in us.

In fact, it seems that this dynamic—that is, the destruction of such critical a subject matter as cheater detection by a form—has been present in literature for some time. Perhaps the best genre to illustrate this dynamic is picaresque, for it is unequivocally built around deception. Since its earlier days, from *Lazarillo de Tormes* (1554) and Mateo Alemán's *Guzmán de Alfarache* (1599–1604) to Miguel de Cervantes's *Rinconete y Cortadillo* (1613), the picaresque novel focused on protagonists who cheated and lied their way to economic survival. Yet the complex embedments arising from the shenanigans of a resourceful picaro were often interlaced with complex embedments arising from the conversation that the narrator was having with readers. Consider the opening paragraph of *Guzmán de Alfarache*:

> I was so desirous, curious reader, to relate to you my own adventures, that I had almost commenced speaking of myself without making any mention of my family, with which some sophist or other would not have failed to accuse me: "Be not so hasty, friend Guzman," would he have said, "let us begin, if you please, from the definition, before we proceed to speak of the thing defined. Inform us, in the first place, who were your parents; you can then relate to us at your pleasure those exploits which you have so immoderate a desire to entertain us with."[131]

There are many different ways to map out this paragraph in terms of its embedded mental states. We can say, for instance, that the implied author wants us to believe that the narrator is afraid of being censored by a pedant; or that the narrator wants his ideal (i.e., "curious") reader to feel superior to an obtuse reader (a "sophist") who is not quite aware of what kind of story he or she is about to hear and thus demands a conventional opening; or that the implied author wants to tease the reader as he defers the actual account of his adventures (which is, presumably, something that the reader is impatient to hear) and instead gives in to the convention of lengthy self-introduction (which, he expects, or pretends to expect, the reader will find tedious).

Swamped by complex embedments, and we haven't even gotten to the story's first official swindle! By the time we do, we will be frequently dealing with two parallel sets of embedments: those involving mental states of liars and their victims and those involving mental states of the narrator, the implied reader, and the various imagined onlookers who are similar in their function to the "sophist" of the opening paragraph. The affect associated with the act of deception as such—for example, with its negative communal repercussions, with private sufferings experienced by the people immediately involved—rubs against the affect arising from our awareness of the playful conversation that we, as implied readers, are having with the narrator.

The presence of complex embedments involving mental states of the critic, the narrator, and the reader may thus be the reason that the picaresque is considered "one of the earliest traditions (perhaps the earliest) in the history of the novel."[132] For take those mental states out, and you will end up with a mere trickster story. I use the word "mere" advisedly, not to downplay this genre's prominence in the world's folklore. In fact, the trickster story often demonstrates the crucial role of complex embedment in the construction of narrative, for the trickster *wants* his victims to *think* that his *intentions* are different from what they really are. Still, literature, as we know it today, happens when authors move beyond straightforward accounts of deception—that is, when they begin to "destroy" that particular "subject-matter by the form"—even while still benefiting from the presence of lying characters.

Let us conclude our conversation about writers simultaneously using liars and moving beyond their lies with another "novel full of deception and self-deception,"[133] also set in Spain, albeit four hundred years later, and written in English: Ben Lerner's *Leaving the Atocha Station* (2011). Lerner's protagonist is an American poet on a fellowship in Madrid. Unlike a picaro, he lies not so much to ensure his economic survival (although he *is* receiving money from a Madrid-based foundation for a "research-driven poem" about the Spanish Civil War, which he has no intention of writing)[134] but to create and maintain a certain image of himself among his Spanish friends and (prospective) lovers. He lies about his parents, about his feelings, about what he is doing now, and about what he plans to do next. Those lies contribute their fair share of complex embedments, yet—a dynamic similar to what we have seen in "When My Love Swears That She Is Made of Truth"—they also don't matter.

This is to say that Adam's lies—even the ones that seem to be quite atrocious—have no real social consequences. Neither his Spanish friends nor his parents take them seriously. For instance, when Adam confesses on the phone to his parents that he has been telling people in Spain that his "mom was dead or gravely ill and that [his] dad was a fascist," his mother and father, both professional psychologists, are "confused, but not upset." They accept his explanation that he has been saying those things in order "to get sympathy" and turn the conversation to other, more pressing, matters.[135]

In fact, it seems that Adam's lying functions primarily as a trigger for self-reflexivity, and that self-reflexivity is what generates the majority of the novel's complex embedments. For instance, when Adam feels disoriented in the foreign social environment (as he does for most of the novel), he responds by faking his emotional reactions and vividly describing involved mental states that he *would* experience were he to actually have those reactions.

Thus, when one of his new acquaintances, Carlos (of whose exact stance toward himself Adam is not sure but whom he is beginning to hate), observes, in the presence of several other people, that it "must be an interesting time to be an American in Spain" and asks Adam what he thinks about "everything," Adam adapts a series of postures that, he hopes, will be read by the onlookers as signaling his sophistication and Carlos's stupidity: "I looked off in the distance as though I was making an effort to formulate my complex reaction so simply even an idiot like him might understand. Then, as if concluding this was an impossible task, I said I didn't know."[136]

Does the shaft hit the mark? Are Carlos's friends now convinced that no reasonably intelligent person would ask the kind of question that Carlos just asked, and do they appreciate Adam's earnest, if ultimately futile, attempt to tackle it? There is no telling. For all that we know, they may be thinking of something else, completely unrelated to Adam's hopeful performance of his emotional complexity.[137]

Yet Adam is not a deluded/unreliable narrator. He is open to revising his perceptions if new evidence presents itself (e.g., if Carlos turns out to be less hostile to Adam than he thought he was), and he can contemplate critically his endeavors to shape other people's impressions of him. This, of course, supplies more grist for the mill of complex embedment.

Here, for instance, is a characteristically funny moment when Adam reaches for his notebook to write down a potentially poetic observation that has occurred to him, only to stop and blush at the realization that he has apparently bought into his own lie (manufactured to impress a current girlfriend) about being the kind of person who writes down potentially poetic observations that occur to him: "Why would I take notes when Isabel wasn't around to see me take them? I'd never taken notes before: I carried around my bag because of my drugs, not because I intended to work on my 'translations,' and the idea of actually being one of those poets who was constantly subject to fits of inspiration repelled me; I was unashamed to pretend to be inspired in front of Isabel, but that I had just believed myself inspired shamed me."[138]

Shame attendant on self-deception has long been a reliable source of complex embedment in the novel. (Think, for instance, of another lonely traveler in a strange land, Robinson Crusoe, who is ashamed when he catches himself thanking God for bringing him to a desert island: "'How canst thou become such a hypocrite,' said I, even audibly, 'to pretend to be thankful for a condition which, however thou mayest endeavour to be contented with, thou wouldst rather pray heartily to be delivered from?'")[139] Lerner's "skeptically postmodern comedy" thus continues to work the rich territory staked by connoisseurs of abashed self-consciousness, from Defoe to Dostoevsky, who used self-deception as a reliable jumping-off point for other mindreading entanglements.[140]

Indeed, there seems to be a good-husbandry aspect to being a writer: Why waste a perfectly expedient, time-tested way to embed complex mental states, such as lying, even if the majority of the text's embedments now come from other social contexts?

6 Embedded Mental States in Children's Literature

6.1 Mental States versus Embedded Mental States in Stories for Children

Children's literature is a particularly fascinating area of study when it comes to complex embedment of mental states. On the one hand, it seems to track certain milestones in the development of children's theory of mind. For instance, as I show in this chapter, complex embedments are mostly absent in stories targeting one- to two-year-olds; they are present, but in a limited way, in those for three- to seven-year-olds; and they increase both in number and variety in literature for nine- to twelve-year-olds.

On the other hand, when it comes to novels written for children of this latter age group, they can range widely, from featuring almost no complex embedment (which is not something we would expect from a novel today!) to displaying an intense "grown-up" pattern of embedment. To explain this range, we have to look at specific historical factors that influence the process of designating some texts as children's literature and/or novels, which reminds us, once again, how tightly cognition and history are bound together in the case of any complex cultural artifact.

We start by revisiting research in developmental psychology that focuses on depiction of thoughts and feelings in books for young children and then see what can be added to this conversation by shifting the focus from mental states as such to *embedded* mental states. You may remember, from the discussion in chapter 3, that the current view of theory of mind is that it develops continuously and can already be studied in preverbal infants. For a long time, psychologists have been especially interested in the changes that occur around the age of four, when children seem to be able to clearly demonstrate their understanding of false belief, that is, the understanding that people may believe something that is not, in fact, the case. (Though see studies by ethnographers working with communities that subscribe to the opacity model for useful qualification of this view).[1] The understanding of false belief may, in principle, be a key condition for appreciating stories that feature embedded mental states, for example, stories whose readers are led to realize that a given character does not know something crucial about another character's intentions.

Yet, to the best of my knowledge, although cognitive scientists have looked at the frequency and types of mental states in children's stories, they have not looked specifically at embedded mental states. Thus, Jennifer Dyer and her colleagues used a sample of ninety books to see if "the information about mental states" present in children's storybooks differed in books for younger preschoolers (three- to four-year-olds) and older preschoolers (five- to six-year-olds), "either in quantity or kind."[2] What they found was that "mental state information in storybooks for young children" doesn't simply increase "with the children's sophistication from 3 [to] 6 years of age"; instead, books for younger and older children are "notably similar in the rates of types and tokens of mental state expressions and the richness of mental state concepts, particularly those expressed by cognitive state terms and situational irony." Yet, at the same time, books for older children contain "more mental state terms [and] varied mental state vocabulary." Additionally, a greater number of the books for older children feature a "variety of references from more of the different categories of mental state."[3]

The textual dynamic described by this study as "situational irony" comes close to what I call "implied embedment." Dyer et al. use this expression to refer to moments when readers are aware of, say, a disjunction between two characters' perspectives, even if it is never explicitly spelled out. Observe, however, the difference between the two terms. "Situational irony" is relatively abstract, while "implied embedment" calls for an articulation of the relationship among the minds involved, which, in turn, allows us to calculate the level of embedment, as in, "the reader is aware that character A doesn't know what character B is thinking" (third level). Such a calculation might not always be an easy task (though it may be more so in children's literature than in literature for grownups), but it would add an important new dimension to the inquiries conducted by cognitive scientists.

Here is another study that also comes close to articulating the educational role of implied embedments in children's literature. Joan Peskin and Janet Wilde Astington wanted to explore further the connection between the acquisition of vocabulary in young children and development of theory of mind.[4] It's been shown that children attending schools in low-income neighborhoods "demonstrate substantial lags in their theory-of-mind understanding" and also that at six years old, they know only half the number of words as do children from higher socioeconomic groups: "Children whose parents do not provide a rich lexicon for distinguishing language about perceiving, thinking, and evaluating might make important gains from hearing and talking such talk in their everyday story reading. . . . A rich vocabulary, more than any other measure, is related to school performance."[5]

Peskin and Astington decided to test whether exposure to an explicit discussion of mental states (they call it metalanguage) "will result in a greater conceptual understanding of one's own and other people's beliefs or whether this understanding develops more implicitly."[6] They rewrote kindergartners' picture books "specially for the study so that the texts were rich in explicit metacognitive vocabulary, such as *think, know, remember, wonder, figure out,* and *guess,* in both the texts and text questions."[7]

Thus, Pat Hutchins's classic *Rosie's Walk* (1968)—which features a chicken on her daily promenade, unaware that a hungry fox is right behind her—was altered to include such descriptions of the chicken's thoughts as, "Does Rosie *know* that Fox has been following her? No, Rosie doesn't *know*. She doesn't even *guess*."[8] The children in this "explicit metacognitive condition were compared with a control group that received the identical picture books, with a similar number of words and questions, but not a single instance of metacognitive vocabulary." (See figure 6.1 for snapshots of typical pages from Hutchins's book, which was what the control group was given.)

What Peskin and Astington found was that "hearing numerous metacognitive terms in stories is less important than having to actively construct one's own mentalistic interpretations from illustrations and text that implicitly draw attention to mental states."[9] Children exposed to explicit metacognitive terms did start using them more, but they used them *incorrectly*.

On the one hand, this study supports findings of psychologists who argue that what parents say in their interactions with their children is less important than how they say it. As Paul L. Harris et al. observe, "Parents elucidate a variety of mental states in conversation with their children. That elucidation is not tied to particular lexical terms or syntactic constructions. Instead it reflects a wide-ranging sensitivity to individual perspectives and nurtures the same sensitivity in children."[10]

Rosie the hen went for a walk

across the yard

Figure 6.1
Pat Hutchins, *Rosie's Walk*.

On the other hand, finding that *explicit* use of metacognitive vocabulary in stories doesn't seem to benefit children's theory of mind led Peskin and Astington to take another look at the *implicit* mentalizing expected of readers. In doing so, they were also prompted by an earlier study by Letitia Naigles, who found that "children exposed to more metacognitive terms of certainty (*think, know,* and *guess*) in a television show later displayed a poorer understanding of certainty distinctions than those exposed to episodes containing fewer of these terms," as well as by the (separate) studies of Deepthi Kamawar and Elizabeth Richner and Ageliki Nicolopoulou, who "compared children whose teachers used more metacognitive vocabulary to those whose teachers used less" and "found superior performance on theory-of-mind tasks for children whose teachers used fewer metacognitive terms."[11]

To explain such counterintuitive findings, Peskin and Astington suggest that "the teaching of information does not automatically lead to learning." What is required instead is a "constructive, effortful process where the learner actively reorganizes perceptions and makes inferences. . . . These inferences lead to an understanding that may be all the deeper because the children had to strive to infer meaning. Ironically, the more direct, explicit condition may have produced less conceptual development precisely because it was explicit."[12]

Crucially for our present argument, Peskin and Astington's main recommendation for fostering constructive learning in children was having them read literature: "Dramatic tension in stories is created when the various characters have disparate knowledge with regard to the action. This may be through error: The reader knows that Romeo does not know that Juliet lies drugged, not dead. Or it may be through deception: Pretending his assigned chore is an adventure, Tom Sawyer tricks his friends into whitewashing the fence."[13]

The examples chosen by Peskin and Astington are prime examples of implied third-level embedments. To stay just with the action that they describe (and thus ignoring complex embedments created by the *tone* of Twain's narrator, which I discussed in chapter 1), Tom *doesn't want* his friends to *realize* that he *hates* whitewashing the fence. Just so, Romeo *doesn't know* that Juliet *wants* some people to *think* that she is dead. Neither Shakespeare nor Twain spells out those mental states for his readers; we have to deduce them ourselves in order to make sense of what we read.

Think about it. Works of literature that do *not* spell out embedded mental states may enrich understanding of mental states,[14] foster the ability for constructive learning, and improve vocabulary in preschool and school-age children. I wouldn't claim that the effect is exactly the same for grown-ups. After all, if theory of mind goes through some important developmental milestones in young children and adolescents, the impact might be more pronounced for those age groups. For older readers, we may want to speak about a different kind of impact, for instance, the one suggested by Kidd and Castano, which is that a long-term exposure to literary fiction may sensitize one to the presence of intentionality cues in one's social environment (including the social environment within a fictional world) and make one more prone to considering ambiguous (rather than clear-cut) intentions.

In fact, we can bring the two kinds of impact together when we look at the history of literary criticism. It appears to be the case, for instance, that experienced readers of literature in the Ming and Qing dynasties (i.e., those *particularly eager to discern the less-than-obvious intentionality cues*) praised the pedagogical acumen of authors who made them work hard to figure out characters' mental states. As David Rolston puts it, although "use of direct psychological description in fiction increased throughout the Ming and Qing dynasties, it was never popular or influential." One reason for this was that "the main justification for reading [literature was] to develop the ability to judge human character; easy access to the inner life of characters would defeat this pedagogical purpose."

As traditional commentators saw it, "the author who is presented as the most subtle in his laying down of . . . clues that raise suspicions about a gap between an inner state of the character's mind and his or her actions or words . . . becomes the author most worthy of praise."[15] Worthy of praise, that is, by a very specific group of readers, ones whose long exposure to literature may have made them seek out and appreciate texts that would provide them with cues for "constructive, effortful [mindreading] where the [reader] actively reorganizes perceptions and makes inferences."

Note that what Peskin and Astington call "disparate knowledge with regard to action" is similar to Dyer et al.'s "situational irony." Once again, we come close to the concept of "implied embedment," particularly with Peskin and Astington's emphasis on texts "that implicitly draw attention to mental states." Let us see, however, if we can go further than simply recognizing some moments in stories for children as instances of "situational irony," or "disparate knowledge in regard to action," if we inquire more minutely into the configuration of mental states involved.

What follows is a preliminary assessment of patterns of embedment in stories for children aged nine to twelve, three to seven, and one to two. These age groupings are taken from the most recent editions of Judy Freeman's, John Gillespie's, and Eden Ross Lipson's guides to children's books and cross-checked with scholastic.com. (Although scholastic.com is by no means immune to the charge of being "primarily a marketing device,"[16] it is a resource widely used by parents and teachers. As long as one is aware of its limitations, it is a good starting point for a conversation about reading "interest levels.")

6.2 Ages Nine to Twelve

Among the books recommended for children aged nine to twelve are Twain's *Tom Sawyer*, Frances Hodgson Burnett's *The Secret Garden*, A. A. Milne's *Winnie the Pooh*, Laura Ingalls Wilder's *Little House in the Big Woods*, Tove Jansson's graphic novel *Moomin Falls in Love*, Jeff Kinney's "Diary of a Wimpy Kid" series, Lewis Carroll's *Alice in Wonderland*, P. L. Travers's *Mary Poppins*, and E. B. White's *Stuart Little*. I list in what follows some examples of third-level embedment more or less in their order of appearance in these stories (with emphases added), leaving out for now *Tom Sawyer* and *Little House in the Big Woods*.

We learn in the first paragraph of *The Secret Garden* that when Mary was born, her nurse was made to *understand* that if she *wanted* to *please* her mistress, she should keep the child to herself. As the narrator explains, Mary's mother "had not wanted a little girl at all, and when Mary was born she handed her over to the care of an Ayah, who was made to understand that if she wished to please the Mem Sahib she must keep the child out of sight as much as possible."[17]

When Mary's mother dies and the little girl is shipped to England, she meets Mrs. Medlock, the housekeeper of her new guardian. Mary instantly dislikes Mrs. Medlock and tries walking farther away from her because she *hates* to *think* that people would *assume* that she belongs to her: "It would have made her angry to think people imagined she was her little girl."

When Mrs. Medlock tells Mary about her new home, Mary listens "in spite of herself," but she *doesn't want* Mrs. Medlock to *think* that she is *interested*: she "did not intend to look as if she were interested."[18]

In the first chapter of *Winnie the Pooh*, Pooh, in his quest for honey, floats up to a bees' nest on his balloon and hopes that the bees will think that he is a small black cloud in the sky. But the honey is still out of reach, and, moreover, he worries that the bees suspect something. So he asks Christopher Robin for help: "'Christopher Robin!' 'Yes?' 'Have you an umbrella in your house?' 'I think so.' 'I wish you would bring it out here, and walk up and down with it, and look up at me every now and then, and say "Tut-tut, it looks like rain." I think, if you did that it would help the deception which we are practising on these bees.' Well, you laughed to yourself, 'Silly old Bear!' but you didn't say it out loud because you were so fond of him, and you went home for your umbrella."[19]

Short as it is, this passage contains several complex embedments: Pooh *doesn't want* the bees to *know* that he *wants* to steal their honey; Christopher Robin *doesn't want* Pooh to *know* that he *thinks* his plan won't work; the narrator *knows* that Christopher Robin *doesn't want* to hurt Pooh's *feelings*.[20]

In *Diary of a Wimpy Kid: Rodrick Rules*, the main protagonist, Greg, observes his parents "acting all lovey in front of [their youngest son] Manny," because they *don't want* Manny to *think* that their arguments mean that they *don't love* each other.[21] (Does the implied author want his grown-up readers to squirm in recognition as they think of the times when they hoped to manipulate their own kids the same way? I leave it up to you to decide if this particular embedment is part of our "mentalistic interpretation" of the action.) On another occasion, Greg reports *thinking* about his father's *feelings* about Greg's older brother's *intentions*: "I'm pretty sure Dad's worst fear is that . . . Rodrick will want to follow in Bill's footsteps."[22]

In *Moomin Falls in Love*, Moomintroll develops a crush on a circus performer, La Goona. His girlfriend, Snorkmaiden, is heartbroken and lonely. As she confides to Mymble, "If you only *knew* how I have *longed* for a friend's *understanding* and advice."[23] Mymble suggests that Snorkmaiden pretend that she doesn't care for Moomin anymore, but when Snorkmaiden follows Mymble's suggestion, she's bitterly disappointed because Moomin's only too happy to learn that he can do anything he wants.[24] Moreover, it transpires that La Goona fancies a circus acrobat who can lift big stones.

Moomin tries to wrench a heavy boulder out of the ground and fails. Little My, who observes his effort, tells him, "I *guess* you must *think* of an entirely different way of *impressing* La Goona."[25]

In *Alice in Wonderland*, Alice "[*thinks* that she] can *remember feeling* a little different."[26] In *Mary Poppins*, Mrs. Banks *wishes* that Mary Poppins wouldn't "*know* so very much more about the best people" than she *knows* herself.[27] (This is an explicitly spelled-out embedment, but an equally interesting implied one is lurking just beneath the surface, involving a grown-up reader's awareness of Mary Poppins's manipulation of her class-conscious employer.) Furthermore, Jane and Michael can't figure out if Mary Poppins only *pretends* to get *angry* at them and not *understand* what they *mean* when they say that her Uncle likes "rolling and bobbing on the ceiling";[28] and Jane "*wonder*[s] if she would ever be able to *remember* what Mrs. Corry *remembered*."[29]

In *Stuart Little*, we learn that Stuart's father, "Mr. Little, was *not at all sure* that he *understood* Stuart's *real feelings* about a mousehole."[30] Later on, the family cat wants everyone to think that Stuart ran down the mousehole while he's actually trapped in a window shade. Stuart knows what the cat had in mind, yet when he is finally found and rescued, he decides not to tell on the cat. Instead, he *wants* his family to "*draw* [*their*] *own conclusions*" about who *might have wanted* them to *think* that he would run down the mousehole and why.[31]

It appears that, in spite of obvious differences in subject matter, the pattern of embedment that one encounters in books for this age group is similar to the one encountered in fiction for "grown-ups." Both feature complex (that is, at least third-level) embedments of mental states, which are either implied or explicitly spelled out and associated with characters, narrators, readers, and authors.

One important difference—at least in this sample—seems to have to do with the frequency of complex embedments. In story after story, from *Alice in Wonderland* to *Stuart Little*, I had to actively search for third-level embedments, sometimes coming up empty for a whole page. This situation would be difficult to imagine in literary fiction for grown-ups, in which the main effort required to find an instance of complex embedment involves opening the book.[32] (There, even when descriptions of mental states are intentionally omitted, to make it seem that characters lack what we may call interiority, embedded mental states are still implied.)

Let me complicate this narrative of difference, if only up to a point. Books in this age bracket (nine to twelve) are sometimes characterized by what Ulrich Knoepflmacher and Mitzi Myers call "cross-writing." That is, they activate a dialogue "between phases of life we persist in regarding as opposites," appealing in different ways to young and to adult readers.[33] And I don't just mean implied embedments, as when adult readers are aware of Travers's intention to show that Mary Poppins knows how to tacitly exploit Mrs. Banks's class anxieties. I also mean subtle interactions between the author and the reader that arise from the parodic feel of the text. As Sandra Beckett observes, to "appreciate parody [of, for instance, Carroll's *Alice* books] the reader must first recognize the intent to parody another work and then have the ability to identify the appropriated work and interpret its meaning in the new context."[34] This recognition of intent is already a complex embedment—I *realize* that the author *wants* me to *think* of text A as I am reading text B—even before we factor in mental states of characters whose motivations we may have to interpret in light of this "new context."

What does a reader's potential awareness of an author's intent do to my present argument about a somewhat less frequent incidence of complex embedment in literature for children aged nine to twelve as compared to literature for adults? Should we say that at least in some of these books, the frequency of complex embedments may approach that encountered in books for grown-ups, but only for those readers who "possess all of the codes necessary to understand all of the parodic allusions"?[35]

In principle, a version of this argument—which is that there are always more implied embedments in a text than meet the casual eye—can be made about many stories. One can say, for instance, that experienced readers bring to *anything* they read the "mastery of the codes of fiction" and a heightened attunement to intentionality cues,[36] while less experienced readers do not. Literary critics, too, may find new ways to read implied embedded mental states into a text by expanding the range of minds associated with it, as we have seen scholars of eighteenth-century literature have done with Haywood's *Fantomina*. Still, even if we allow that, with enough effort, we can import more complex embedments into just about any work of literature, some of them do clearly require less of this kind of effortful importation than others, for example, *Howards End* less than *Mary Poppins* and *Dream of the Red Chamber* less than *The Secret Garden*.

5.3   Young Adult Fiction (Thirteen to Eighteen): Preliminary Notes

One category of books that I did not explore systematically for this study but that may be an interesting one to watch is young adult fiction, or YA. Were I to venture some general observations about YA novels, I would say that they tend to combine features that we encounter in the nine-to-twelve age bracket with those that we encounter in so-called popular fiction. That is, they embed complex mental states somewhat less frequently (i.e., similarly to books for nine- to twelve-year-old readers), and they also tend to associate their complex embedments with characters and spell those out (i.e., a pattern similar to that we find in works of "popular" fiction). At the same time—and this is why I want you to take what I just said with a grain of salt—some YA books also experiment with forms of embedment that arise from the interaction between the implied reader and the implied author and, as such, develop a more complex and less predictable sociocognitive profile.

For instance, Mariama J. Lockington's *For Black Girls like Me* (2019) features an eleven-year-old narrator, Makeda, who is missing one crucial bit of information about her environment, namely, that her adopted mother is bipolar and that while she is taking her daughters on an exciting cross-country trip, she is actually having a manic episode. Once we figure out that our first-person narrator is unreliable, we start communicating with the implied author behind Makeda's back, as it were. For instance, when Makeda's father (who is currently away, performing at a concert in Japan) is talking to his family on the phone, making Makeda wonder why he is using "that fake cheery voice again,"[37] we *know* that Makeda *doesn't know* that her father is terribly *worried* that his wife's life is in danger and that, even though he does tell his daughters, when he gets a chance, to watch her, he cannot communicate to them the full extent of his fear.

In another, richly suggestive episode, the father gets to talk to Makeda on the phone in the middle of the night, while her mom can't hear them, and he asks her if she thinks that he should come home early. Again, we know that Makeda doesn't know how worried he really is and how worried she, herself, should be. But then we also realize that Makeda *is* worried but doesn't want to acknowledge it and that she thus seems to be learning from her mom—without being aware of it!—the habit of refusing to deal with scary feelings:

"Makeda. Wait. Do you think I should come home early? I think maybe this trip is too long."

"Honestly." I hear myself saying. In what sounds like someone else's voice. "We don't need you. We're having fun. Just enjoy your tour. I gotta go. Talk tomorrow."

Before he can say anything else I hit END CALL. I shake the dark of the bathroom off me. I shut the front door and lock it. I slip into my bed. Then I keep my eyes open until the sun comes up.[38]

It is not a given, moreover, that the novel's primary audience will grasp the extent of its narrator's unreliability upon first reading. Thirteen-year-old readers may know something about the Bipolar II mental disorder (and, arguably, they are more likely to know about it in the 2020s than their counterparts a decade or two ago would have). But, then again, they may not, in which case they will have to learn about it together with Makeda, almost at the end of the book. If they then *reread* the story, armed with that new knowledge, they may end up processing an additional set of implied complex embedments. But if they do not reread it, it is hard to say how many of the embedments arising from the unreliability of Lockington's narrator will have registered with them.

The potential fluidity of the audience for YA books—owing to their pronounced cross-writing tendencies—is what makes this category such a fascinating subject for cognitivist inquiry. Will their patterns of embedment gradually come to replicate our present division between "literary" and "popular" fiction, that is, with some of them featuring mostly explicitly spelled-out mental states of their characters and others featuring mostly implied as well as spelled-out mental states of narrators and implied readers as well as those of characters? Or will YA books evolve their own distinct profile, which may be characterized, for instance, by less frequent instances of complex embedment than we encounter in adult literature yet also by more active use of complex embedment associated with implied narrators and implied readers than we encounter in popular literature?

6.4 History and Cognition: Case Study 1 (*Tom Sawyer*)

Here is the reason I set *Tom Sawyer* aside when dealing with literature for nine- to twelve-year-olds. Although it is typically placed on the same reading level as *Mary Poppins*, *Alice in Wonderland*, and *Stuart Little*, its pattern of embedment differs from that prevalent in those books. That is, even if we take into account those books' cross-writing tendencies and say that an experienced/adult reader intuits more intentionality in them than does a less experienced/child reader, they still do not live up to the furious rate with which complex embedments (especially implied ones, involving the narrator and the implied reader) present themselves in *Tom Sawyer*. When it comes to the frequency of such embedments, Twain's novel is on par with unambiguously "grown-up" texts that which I have looked at throughout this study (e.g., novels by Murasaki Shikibu, Cao Xueqin, Frances Burney, Jane Austen, Alexander Pushkin, E. M. Forster, and Zadie Smith).

Why, then, is *Tom Sawyer* considered to be a book for children? Several factors seem to have made it so. First, as Beverly Lyon Clark has shown in her study of the history of children's literature in America, Twain "himself notoriously vacillated about the intended audience for what are now sometimes called his boy books."[39] In July 1875, he wrote to William Dean Howells that *Tom Sawyer* was "not a boy's book at all," that it was "only written for adults" and would "only be read by adults."[40] When Howells suggested that it should rather be (to use our present term) a cross-writing novel, Twain responded by "toning down [its] satire and strong language."[41] In January 1876, he was able to assure Howells that *Tom Sawyer* was now "for boys and girls."[42] In the preface to the published novel, he evokes both audiences, hoping that, though "intended mainly for the entertainment of boys and girls, . . . it will not be shunned by men and women on that account."[43]

And nineteenth-century men and women did not shun *Tom Sawyer*. It was said to "appeal to all ages," reflecting, among other things, the perspective of a culture "in which the [grown-up and children] audiences were not yet fully discrete."[44] In that culture, a review of books titled "For the Young" could still appear in the *Atlantic* (a practice apparently discontinued after 1903), stating that, although a child "will devour tales like Tom Sawyer or Huckleberry Finn, . . . he cannot understand their real merit. . . . The adult intelligence is necessary to understand them."[45]

But, although both "tales" were initially thought to demand the "adult intelligence," that perception did not last. Over the course of the twentieth century, *Huckleberry Finn* was gradually elevated to the "great American novel," an elevation that depended, Clark argues, on the simultaneous relegation of *Tom Sawyer* to "kiddie lit." As she puts it, the construction of *Huckleberry Finn*'s greatness "at the expense of *Tom Sawyer*" entailed erosion "of a fundamental respect for childhood and children's literature."[46]

Here is what a cognitivist perspective may contribute to this kind of historicist reconstruction. If we consider the difference between the two novels' patterns of embedment, we can suggest that it was this difference that may have made easier—though not necessarily determined!—the elevation of one book at the expense of the other.

Let us take as our starting point James Phelan and Peter Rabinowitz's contrast between the respective implied authors of *Tom Sawyer* and *Huckleberry Finn*. As they put it, Twain of *Tom Sawyer* speaks in the "avuncular" voice—"one that sold well in the public marketplace" but that may have demanded less work from his readers than the voice behind *Huckleberry Finn*, which is characterized by a "multilayered" ethical consciousness.[47] Thus, in one of the passages used by Phelan and Rabinowitz to illustrate their point,

> Huck describes the widow Douglas's response to his return to her home this way:
> "The widow she cried over me, and called me a poor lost lamb, and she called
> me a lot of other names, too, but she never meant no harm by it." . . . Huck
> misinterprets the widow's joyous religious references as name-calling because he
> doesn't recognize the New Testament source—and that misinterpreting leads him
> to undervalue the ethical quality of her response. Yet this comic failure of under-
> standing simultaneously reveals a moral strength. Although Huck's ignorance
> means that he fails to grasp both the extent of the widow's joy and her beliefs
> about what his return means, Twain demonstrates that Huck's ethical compass is
> sufficiently sensitive for him to appreciate that she "never meant no harm." . . .
> The overall effects are to bring us affectively and ethically closer to Huck even as
> we continue to register our interpretive difference from him.[48]

To translate Phelan and Rabinowitz's analysis into our "cognitivist" one, focusing on high-level embedments, the implied author *wants* us to *know* that Huck *doesn't understand* the widow's *motivations* (i.e., he "undervalues the ethical quality of her response"). At the same time, he *wants* us to *know* that Huck *understands* the widow's kind *intentions*. What I find particularly interesting is that it seems that to experience the full rhetorical and emotional impact of the passage—which brings us "closer to Huck even as we . . . register our interpretive difference from him"—we have to process both of these complex embedments simultaneously.

I actually don't know what this kind of dual ethical processing entails in terms of mindreading. I strongly believe that it *does not* simply ratchet up the overall level of embedment, adding up, say, to the seventh or eighth level. Still, something peculiar is happening here, something that cognitive scientists who study complex embedments in laboratory and in real-life social interactions don't tend to encounter.[49] At the very least, it shows that, while remaining inextricably bound with the social, literature has run away with it, "having amassed a repertoire of extremely nuanced stylistic tools for embedding mental states,"[50] as well as having cultivated cultural niches in which the capacity for this kind of somewhat "ecologically implausible"[51] mindreading is prized and rewarded.

Thus *Huckleberry Finn*. I do not mean to say that *Tom Sawyer* never once demands such dual ethical processing from our theory of mind but that such demands are more frequent in *Huckleberry Finn* and central to the development of its main character, that is, to "the wisdom and understanding [Huck gains] during the trip down the River."[52] Huck's reaction to the widow's response comes early and, as Phelan and Rabinowitz put it, is a "fairly simple" case of split ethical evaluation. The "same kind of interplay," only "with more subtlety and greater consequences," will mark Huck's "self-examination" later, when he decides "to go to hell rather than inform" the owner of Jim (i.e., of the runaway slave and Huck's friend) of Jim's whereabouts.[53]

In fact, so integral is this pattern of "multilayered communication" with the reader to the voice of this novel that when, at one point (i.e., when Tom plots to arrange Jim's escape from Silas Phelps), Twain abandons it, lapsing into the broad humor familiar to the readers of *Tom Sawyer*, the change feels like "a serious come-down."[54] The story still gets told through a series of complex embedments—what with all the lies that Tom is feeding the Phelpses and with the implied author winking to the reader as he parodies the chivalric romance—but the dual ethical processing is notably absent.

Where does it all leave us in the conversation about the twentieth-century designation of *Tom Sawyer* as "kiddie lit"? Looking at the dual ethical processing expected from readers of *Huckleberry Finn*—which marks some of its third- and fourth-level embedments as qualitatively different from the third- and fourth-level embedments in *Tom Sawyer*—we may speculate that had Twain never written *Huckleberry Finn*, the frequency of such embedments in *Tom Sawyer* would have made its relegation to children's literature less certain. But with *Huckleberry Finn* next to it, the intuitive awareness of a different kind of sociocognitive complexity underlying the latter's affective charge may have contributed to this cultural phenomenon.

Still, the main payoff of factoring the cognitive perspective into the historicist explanation of this process offered by Clark may be a more nuanced understanding of why the designation of *Tom Sawyer* as "kiddie lit" remains troubling enough for critics to keep wanting to account for it. The cascading frequency of complex embedments expected from the reader of *Tom Sawyer*—a frequency that, though not inconceivable in a book for children,[55] is nevertheless rare—may be the reason why this novel does not stay meekly put in the category of kiddie lit. For as long as we place in that category texts that embed complex mental states of characters, narrators, and implied readers, but not at the same high rate that we've come to expect from a work of "grown-up" literature, *Tom Sawyer* shall remain an outlier.

6.5   History and Cognition: Case Study 2 (*Little House in the Big Woods*)

Tom Sawyer is not the only outlier that I found in the nine-to-twelve age group. An even more striking case, though for the opposite reason, is Laura Ingalls Wilder's *Little House in the Big Woods*. It contains very few embedded mental states and practically no third-level embedments. Though a highly compelling narrative in its own right, it has, as its readers observe, "no plot."[56] Instead, we learn details of life on the frontier: how bullets were made, how butter was churned, and how meat was cured. The near-total absence of social situations that would call for attribution of complex mental states is, one can safely say, extremely unusual for a text considered to be a novel. To see how this classification came to pass, we have to inquire once again, into the circumstances of its writing and publication.

The original version of the "Little House" series was called *Pioneer Girl*. It was an autobiographical account of Wilder's "family pioneering experiences in the American West," intended for adults. As Wilder's biographer Pamela Smith Hill puts it, it was "nonfiction, the truth . . . as only Wilder remembered it."[57]

What happened then was that *Pioneer Girl* could not find a publisher. A typical rejection, from *Country Home* magazine, praised it for "some very interesting pioneer reminiscences" yet explained that they had "no place for non-fiction serials."[58] Wilder then turned her autobiographical manuscript into a book of fiction for children, with the assistance of her daughter, the established writer Rose Wilder Lane. As Hill puts it, "Lane not only switched audiences, she switched genres—from nonfiction to fiction. When she replaced Wilder's intimate first-person voice, her 'I' narrator, with a third-person narrative, the juvenile manuscript instantly became fiction."[59]

Did it? If we think of fiction in a broader sense of the term, as something fabricated rather than factual, we can say that Wilder's manuscript "became fiction" even earlier, when, for instance, to make *Pioneer Girl* more dramatic, Lane adjusted the timing of the Ingallses' move to Wisconsin to bring them into contact with a notorious family of Kansas mass murderers.[60] Or we can say that the fictional status of the Little House books was clinched when, as staunch opponents of the New Deal, Wilder and Lane took "serious liberties with the facts of the Ingallses' lives" to portray the US government as "nothing but destructive to the enterprising individual." Or that it happened when they "entirely made up or altered in fundamental ways" scenes that testified to Laura and her sisters' schooling "in emotional and physical stoicism" and to their family's socioeconomic self-reliance.[61] As far as historical accuracy goes, the series is certainly fiction: a heady blend of libertarian ideology and emotional warmth, mythologizing life on the frontier.

Yet we have also come to intuitively expect something else from fiction/literature, particularly with the novel as its flagship genre. While the presence of complex embedments alone does not determine if a given text is considered fiction, the near absence of such embedments in *Little House in the Big Woods* makes one wonder just how those joint appellations—that of fiction and that of novel—came to stick. To see how it happened, we retrain our attention on its cultural reception.

And what we learn when we look at the history of that reception is that readers have always seen *Little House in the Big Woods* in the context of other books in the series, which are more "novelistic" in their outlook. For, as Wilder continued to draw on *Pioneer Girl* for her subsequent volumes, she went further than merely substituting "I" with "Laura." As Smith Hill observes: "[As] Wilder transformed her original material into fiction for young readers, she grew both as a writer and ultimately as an artist, creating dynamic characters, building more suspenseful stories, and manipulating her themes more masterfully."[62]

From a cognitive literary perspective, we can see the evidence of this transformation in a gradual increase of the number of situations calling for third-level embedment. Take *Little Town on the Prairie*, "the best-selling of the Little House books," which serves for many readers as the gateway into the series. It turns out to owe very little to the original manuscript: the

"comparable segment" of *Pioneer Girl* is "only six and a half pages long."
In this "product of . . . Wilder and Lane's imaginations," Laura *feels shocked*
when her sister Mary tells her that she *knows* why Laura used to *want* to
slap her and that she thinks she deserved being slapped.[63] She also feels bad
about reading a poem in a fine book that she finds in a drawer because she
realizes that her mother *wanted* that book to be a *surprise* gift for her.[64]

Similarly, in *These Happy Golden Years*, older Laura is "furiously angry" at
her student Clarence and trying to conceal her anger, for "as her eyes met
his she *knew* that he *expected* her to be *angry*."[65] When Laura goes for a ride
with Almanzo and her potential rival, Nellie, Laura is thinking that her
acquaintance Mr. Boast knows that she intends to take Nellie down a road
that she won't like: "His eyes laughed at Laura. She was *sure* he *guessed* what
was on her mind." Later on, Laura is having a similar exchange of glances
with Almanzo: "She let her eyes twinkle at him. She *didn't care* if he did
know that she had frightened the colts to *scare* Nellie, on purpose."[66]

This is very different from the inaugural volume, which focuses on how
things are made as opposed to what people think and feel. Still, because
the Little House books are treated as one continuous narrative—a story of
Laura's "transition from a tomboyish girl to a marriageable woman"[67]—it's
possible that the sociocognitive complexity of the later volumes colors our
perception of the first. Had those later volumes been constructed similarly
to *Little House in the Big Woods*—that is, had they focused on objects and
processes to the exclusion of complex social dynamics—perhaps *Little House
in the Big Woods* wouldn't have been considered a novel today. Instead, it
might have been viewed as an arresting description of a child's experience
on the frontier—for remember that expository nonfiction does very well
with lower (i.e., first and second) levels of embedment!—perhaps something along the lines of Susan Sinnott et al.'s *Welcome to Kirsten's World,
1854: Growing Up in Pioneer America*.

To see how the perception of the Little House books as one continuous
narrative has become entrenched in US popular culture, we can inquire
into the role of the 1974–1983 television series, which didn't follow the
original's division into volumes (indeed, didn't follow the original at all).[68]
I prefer, however, to look at another, subtler factor, one that has to do with
Little House's career as a mainstay of basal readers used by US elementary
school teachers from the 1930s until the 1990s. The original inclusion in
basal readers owed to the fact that Wilder's book seemed to fit several diverse

criteria articulated by 1920 research studies, which called for more "adventure stories (boys) and home-and-school stories (girls)" as well as for more "informational books." The criteria changed by the 1970s—with stress on the emotional security provided by family and on the child's ability "to master environment without adult help."[69] Once again, Little House books met those criteria because they have long been perceived—and taught!—as a story of Laura's personal journey toward maturity and independence, made possible by her warm, supportive family.

So here we have *Tom Sawyer* classed with "kiddie lit" even as the frequency of its complex embedments makes it stand out among other books in its designated cohort and *Little House in the Big Woods* considered a novel in the absence of any complex embedments. What these two outliers tell us (besides illustrating the importance of historical inquiry for a cognitive literary analysis) is that patterns of embedment don't always determine genre designations even if, at present, our "grown-up" literature, and novels in particular, are dominated by subjectivity arising from complex embedment of mental states.

6.6 Ages Three to Seven

Three to seven is an extremely interesting age when it comes to embedment, because this is when children are more consistently found to be aware of first- and second-order false beliefs in themselves and others. Although the boundary between books for seven-year-olds and nine-year-olds is porous, here is one intriguing pattern found in stories signposted specifically for the younger age group.

Some books marked for ages three to seven contain just one third-level embedment, although it can be repeated several times either with different characters or in slightly different settings. This embedment is central to the story, constituting, in effect, its punch line, its raison d'être. It is typically structured as a dawning awareness, on the part of young readers, that they know something about one character's thoughts that another character doesn't know. (Literary scholars may recognize this as a preschool version of dramatic irony and thus talk of cultural scaffolding involved in shaping children into future mature readers,[70] while developmental psychologists may note its similarity to their made-up scenarios used in double-embedment false-belief tests with six-year-olds.)

Thus, Jon Klassen's *This Is Not My Hat* follows the path of a small fish who has stolen a big fish's hat. Young readers gradually *realize*—and presumably delight in their realization[71]—that the small fish erroneously *believes* that the big fish *doesn't know* who has stolen his hat.[72] Julia Donaldson's *Gruffalo* tells a story of a big scary monster who believes a mouse's claims that the mouse is the most powerful animal in the forest. Once more, preschoolers are "in" on the joke: they *know* that the Gruffalo *doesn't realize* that when she[73] is walking behind the mouse in the forest, other animals are scattering because they are *afraid* of her and not of the tiny mouse.[74]

Similarly, reading Pat Hutchins's *Rosie's Walk*, children *know* that Rosie the hen *doesn't know* that the hungry fox *wants* to devour her and that she has one lucky escape after another. In Gene Zion's *Harry the Dirty Dog*, the premise of the story is that Harry's owners don't recognize Harry, a white dog with black spots, because running around the city and getting dirty has turned him into a black dog with white spots. The young readers thus *know* that Harry's owners *don't suspect* that the reason this strange dog brings them a scrubbing brush (a hateful implement, which Harry earlier buried in the backyard) is that he *thinks* that, once they wash him, they'll *recognize* him as their beloved pet.

The positive affect presumably elicited in young readers by such embedments is a fascinating phenomenon. One may argue that it derives from identification with the characters,[75] particularly those who get to have their way, such as Rosie, Harry, the big fish, and the little mouse. I tend to think that it comes from the perception of social mastery fostered by the plot. Children know—and they know that they know!—that the small fish doesn't realize that the big fish has already figured out who has stolen his hat and is on the way to catch the thief. So the big fish may end up eating the little fish, but it's the young reader who is having the satisfying experience of being on top of the epistemological food chain.

We may do well to remember here that contemporary writers for young children didn't invent the concept of a triply embedded punch line and that it has been long present in "trickster" stories worldwide. Thus, the premise of *Gruffalo* is based on a classic Chinese tale of a tiger and a fox. (The fox wants the tiger to think that, when they walk together, the fox slightly ahead, other animals run away because they are afraid of the fox.) We find triply embedded mental states in West African folklore (e.g., Brer Rabbit wants Brer Fox to think that he's afraid of the briar patch); in Native

American legends (e.g., Badger knows that Coyote thinks that Badger is lying to him when he says that there is no food in the sack that the Badger is carrying on its back); in Bornean folktales (e.g., a mouse-deer wants a crocodile to think that the mouse-deer doesn't know if the body in the water is the crocodile or just a log); and in Russian fairy tales (e.g., an exhausted old house cat wanders into the forest, where he meets a fox, who promptly offers to marry him; once married, the fox has to figure out how to protect and feed her new husband; she decides to make a bear and a wolf think that the cat is an important government official who'll be angry at them if they come to see him without substantial gifts).

Of course, not all trickster tales feature triply embedded mental states. Just so, not all are geared toward children. Still, if we only consider those that do and are, it is an extremely suggestive sociocognitive phenomenon. It seems that many cultural traditions offer young children stories centering around doubly embedded false beliefs just at the time when children go through a developmental stage that makes them particularly attuned to such beliefs.[76] In this particular case, the "cultural" and the "cognitive" appear to form a feedback loop, shaping and reinforcing each other.

6.7   Ages One to Two

Recall that in the study of children's books by Dyer et al. (which found that books for younger and older children are similar in their "richness of mental state concepts"), the youngest subjects were three years old. I wonder if, at three, children are already too far advanced on the developmental trajectory that leads to awareness of (first-degree) false belief. For that awareness is not achieved suddenly once the child turns four. It is being continuously built up, in conjunction with other "maturational factors," such as language ability.[77]

This is why I believe it's worth our while to take a closer look at books for toddlers.[78] (This age group, as you remember, is now a subject of controversy: it used to be assumed that they have not yet reached the theory of mind milestone of appreciating false beliefs, but now experimental evidence suggests that one may elicit such appreciation from them.) What I found after a preliminary study of books in this group is that they do demonstrate a significant drop in third-level embedment. This is not to say that they don't contain references, both explicit and implied, to mental states: they do. (This is a key difference between my approach and that of developmental psychologists studying children's theory of mind: they look at mental states; I look at embedded mental states.) What they don't seem to contain—at least those that don't function as crossovers that appeal both to toddlers and to older readers—is third-level embedment.

In looking at books geared toward children aged one to two, I focus on those that lay a claim to telling a story, as distinct, that is, from books of colors, numbers, body parts, and so on, which don't.[79] There is, for instance, *Curious George at the Zoo: A Touch and Feel Book* (not to be confused with the original Margret and H. A. Rey's "Curious George" stories and their more recent versions: the touch and feel books do not reproduce any of their plots; indeed, the only thing they seem have in common with the "real" Curious George series are the two main characters.)

We learn on the first page that the "man with the yellow hat is taking George to the Zoo today. There are so many things to see and do and touch." Most of the pages that follow focus on the sensory: "Feel the black and white penguin's thick coat," "Feel the smooth shiny water," "Feel the rhino's rough skin." The book does contain references to mental states (e.g., "Where has George gone? He would love to watch the pink flamingo standing on one leg"),[80] but it has no complex embedments.

Note that *Curious George* currently has 175 reviews on Amazon, and 62 of them mention explicitly the age of the young reader (another 10 merely say that the reader is a "toddler"). Out of these 62, 58 cluster between the ages of four and twenty-four months. While we may not want to put too much emphasis on this bit of digital data mining, it offers a useful glimpse at the perspective of caregivers who actually buy these books and judge their appropriateness for their young charges.

Here is another example: Disney's *Pooh's Honey Trouble*, based, loosely, on the first chapter of Milne's *Winnie the Pooh*. That's the chapter in which Pooh hopes to fool the bees into thinking that he is a black cloud and not a honey-stealing bear floating on a balloon and in which Christopher Robin doesn't want to hurt Pooh's feelings by telling him that his plan won't work. In Disney's version, Pooh wakes up in the morning feeling hungry and goes out in search of honey. He comes across several of his friends, busy doing what they like to do. Then Christopher Robin finds out that Pooh is hungry and gives him a balloon, with which he finally manages to get some honey:

Winnie the Pooh awoke one morning with rumbly in his tumbly. "Oh, bother," he said, finding his honeypots not at all full. The trouble with empty honeypots, thought Pooh, is that they're so very empty. Pooh went to see Piglet who was busy gathering haycorns. Pooh helped his friend for a bit, but picking haycorns didn't help to take his mind off his rumbly tummy, so he continued on. . . . "Hello, Pooh Boy!" said Tigger, bouncing his way through the forest. "Tiggers love bouncing." "And bears love honey," Pooh replied in a rumbly voice. . . . When Christopher Robin heard of Pooh's honey trouble, he gave him a balloon. The balloon was very nice, in a balloonish sort of way, but Pooh was quite sure it wouldn't make his tummy any less rumbly. "Silly old bear," said Christopher Robin, watching Pooh float up, up, up, up to the spot where the honey was. And, at last, Pooh's tummy wasn't rumbly anymore.[81]

What kind of embedments do we have here? Most of them are first level, such as "Pooh wants honey," "Tiggers love bouncing," "Rabbits like carrots," "Piglet likes haycorns," although there are also some implied second-level ones, such as "Pooh knows that Piglet likes haycorns" or "Christopher Robin knows that Pooh doesn't understand what the balloon is for."

There are currently seventy-two reviews of this book available on Amazon,[82] and twenty-nine of them explicitly mention the age of the child for whom the book was bought. Out of these twenty-nine, twenty-eight fall between the ages of eight and twenty-four months, making it, as one reviewer puts it emphatically, a "book for toddlers."[83]

The development of theory of mind is intertwined with the acquisition of vocabulary, but it's not a simple vocabulary that makes *Pooh's Honey Trouble* "a book for toddlers." Take another look at *Rosie's Walk*, mentioned earlier (a book targeting children aged three to seven). *Rosie's Walk* contains fewer words than either *Curious George at the Zoo* or *Pooh's Honey Trouble* does, and, unlike them, it has no explicit references to mental states.[84] Nevertheless, it does embed mental states on the third level—via illustrations!—and the reviews on Amazon testify to its popularity with parents of preschoolers, with kindergarten teachers, and with beginner readers themselves.[85]

Still, although I am encouraged by early findings about the relative scarcity of third-level embedments in books for one- to two-year-olds, I would be cautious about simply concluding that they signal intuitive awareness on the part of authors and caregivers of the stages in the development of theory of mind.[86] For the excision of complex mental states from such books must also have its own history, bound up with the emergence of what Alan Richardson calls "the children's book industry," which in Eng-land, for instance, goes back to at least 1744.[87]

Complicating the issue even further are recent experiments of cogni-tive scientists that demonstrate some awareness of false beliefs in fifteen-month-olds. Given these experiments, one would think that it may be good for one-year-olds, now and then, to hear a story that is "above their head"—that is, a story that embeds mental states on the third level—especially if their parents make a point of talking with them about the characters' thoughts and feelings. Benefits of this practice are borne out by research of the developmental psychologist Paul L. Harris and his colleagues, who have shown that parents "who talk about psychological themes promote their children's mental state understanding," especially when their elucidation of mental states "is not tied to particular lexical terms or syntactic con-structions, . . . [reflecting instead] a wide-ranging sensitivity to individual perspectives and [nurturing] the same sensitivity in children."[88]

Of course, to extrapolate from Peskin and Astington's study, there may be a delicate balance between letting toddlers *infer* implied mental states of characters in a children's book and *talking* to them about those mental states. This, moreover, is the point at which our current state of knowl-edge makes me cautious about speculating any further, calling (predictably) for more research into historical and cognitive-developmental aspects of embedment in stories for toddlers.

Crossovers

It's fitting to conclude this chapter with a discussion of crossovers: books that appeal to toddlers *and* to their parents, such as Marla Frazee's *Hush, Little Baby.*[89] The "story" told by this board book is an old folksong, "Hush little baby, don't say a word," transcribed verbatim. There are no third-level embedments in the song. In fact, there are no references to mental states at all, although we may come up with a couple of implied embedments, such as, papa and mama are *willing* to buy anything to make their baby *happy* ("If that billy goat don't pull, / Papa's gonna buy you a cart and a bull"), and papa and mama *love* the baby ("If that horse and cart fall down, / You'll still be the sweetest little baby in town").

Frazee's illustrations, however, tell a different story. Its protagonist is an older sister, who is about eight and jealous of the attention that the new baby gets. So when the baby's peacefully asleep and the parents are looking the other way, the girl pushes the cradle roughly. The baby wakes up screaming, and the girl pretends to be concerned and eager to calm it down ("Hush, little baby, don't say a word"), while the startled and bleary-eyed parents look on. The girl then convinces the father that they should go visit a village peddler, because a mockingbird in a cage would surely console the baby. Frazee's drawings seem to imply that the girl has wanted the bird for some time and that she is thrilled to get some time alone with her daddy. And so it goes. The baby keeps crying, while the older sister keeps accumulating one treasure after another (a diamond ring, a looking glass, a puppy), delighting in her important role in the common project of calming down the baby and, in fact, gradually warming up to the little interloper.

There are numerous third-level embedments in the story told by the pictures. At first, we are encouraged to think that the parents don't suspect that the girl is jealous, just as they don't suspect that she only wants them to think that these toys are for the baby while, in reality, they are for her. But toward the end of the narrative, we begin to wonder if the parents are indeed as clueless as the girl thinks they are. In fact, when she gets the puppy, the father's facial expression seems to imply that he has understood all along more than his daughter thought he did (figure 6.2). His glance breaks the fourth wall and draws us in: he wants us to know that he knows what's going on. (Or, given that the narrative thus foregrounds the

Figure 6.2
"If that dog named Rover don't bark." (Marla Frazee, *Hush, Little Baby: A Folk Song with Pictures Board Book*)

relationship between the implied reader and the implied author, another way to map out this scene would be to say that *Frazee* wants us to know that the father knows what's going on.)[90]

Eden Ross Lipson as well as Amazon put the age of the reader for *Hush, Little Baby* at two to three years old, which is reasonable, given that the original folksong has no third-level embedments.[91] Freeman, Gillespie, and Scholastic, however, estimate the age of the reader as pre-K to second grade.[92] The difference between two to three and pre-K to second grade appears striking unless we assume that Freeman, Gillespie, and Scholastic respond to the story told by the book's illustrations. The level of embedment in that story, indeed, makes it appropriate for readers who can appreciate the first- and even second-order false beliefs, that is, for four- to seven-year-olds. Moreover, responses accumulated on Amazon show that parents and grandparents are intuitively aware that Frazee's book contains two stories under one cover, one geared (we can say) toward a more mature theory of mind, and another, toward a theory of mind early in development.[93]

What I have hoped to show throughout this chapter is that embedded mental states are richly present not just in "grown-up" fiction but also in children's literature and that a critical inquiry into patterns of embedment in children's literature draws on close reading, cultural-historical analysis, research in cognitive science, and even some occasional digital data mining. As such, it makes a practitioner of the cognitive approach to literary criticism accountable to several different fields and, moreover, aware of the provisional state of one's conclusions. This may imply more uncertainty than our discipline is used to, but, then, one doesn't turn to interdisciplinary work seeking certainty and familiarity.

Conclusion: On the Future of the "Secret Life" of Literature

> Then Ea opened his mouth and said to me, his servant, "Tell them this: I have learnt that Enlil is wrathful against me, I dare no longer walk in his land nor live in his city; I will go down to the Gulf to dwell with Ea my lord. But on you he will rain down abundance, rare fish and shy wild-fowl, a rich harvest-tide. In the evening the rider of the storm will bring you wheat in torrents."
> —*The Epic of Gilgamesh*, 2100 BC

> It is interesting that we either fictionalize or become tongue-tied when it comes to personal matters. We may have good reasons to hide from ourselves (at least to hide certain aspects—which amounts to the same). But even if there is little hope of an eventual self-acquittal, it would be enough to withstand the lure of silence, of concealment.[1]
> —Christa Wolf, *Patterns of Childhood*, 1976

There are four thousand years and several worlds of difference between the promise of abundance that the god Ea dangles in front of the people of Shurrupak, just before they are all swept to their death by a giant flood, and the painful self-searching awareness of Wolf's autobiographical novel about growing up in Nazi Germany. Yet to make sense of either situation, we engage in a very particular kind of social reasoning. We navigate, without being consciously aware of it, the multilayered intentionality of the text. That is, we recursively embed—mostly on the third level—thoughts, feelings, and wishes of its characters, as well as (if we are *that* kind of readers) of its narrators and implied audiences.

Thus, we may recognize that Ea *wants* the citizens of Shurrupak to *believe* that Enlil is *angry* at Utnapishtim. We may also surmise that, with all the talk about "a rich harvest-tide," Ea is enjoying his cruel joke, as befits a trickster deity—which is to say that the narrator of *Gilgamesh wants* to draw his audience's *attention* to Ea's *intention* to mock the doomed Shurrupakians.

When it comes to Wolf, her narrator *knows* that she *may not like* much of what she will *learn* about herself when she *starts thinking* about her childhood. Yet she also *intuits* that there is some *hope* that she may *forgive* her past self. She *thinks* that her *awareness* of that *hope*, however small, should help her to keep going even when it would feel so much easier to stop and keep her memories hidden from herself and others.

Moreover—again, if we are that kind of readers—we may start reading additional intentionality into the present juxtaposition of the two passages. After all, the child protagonist of Wolf's novel is no more aware of what kind of deadly "harvest-tide" lies in wait for her and her countrymen than are the people of Shurrupak. Although I did not intend any such conversation between the two passages when I selected them—indeed, my goal was to use works of literature as distinct from each other as possible—I now can't help *wondering* if some of my readers will see the connection and *think* that I *meant* for it to be there.

(Herein lies an object lesson in what happens when you put two random literary passages in front of a person who makes her living by reading complex intentionality into cultural artifacts: "Hey, what do you mean 'two random passages'? I see a connection here!")

And now I also wonder if you will take this emerging conversation between *Gilgamesh* and *Patterns of Childhood* as me saying that nothing much has changed in the depiction of literary subjectivity over the past four thousand years. In fact, I am saying the opposite. I want you to see how different literary subjectivity has become as it has moved from the occasional reliance on complex embedment of mental states (e.g., in *Gilgamesh*) to the constant one. For, to find my *Gilgamesh* example, I had to comb the text; to find my *Patterns of Childhood* one, I had to merely open the book. The challenge of casting about for social situations conducive to incessant complex embedment of mental states has been shaping literature as we know it for several centuries. Without being consciously aware of it (which is a good thing, too, as my experience in the writing workshop confirms), authors keep inventing new and tweaking old ways of recursively embedding thoughts and feelings. To quote Lewis Carroll's *Through the Looking-Glass*, just to stay in place—here, on the third level of embedment—they have to run as fast as they can.

It remains an open question whether literature will ever be able to break free of this relentless gravitational pull of complex embedment. Writers who *seem* to attempt such a break, driven by a wide variety of personal, political, and aesthetic motivations (e.g., Evgeny Zamyatin in *We*, Alain Robbe-Grillet in *Jealousy*, Muriel Spark in *The Ballad of Peckham Rye*, Cormac McCarthy in *Blood Meridian*, and Fedor Gladkov in *Cement*), manage it only to a point. The odds are stacked against them. Reading complex intentionality into a literary text—which is to say, intuitively expecting literary subjectivity to be constructed as a series of complex embedments, explicit or implied—has become our standard experience of literature.

This expectation/experience is buttressed by several cultural factors. First, there is a vast ocean of popular fiction that embeds complex mental states of (mostly) characters. Though differing from literary fiction (which embeds mental states of narrators and implied readers, as well as characters), such books nevertheless contribute to making their readers experience complex embedment as a default mode of engagement with fictional imagination. Second, this "induction" into the association between fictional stories and complex embedment begins quite early—with books targeting three- to seven-year-olds. Third, cultural institutions—from college literature departments to critical reviews—implicitly train their adepts to think in terms of embedded motivations of characters, writers, and readers and reward them for compellingly articulating such motivations.

Fourth, there is also the possible impact of moving images, which I mention here only briefly, not having addressed it in this book. Feature films and television series use medium-specific methods to generate complex embedments of mental states. Moreover, critics (as in my Susan Sontag example, in chapter 4) depend on complex embedments to talk about films, which means that institutional structures that reward thinking about moving images in terms of complex intentionality have been in place for some time. Whether the experience of watching certain films and TV series and reading reviews of such films/series sensitizes viewers to cues of intentionality in their social environment and whether such a sensitivity translates between media, influencing reading practices, are open and intriguing questions.[2]

Finally, consider that we tend to view as ethical and prosocial the practice of rendering minds transparent—which is to say, of talking publicly about one's own and other people's feelings, even when (in fact, sometimes especially when) we think that we can articulate other people's true motivations better than they themselves can articulate them. Though adapting the rhetoric of opacity when it is expedient, our culture inclines, on the whole, toward the transparency end of the opacity-transparency spectrum. This means that representations of and conversations about complex intentionality of fictional characters, their creators, and their audiences are entrenched in our public discourse and, indeed, in our current cultural perception of how the social mind works.

Imagine, then, an author who is firmly committed to writing a novel that will transcend the pull of embedded subjectivity. (Not that they themselves would put it that way; they may think of it as "antipsychological" or "surface based," or "a story without interiority"—you name it.) That writer will face an uphill battle at every step of their interaction with their audience. Readers will come to that novel intuitively expecting to encounter recursively embedded subjectivity either of characters or of characters, narrators, and implied readers. They will force-read as much of that kind of subjectivity into the story as the text itself and their own past reading history will allow them. Critics, too, will find ways of talking about embedded thoughts and feelings, by speculating about *the writer's* intentions and describing *their* reactions. If the novel is adapted for screen, social situations and/or shots calling for complex embedment of mental states will be introduced, and that will, in turn, influence the experience of readers who will come (or return) to the original text after watching the film. Can our experimental novel survive this onslaught of embedded mentalizing and even start a new literary trend of embedment-free writing? Perhaps it still can, but it won't be easy.

This is not to say that complex embedment of mental states is an inevitable feature of the literary landscape of a mindreading species—merely that it has been around for a while and is still going strong. Contributing to its longevity is its integration with our ideology of mind: we believe that mental states are knowable and can be discussed in public, and we have cultural institutions that reward elaborate forms of such discussions, be they about real or fictional minds. But while there is no way of knowing what the future holds either for such institutions or for the secret life of literature, we can follow, with new awareness, the remarkable current career of this inconspicuous yet pervasive phenomenon: watch it as it adapts to new media and reinvents itself in the old ones.

Notes

Preface

1. For an introduction to the field, see Zunshine, "Introduction to Cognitive Literary Studies." For a representative bibliography, see Zunshine, "May 2020 Bibliography," as well as its more frequently updated counterpart at my Academia page: https://uky.academia.edu/LisaZunshine.

2. B. Schieffelin, "Found in Translation," 143.

Chapter 1

1. Twain, *Mississippi Writings*, 20.

2. For a discussion, see Fernyhough, "Metaphors of Mind." Note that plenty of cognitive scientists use "theory of mind," "mindreading," and "mental states" (or "internal states") in a literal sense. Indeed, for the purposes of studying the phenomena referred to by these terms, it does not seem practicable to be always carefully foregrounding their metaphoricity.

3. For a review of mindreading, see Apperly, *Cognitive Basis of "Theory of Mind."*

4. See, for instance, Tomasello, *Origins of Human Communication*, 173, 189.

5. Of course, I can imagine a context in which this statement will contain mental states. For instance, if I am standing in a long line and an authority figure comes over and tells us that this is the line only for people whose last names begin with a *Z* and that we should thus disperse, I may call out with some strong feelings, "My last name begins with a *Z*!" This is to say that my present examples are synthetic constructs designed to make a point rather than to represent accurately a range of real-life situations.

6. Mercier and Sperber, *Enigma of Reason*, 81. See also Martins and Fitch, "Do We Represent Intentional Action as Recursively Embedded?"

7. Tomasello, *Origins of Human Communication*, 173.

8. Miller, Kessel, and Flavell, "Thinking about People Thinking," 622.

9. On the difference between the effect on theory of mind of reading fiction and expository nonfiction, see Mar et al., "Bookworms versus Nerds."

10. In prose fiction, sentence- and paragraph-level complex embedments may be particularly predominant. Compare to Auyoung's observation that "the prosaic organization of text across sentences and paragraphs emerges as a crucial scale at which narrative information can be strategically arranged" (*When Fiction Feels Real*, 63).

11. See Zunshine, "Commotion of Souls"; Whalen, Zunshine, and Holquist, "Increases in Perspective Embedding."

12. From this point on, I frequently omit the term "implied" as a modifier for reader/audience. Although the narratologist in me would strongly prefer to speak of implied readers as opposed to just readers, I find useful the distinction between the literary-critical (in my case, narratological) and empirical perspectives, recently outlined by Andrew Elfenbein. As he puts it, "Literary scholars may at times strive to occupy a position as close as possible to their understanding of the implied reader. . . . While I am comfortable with the 'implied reader' as a literary critical construct, I have seen no psychological evidence that actual readers envision an implied reader as they read or use an implied reader to gauge their own performance" (*Gist*, 199).

13. Compare to arguments developed by narratologists, such as Henrik Skov Nielsen, James Phelan, and Richard Walsh, who contend that "the rhetoric of fictionality is founded upon a communicative intent" ("Ten Theses about Fictionality," 64); by philosophers, such as Gregory Currie, who observes that a "narrative is an artefact, wherein the maker seeks to make manifest his or her communicative intentions" ("Framing Narratives," 18); by cognitive literary scholars, such as Andrei Ionescu, who notes that the relationship between reader and writer can in itself be "a very complex form of intersubjectivity" ("Manifesto," 9); and by cognitive linguists, such as Yanna Popova, who sees literature as "framing an interactive engagement with a reader" (*Stories, Meaning, and Experience*, 71).

14. Phelan and Rabinowitz, "Authors, Narrators, Narration," 37.

15. As Elfenbein puts it, "As soon as a reader can recognize that paradoxes, ambiguities, and uncertainties are intentional, at whatever level of agency intention is understood, representation becomes coherent" ("Mental Representation," 251).

16. See Bowes and Katz, "Metaphor Creates Intimacy."

17. Twain, *Mississippi Writings*, 24 (emphasis added).

18. Twain, 25 (emphasis added).

19. Gavaler and Johnson, "Genre Effect," 86, 91. For an analysis of the "interaction effect between genre and mentalizing" in case of espionage stories as compared to relationship stories, see also Carney, Wlodarski, and Dunbar, "Inference or Enaction?"

20. Ferrante, *Story of the Lost Child*, 250.

21. Rooney, *Conversations with Friends*, 242.

22. Cusk, *Transit*, 174.

23. Dangarembga, *Nervous Conditions*, 116.

24. Al Harthi, *Celestial Bodies*, 97.

25. Lerner, *Leaving the Atocha Station*, 84.

26. Williams, *Ninety-Nine Stories of God*, #61.

27. Gavaler and Johnson, "Genre Effect," 79–108.

28. For a related critique of this stance, see Savarese's *See It Feelingly*, in which he objects to its "very narrow conception of the social, as if the social were something that only human did with each other" (111).

29. Dick, *Do Androids Dream of Electric Sheep?*, 4.

30. Savarese, *See It Feelingly*, 101.

31. Jackson, "Beautiful Stranger," 79.

32. Forster, *Howards End*, 254.

33. Tolstoy, *Anna Karenina*, 413.

34. Austen, *Pride and Prejudice*, 180.

35. Cao, *Story of the Stone*, vol. 1, 438.

36. And if we agree with Elaine Scarry's argument that Shakespeare's "beautiful young man" (*Naming Thy Name*, 4) was Henry Constable, then we have a poetic rejoinder written by Constable, who casts about for illusory explanations that may soothe *his* pain. The speaker of this sonnet (number 8 in Constable's "Diana" cycle) suspects that his beloved (i.e., Shakespeare) placed him in harm's way—by asking him to keep company with his mistress in his absence—on purpose. But if that's the case—and here comes the complex embedment—then the speaker can make himself feel better by *imagining* that the beloved *wanted* him to *feel* this pain:

> So when this thought my sorrowes shall augment,
> That mine owne folly did procure my paine;
> Then shall I say, to give my selfe content,
> Obedience only made me love in vaine:
> It was your will, and not my want of wit;
> I have the paine—beare you the blame of it.

Or, as Scarry explains in her own tour-de-force of complex embedment, "I am in torment, says the speaker, a torment made worse by knowing my own folly brought this about; the only explanation that would make me gladly accept my pain would

be to know you so take pleasure in my torment that you scripted the entire event; i
my pain gives you pleasure, I can accept my pain" (*Naming Thy Name*, 68).

37. Nizami, *Story of Layla and Majnun*, 10.

38. Anonymous, "The Wanderer," n. p.

39. Petronius, *Satyricon*, 50.

40. Homer, The Odyssey, 219.

41. Anonymous, *Epic of Gilgamesh*, 108.

42. Furlanetto et al., "Through Your Eyes." Note that in this case the word "actor" a
used by the authors of this essay refers not to an actor onstage but to a person whose
actions are observed by others.

43. Furlanetto et al.

44. See, for instance, Noel, "What Do We Actually See on Stage?"

45. Shakespeare, *Measure for Measure*, 5.1.

46. Hogan, *Sexual Identities*, 141.

47. Shakespeare, *Twelfth Night*, 3.4.97–98.

48. For a detailed discussion, see Zunshine, "Why Jane Austen Was Different."

49. Nizami, *Layla and Majnun*, 10.

50. See Whalen et al., "Validating Judgments," 293.

51. Chekhov, "Skripka Rotshil'da," n.p., translation mine.

52. Lu, *Madman's Diary*, 19. In the original: 我忍不住, 便放聲大笑起來, 十分快活。自己曉
得這笑聲裏面, 有的是義勇 和正氣。老頭子和大哥, 都失了色, 被我這勇氣正氣鎮壓住了。

53. Pittard, *Listen to Me*, 2, 3.

54. Z. Smith. *On Beauty*, 3.

55. Forster, *Howards End*, 3.

56. Note the specific meaning of the word "intentionality" when used interchangeably with "mental state." As Mauricio D. Martins and W. Tecumseh Fitch observe,

> It is important before going further to identify a potential source of confusion concerning
> "intention" and "intentional" stemming from the specialized interpretations of these terms
> as traditionally used by philosophers, that differ considerably from their ordinary English
> meanings. In ordinary English, "intentional" means "on purpose," but philosophers use
> "intentionality" to designate a particular characteristic of mental states. . . . In this sense
> intentionality is a pervasive and fundamental feature of mental states like beliefs or desires
> but including a wide range of other states including memories, hopes, knowledge, love—or
> intentions (in the ordinary sense). Thus, from the philosophers perspective, intentions are

just one among many different forms of intentional state. . . . [Thus] it is important to note that "intentionality" does not imply an "intention to do something." ("Do We Represent Intentional Action as Recursively Embedded?,"18)

57. Williams, *Ninety-Nine Stories of God*, #61.

58. Kulpa, "Review of *Ninety-Nine Stories of God*."

59. Apuleius, *Golden Ass*, 112.

60. Ruden, "Translator's Preface," xv.

61. Peter Stockwell, MIT Press reader report, 2020.

62. Anonymous, *Epic of Gilgamesh*, 108.

63. Kidd and Castano, "Reading Literary Fiction and Theory of Mind," 8. For a discussion of a controversy involved in the replication of the findings from the original 2013 study (i.e., Panero et al., "Does Reading a Single Passage of Literary Fiction Really Improve Theory of Mind?"), see also Kidd and Castano, "Panero et al. (2016)"; and van Kuijk et al., "Effect of Reading a Short Passage." For a useful metastudy that reviews recent research on fiction's effects on social cognition, see Dodell-Feder and Tamir, "Fiction Reading Has a Small Positive Impact." As its authors summarize citations omitted throughout),

> The effect of fiction on social cognition was larger when compared to no reading versus nonfiction reading. Indeed, if fiction's causal impact depends on the extent to which a text provokes readers to consider mental states, then many forms of nonfiction (e.g., memoir) may likewise improve social cognition. Individual difference factors may also moderate the causal relation between fiction and social cognition. Given the same text, some readers may be more likely to benefit from fiction than others. Reading is an active experience, requiring willful participation by the reader. Thus, the benefits to social cognition may depend on the quality of a reader's engagement with a text and motivation to understand the characters. For example, fiction's impact may depend on a reader's propensity to be transported into narratives, generate imagery while reading, or to simulate other minds. In the absence of this type of reader engagement, fiction is unlikely to effect any change at all. Furthermore, one's existing knowledge base, expertise, or age of exposure may determine how likely one is to benefit from fiction reading. If so, prior social-cognitive ability would also moderate fiction's impact. While we were not able to test these factors here, we recommend that future studies measure the role that individual differences play in moderating the effect of fiction reading on social cognition. While we show here that fiction effects a small causal improvement of social cognition, it is also likely the reverse causal relation exists. That is, fiction reading and social cognition might form a mutually facilitating and reinforcing pathway, akin to a "Matthew Effect." Socially skilled individuals may gravitate toward fiction due to its social content more than less-skilled individuals. In doing so, readers further differentiate their social-cognitive skills from nonreaders as part of a self-reinforcing cycle. In summary, we find that fiction reading leads to a small improvement in social cognition. (1725)

64. As Phillips puts it, "We define close reading . . . as a style of focus—a mode of noticing details about literary form—that serves as a springboard for later analysis, writing, and criticism" ("Literary Neuroscience and History of Mind," 58).

65. I have more to say about this in section 1.19.

66. See Zunshine, *Why We Read Fiction*.

67. Elfenbein, *Gist*, 59, 139.

68. Elfenbein, 58. Compare to H. Porter Abbott's discussion in *Real Mysteries*, 10. Also, for a related discussion of the "online/offline" experience of reading and "promiscuous inference generation," see, respectively, Elfenbein, *Gist*, 83–84 and 86. Finally, for a critique of "decoupling reading from interpretation, a linkage so common in literary criticism that the claim 'there is no reading without interpretation' has become a truism, though it rests on a host of unexamined assumptions about both," see Elfenbein, *Gist*, 214.

69. Elfenbein, *Gist*, 19.

70. Elfenbein, 2. Compare to Anezka Kuzmičová's argument that "trained readers' minds may . . . take up various higher-order and formal aspects of the text in addition to the basic gist" ("Consciousness," 272).

71. Elfenbein, *Gist*, 2.

72. See Zunshine, "Who Is He to Speak of My Sorrow?"; and Steven Feld, *Sound and Sentiment*.

73. I have more to say about this in chapter 2, on mindreading and social status.

74. The full list of vignettes can be found here: https://yale.app.box.com/s/qvk12d3vwrppimedrrdkkj5hgrdq5p76.

75. Whalen et al, "Validating Judgments," 287.

76. Whalen et al., 288.

77. See Whalen, Zunshine, and Holquist, "Increases in Perspective"; Whalen, Zunshine, and Holquist, "Theory of Mind and Embedding of Perspective"; and Whalen et al., "Validating Judgments."

78. For details, see Zunshine, "Style Brings In Mental States."

79. To quote Elfenbein again, such stylistic nuances "could create a heightened textural density in reading, a sensation that does not produce paraphrasable meaning but a phenomenological feeling" (*Gist*, 35).

80. Hogan, *Sexual Identities*, 130.

81. While there are different ways to control for subjects' level of expertise, one may want to discuss with them beforehand situations that can lead to an overgenerous counting of embedded mental states. For instance, in my tutorial sessions, I used sentences similar to those I offered to you earlier in this chapter , such as "My last name begins with a *Z*" (no mental states), "I'm glad that my name begins with a *Z* because

the teacher may not get to the end of the list today" (two mental states), and so forth. Were I to do it again now, I would also point out that some sentences may look like complex (i.e., third- and fourth-level) embedments, when in fact they are just parallel sets of low-level embedments. Thus, "I *hope* he *doesn't realize* what I did last week because I am *scared* that he would be *angry* at me" contains two sets of second-level embedments connected by "because" and not one four-level embedment. Here is what a four-level embedment may look like, based on the same material: "I *hope* that he *doesn't realize* that I am *scared* of him being *angry* at me if he finds out what I did last week." Note the very different emotional tenor of this second sentence. Before, the speaker was afraid that "he" would be angry at her for what she did. Now she is worried that he would realize that he holds quite a bit of emotional power over her. It's a rather more interesting feeling (perhaps there is even a whiff of a story to it).

82. This would bear out the findings of the social psychologist Raymond Mar and his colleagues, who have shown that, contrary to the conventional belief that "bookworms" must be antisocial, there is actually a positive correlation between reading fiction and having good social skills ("Bookworms versus Nerds"), as well as the historical perspective offered by Elfenbein, who observes that "although reading is sometimes imagined as a withdrawal from social relationships, for many [eighteenth-century English] readers, reading novels sustained friendship and occasioned new ones" (*Gist*, 145).

83. This, of course, supports the aforementioned finding of Kidd and Castano. A similar argument has been made by social psychologists studying "attributional complexity"—a construct that was "proposed as an integration of several perspectives on cognitive complexity and includes the motivation to understand human behavior, along with the preference for complex explanations of it." It has been suggested that people who are already "interested in explaining human behavior," such as students who decide to major in psychology, are likely, in the process of their studies, to develop a "more complex explanatory schemata for human behavior as compared with other groups of students, such as students majoring in the natural sciences" (Fletcher et al., "Attributional Complexity," 880).

84. Whalen et al, "Validating Judgments," 293.

85. Whalen et al, 292, 294.

86. De Jaegher and Di Paolo, "Participatory Sense-Making," 490.

87. Forster, *Howards End*, 254.

88. For a foundational work on free indirect discourse, see Pascal, *Dual Voice*. See also Gunn, "Free Indirect Discourse." For a recent discussion of FID from a cognitive perspective, see Mäkelä, "Possible Minds."

89. Forster, *Howards End*, 241.

90. Forster, 253.

91. For a discussion of "the disparity between to kinds of engagement with fiction—the experience of the text and its interpretation," see Abbott, *Real Mysteries*, 10.

92. The enactive perspective on the extent to which embedded mental states are already in a work of fiction can be illustrated by the use of the classical example of a sponge, i.e., an object that changes depending on how it is acted upon. To quote De Jaegher and Di Paolo,

> Traditional distinctions between action and perception arise only as the specialization of phases in an act of sense-making. Several examples that illustrate this point have been dis-cussed in the enaction literature, but perhaps the simplest and clearest one is that of perceiv-ing the softness of a sponge. . . . The softness of a sponge is not to be found "in it" but in how it responds to the active probing and squeezing of our appropriate bodily movements (e.g., with the fingers or the palms of the hand). It is the outcome of a particular kind of encounter between a "questioning" agent with a particular body (sponges are solid ground for ants) and a "responding" segment of the world. ("Participatory Sense-Making," 489)

93. See also Troscianko, *Kafka's Cognitive Realism*, 172.

94. De Jaegher and Di Paolo, "Participatory Sense-Making," 502.

95. One may speak here of the guidance provided by the author or, even more spe-cifically, of what Gregory Currie calls the "guided attention" cultivated by the text ("Framing Narratives," 25). For a discussion of how a reader may be made to share "the dispositions, preferences and knowledge that make [one character's] response to [what is going on] a natural one," see Currie, 22. Compare, also, to a recent com-pelling discussion, by Auyoung, of "postdictability" of *Pride and Prejudice*, that is, of the availability of "locally surprising information" that makes readers experience as "inevitable" certain emotional responses of the novel's characters (*When Fiction Feels Real*, 47). Finally, compare to the argument by Scarry, who suggests that texts offer sets of "instructions" to their reader (*Dreaming by the Book*, 13).

96. Kuzmičová, "Consciousness," 275.

97. Kuzmičová, 277. See also Kuzmičová and Bálint, "Personal Relevance in Story Reading."

98. Kuzmičová, 277.

99. Compare to Auyoung: "No two occasions of reading are ever exactly the same, not just for different readers within the same interpretive community but even for the same reader, who may approach a single text with a variety of reading goals, fluctuating levels of motivation to pursue those goals, and newly acquired domains of background knowledge, to say nothing of the reader's variable moods, prefer-ences, and physical surroundings" (*When Fiction Feels Real*, 6). See also Elfenbein, *Gist*, 41–44; Graesser, Singer, and Trabasso, "Constructing Inferences"; and Cook, "4E Cognition and the Humanities," 879.

100. Kuzmičová, "Consciousness," 275.

101. One way to talk about this process is to rely on yet another series of metaphors and say that the secret life of literature is enactive, emergent, and embodied. There is, currently, a rich body of criticism that applies the enactivist approach to literature. See, for instance, Kukkonen, *4E Cognition and Eighteenth-Century Fiction*; Spolsky, *Contracts of Fiction*; Zunshine, "Embodied Social Cognition and Comparative Literature"; Tribble and Sutton, "Cognitive Ecology"; Polvinen, "Sense-Making and Wonder"; Polvinen, "Enactive Perception and Fictional Worlds"; Popova, *Stories, Meaning, and Experience*; Troscianko, *Kafka's Cognitive Realism*; and Garratt, *Cognitive Humanities*.

102. M. Johnson, *Embodied Mind, Meaning, and Reason*, 34. But see also Nikola A. Kompa's useful critique of "embodied accounts of language comprehension," 27.

103. De Jaegher and Di Paolo, "Participatory Sense-Making," 495. Also: "Overemphasis on [explanation and prediction involved in mindreading] has led most of the comparative social cognitive science to paint a picture of individuals who have to work out each other's minds much like they do with scientific problems" (De Jaegher and Di Paolo, 486).

104. This description of the process of reading is explicitly modeled on De Jaegher and Di Paolo's definition of "social interaction as a regulated coupling between at least two autonomous agents, where the regulation is aimed at aspects of the coupling itself so that it constitutes an emergent autonomous organization in the domain of relational dynamics, without destroying in the process the autonomy of the agents involved (though the latter's scope can be augmented or reduced)" "Participatory Sense-Making," 493). Compare to Auyoung's discussion of readers' experience with fictional characters as "a form of social connection in which their i.e., readers'] autonomy is preserved" (*When Fiction Feels Real*, 120; see also 109).

105. For a discussion of challenges involved in this kind of study, see Elfenbein, *Gist*, 101.

106. Natalie Phillips reports a similar experience when she writes about conducting an interdisciplinary study involving cognitive scientists and literary scholars: "One of my favorite moments in the process, however, was when our group—three humanists and two scientists—met one evening to discuss the project. Something happened: the literary critics got excited about experimental variables; the scientists started waxing poetic about Jane Austen's style. Now, this kind of crosstalk has become part of our everyday lives" ("Literary Neuroscience and History of Mind," 57).

107. Elfenbein, *Gist*, 3–4.

108. If that is impossible, the next best thing to do is to "read actual articles rather than overviews popularizing scientific findings" (Elfenbein, 6).

109. Jackson, "Beautiful Stranger," 79.

110. As Ottessa Moshfegh sees it, the man in "The Beautiful Stranger" is "not quite sly enough to convince his wife that he's the same person he was before he left" (foreword to Jackson, *Dark Tales*, viii).

111. Vapnyar, *Still Here*, 168–169.

112. To adapt Stanley Cavell's term used to describe a particular Hollywood genre, this would make Vika's story a variation of "the comedy of remarriage" (*Pursuits of Happiness*, 1).

113. Spark, *Girls of Slender Means*, 71.

114. Van Duijn, Sluiter, and Verhagen, "When Narrative Takes Over," 148; italics in the original.

115. Van Duijn, Sluiter, and Verhagen, 151. Note that Van Duijn et al. use the term "multiple-order intentionality" (149) rather than "embedded mental states."

116. Dunbar, *How Many Friends Does One Person Need?*, 180. See also Whalen, Zunshine, and Holquist, "Increases in Perspective Embedding" and Whalen et al., "Validating Judgments."

117. Van Duijn, Sluiter, and Verhagen, "When Narrative Takes Over," 149, 153 Compare to Ralf Schneider's useful description of various sources of information involved in constructing a mental model of a character ("Cognitive Theory of Character Reception," 122–123).

118. Quoted in Schenkar, *Talented Miss Highsmith*, 270.

119. Highsmith, *Price of Salt*, 363.

120. See Zunshine, *Why We Read Fiction*; Palmer, *Social Minds in the Novel*; and Vermeule, *Why Do We Care about Literary Characters?*

121. As Highsmith wrote later, the appeal of *The Price of Salt* "was that it had a happy ending for its two main characters, or at least they were going to try to have a future together. Prior to this book, homosexuals male and female in American novels had had to pay for their deviation by cutting their wrists, drowning themselves in a swimming pool, or by switching to heterosexuality (so it was stated), or by collapsing—alone and miserable and shunned—into a depression equal to hell" (afterword to *Selected Novels and Short Stories*, 579).

122. Isabelle Johnson, MFA workshop, University of Kentucky, March 22, 2019.

123. Hagan Smith, MFA workshop, University of Kentucky, March 25, 2019.

124. Taylor Sarratt, MFA workshop, University of Kentucky, April 1, 2019.

125. Auyong, *When Fiction Feels Real*, 121.

126. Dunbar, "Why Are Good Writers So Rare?," 7.

127. Dunbar, 17.

128. Miller, Kessel, and Flavell, "Thinking about People Thinking," 622. See also my argument in section 1.1.

129. Dunbar, "Why Are Good Writers So Rare?," 18.

130. Dunbar, 18.

131. As Hogan puts it (citations omitted):

> It is not only unnecessary for universals to apply to all works; they need not apply to all traditions. Linguists use the term universal to refer to any property or relation that occurs across (genetically and areally unrelated) languages with greater frequency than would be predicted by chance alone. An absolute universal is merely a special case—a property or relation that occurs across traditions with a frequency of one. Universals with a frequency below one are referred to as statistical universals. On the whole, we should expect to find a limited number of hierarchies of statistically universal properties and relations, ordered according to abstraction and thus according to frequency (again, as abstraction increases, frequency can only increase or remain the same), with a few absolute universals at the apex of these hierarchies.
>
> This extension of "universal" to statistically unexpected properties may seem odd, even misleading. However, it is perfectly in keeping with standard practices and definitions in all sciences, and is inconsistent only with common prejudices about the nature of literary or, more broadly, cultural universals. An example from the field of medicine may help to clarify things. It is a universal principle of medicine that secondhand smoke causes lung cancer, despite the fact that most people who have inhaled secondhand smoke never develop lung cancer. It is a universal principle because there is a statistically significant correlation between inhaling secondhand smoke and developing lung cancer (or, rather, there is a statistically significant correlation that cannot be explained by other factors--obviously it is important to distinguish between correlations that are primary or causal and those that are derivative or noncausal). Statistical universals of literature, as well as linguistics, anthropology, etc., are no different. ("Literary Universals," 42–43)

132. Zamiatin, Мы, 7; translation mine. Original: "Все это без улыбки, я бы даже сказал, с некоторой почтительностью (может быть, ей известно, что я—строитель "Интеграла"). Но не знаю—в глазах или бровях—какой-то странный раздражающий икс, и я никак не могу его поймать, дать ему цифровое выражение."

133. McCarthy, *Blood Meridian*, 3.

134. Proust, *Remembrance of Things Past*, 34.

135. Isabelle Johnson, MFA workshop, University of Kentucky, March 22, 2019.

136. Defoe, *Robinson Crusoe*, 111.

137. Defoe, 148.

138. Wharton, "Xingu," 25.

139. Defoe, *Robinson Crusoe*, 97.

140. Given the importance of embedded thinking for Defoe's novel, we should no hurry to conclude that, in contrast to twentieth-century introspective characters eighteenth-century "heroes like Robinson Crusoe . . . did things in the externa world that declared their beliefs and character" and that twentieth-century writer. "replaced these kinds of heroes with heroes like Mrs. Dalloway and Stephen Dedalus heroes whose reflective consciousness and inner lives supplied the novel's action" (Lantos, "Reconsidering Action," 157; emphasis in the original).

141. Wordsworth, "Lines Composed a Few Miles above Tintern Abbey," 1493, line. 58–65.

142. Herman, "Multimodal Storytelling," 204.

143. Palmer, "Storyworlds and Groups." For an important related analysis, see Palmer's *Social Minds in the Novel*, in which he shows how a town such as Middlemarck (2006) or Santa Dulcina delle Rocce of Evelyn Waugh's *Men at Arms* can "actually and literally does have a mind of its own" (74).

144. Defoe, *Robinson Crusoe*, 115.

145. As pointed out earlier, research by Dunbar and his colleagues strongly suggests that fifth-level embedment of mental states represents "a real upper limit for mos: people" (*How Many Friends Does One Person Need?*, 180). Moreover, in one of the studies dealing with embedment of mental states that my colleagues and I conducted jointly with cognitive scientists, the question of how to process mental states shared by several people came up. For instance, while counting levels of embedment, some experiment participants felt that when two people experience the same doubly embedded mental states, the total number of embedments adds up to four. To avoid ambiguity on this count, we felt that, for the purpose of future studies, i would be useful to introduce subjects early on to Palmer's concept of "intermenta unit" (Whalen et al., "Validating Judgments," 291).

146. Eliot, *Middlemarch*, 506.

147. Eliot, 505.

148. Henry, *Life of George Eliot*, 38.

149. The Middlemarch crowd knows its mind because George Eliot knew hers. That is, she knew where she stood on the subject of a rich landowner who dabbles in politics without caring about progress. But consider a different scenario. In some works of fiction, crowds don't know what they want. As Monika Fludernik observes fictional crowds and, in particular, rioting mobs can be portrayed as dangerous precisely because they lack in consistency: the "monstrosity of the crowd consists in its magnitude and its divisibility into constituent groups with their own agendas"

("Collective Minds," 702); see also Fludernik, "Many in Action and Thought." And, of course, the crowd's dispersal "into many different viewpoints, [thwarts] the access of the multitude to political impact." Just as in *Middlemarch*, this outcome may be a reflection of a particular ideology on the part of the author, and it is up to literary critics to figure out what "ideological premises and rhetorical strategies of naturalization, defense, or resistance" may underlie such unflattering representations of 'collective minds" (Fludernik, "Collective Minds," 710).

150. Apuleius, *Golden Ass*, 167.

151. Heliodorus, *Ethiopian Romance*, 73.

152. Murasaki, *Tale of Genji* (trans. Tyler), 8.

153. Murasaki, 8. Compare to other translations, e.g., Edward G. Seidensticker's: "Ashamed before the Takasago pines,/I would not have it known that I still live" (Murasaki, *Tale of Genji* [trans. Seidensticker], 9; or Arthur Waley's: "Though I know that long life means only bitterness, I have stayed so long in the world that even before the Pine Tree of Takasago I should hide my head in shame" (Murasaki, *Tale of Genji* [trans. Waley], 9).

154. Irving, *158-Pound Marriage*, 38.

155. Cao, *Story of the Stone*, vol. 1, 124. Original: "不想如今忽然來了一個薛寶釵，年歲雖大不多，然品格端方，容貌豐美，人多謂黛玉所不及" (紅樓夢，第五回).

156. Note that in the original, even that mental state is not present in this explicit form. The word 謂 implies a verbal agreement rather than a mental state. 人多謂 is "all said" rather than "all agreed"—although this is a situation in which the boundary between the two is blurry.

157. Castano, Martingano, and Perconti, "Effect of Exposure to Fiction." The authors consider this finding important for several reasons (citations are omitted): "First, it contributes to the ongoing discourse surrounding the role of fiction in shaping social cognition . . . and is consistent with theory and research on the characteristics of literary fiction and how they may impact cognition and cognitive style. . . . Second, understanding correlates and especially possible predictors of attributional complexity is of importance and has far ranging potential consequences because high attributional complexity attenuates racism . . . and plays a role in attitudes about important policy-related opinions." See also Kidd and Castano, "Reading Literary Fiction and Theory of Mind"; and Wulandini, Kuntoro, and Handayani, "Effect of Literary Fiction."

158. See Fletcher et al., "Attributional Complexity," 880.

159. As Castano et al. put it, "For one thing, the variables that are specifically associated with exposure to literary fiction may be desirable from one perspective, but

problematic from other perspectives. Literary fiction is associated with greater attributional complexity, which seems a valuable cognitive style from a societal perspective.
Yet, attributional complexity may also delay or derail decision-making and it has been
shown to be negatively related to mental health" ("Effect of Exposure to Fiction").

160. See also Castano, "Art Films Foster Advanced Theory of Mind."

161. van Kuijk et al., "Effect of Reading a Short Passage."

162. See, for instance, the discussion by Ralph James Savarese, which complicates
the distinction between genres that involve "understanding characters" and those
"imagining different realities." Specifically, Savarese objects to "the distinction
between literary and science fiction, as if the latter can't be 'literary,'" as well as to
"the claim that science fiction isn't character-based" (*See It Feelingly*, 111).

163. Gavaler and Johnson, "Genre Effect," 82. As they point out, citing Russian
Formalists' analysis of Jonathan Swift's *Gulliver's Travels* and Laurence Sterne's *Life
and Opinions of Tristram Shandy, Gentleman*, a text may straddle categories and also
migrate, in time, between them.

164. Peskin and Astington, "Effects of Adding Metacognitive Language."

165. Culler, "Closeness of Close Reading," 22.

166. Culler, 23.

167. B. Johnson, "Teaching Deconstructively," 141. Quoted in Culler, "Closeness of
Close Reading," 23.

168. For a discussion of translation as a form of close reading, see Culler, "Closeness
of Close Reading," 24.

169. For a discussion of nuances of mindreading attributions that depend on the
translator's understanding of the emotional and kinesic meaning of the scene, see
Bolens, *Kinesic Humor*, 95–105.

170. Zunshine, *Getting Inside Your Head*.

Chapter 2

1. De Jaegher and Di Paolo, "Participatory Sense-Making," 498.

2. Mercier and Sperber, *Enigma of Reason*, 100. See also Andrew Ionescu on how
"folk psychology grounded in a folk sociology" may cause "erroneous interpretations of the others" in works of literature ("Manifesto," 9).

3. Hirschfeld, "Myth of Mentalizing," 101.

4. Sperber, *Rethinking Symbolism*, 115. What used to serve as a radical corrective
to the view that mental life is "for the most part . . . conscious, or at least open

to introspection"—that is, Freud's notion of the "unconscious"—is now outdated because it's not radical enough: "Not some, but all mental processes, affective and cognitive, are now seen as largely or even wholly unconscious" (Sperber, 114). On the useful distinction between the cognitive and the traditional psychoanalytic unconscious in the context of cognitive literary studies, see Crane, "Cognitive Historicism," 18.

5. Hogan, *Sexual Identities*, 232.

6. Snodgrass, "Women's Intuition," 149; see also Vignemont, "Frames of Reference in Social Cognition." For a recent review, see Santos, Grossmann, and Varnum, "Class, Cognition and Cultural Change."

7. Miller, *Losing It*, 180–181.

8. See, for instance, Baum, Garofalo, and Yali, "Socioeconomic Status and Chronic Stress."

9. Simon Stern, email communication, March 8, 2018. For fascinating fictional correlatives—for example, when characters refuse to read mental states of their presumed social inferiors (e.g., as Nelly does Heathcliff's, in *Wuthering Heights*) or when the "unreadable character" is a "socially marginalized" colonial "other" (as in Coetzee's *Waiting for the Barbarians*), see Abbott, *Real Mysteries*, respectively, 125 and 143–144.

10. Sedgwick, "Privilege of Unknowing," 23. I am grateful to Simon Stern for this reference.

11. Solnit, "Nobody Knows," 5. Compare, too, to Fritz Breithaupt's comment that "Nietzsche tells us that we should not expect rulers to have any capacity for self-observation" (*Dark Sides of Empathy*, 150).

12. Note that this is still the same old me who, under different circumstances, feels compelled to carefully parse the nuances of possible mental states of my dean and who, under yet different circumstances (that is, in classroom, analyzing complex mental states of characters, readers, and other critics), may find the process genuinely delightful. As Fletcher et al. observe, "The condition under which people with complex schemata revert to the use of simple schemata or heuristics, or the extent to which they do, are important questions for future research and theorizing" ("Attributional Complexity," 883).

13. Ochs, "Clarification and Culture," 329.

14. Ochs, 333.

15. Tobar, "Assassin Next Door." I thank Doug H. Whalen for bringing this passage to my attention.

16. Vermeule, *Why Do We Care about Literary Characters?*, 86.

17. For a discussion, see Phillips, *Distraction*, 179–180; and Zunshine, "Bakhtin," 118. See also Auyoung's observation that if "even Mrs. Elton and Mr. Collins can feel real to Austen's readers [i.e., "capable of rotundity," as E. M. Forster puts it], the claim that fictional characters seem lifelike [i.e., not flat] is not necessarily a function of how much psychological depth they display" (*When Fiction Feels Real*, 40). Compare to Vermuele, *Why Do We Care about Literary Characters?*, 83.

18. Tolstoy, *Война и Мир*, 331. Original: "Une ville occupée par l'ennemi ressemble à une fille qui a perdu son honneur."

19. Tolstoy, 279. Original (emphasis in the original):

> Лицо Кутузова становилось все озабоченнее и печальнее. Из всех этих разговоров Кутузов видел одно—защищать Москву не было *никакой физической возможности*, в полном значении этих слов, то есть до такой степени не было возможности, что ежели бы какой-нибудь безумный главнокомандующий отдал приказ о даче сражения, то произошла бы путаница, и сражения все-таки бы не было; не было бы потому, что все высшие начальники не только признавали эту позицию невозможной, но в разговорах своих обсуждали только то, что произойдет после несомненного оставления этой позиции. Как же могли начальники вести свои войска на поле сражения, которое они считали невозможным? Низшие начальники, даже солдаты (которые тоже рассуждают), также признавали позицию невозможной и потому не могли одни драться с уверенностью в поражении. Ежели Бенигсен настаивал на защите этой позиции и другие еще обсуживали ее, то вопрос этот уже не имел значения сам по себе, а имел значение только как предлог для ссоры и интриги. Это понимал Кутузов.

20. For an analysis of the difference between the complexity of Kutuzov's thinking as compared to Napoleon's, see Allakhverdov, *Psichologia Iskusstva*, 76–78.

21. Austen, *Mansfield Park*, 197.

22. Austen, 310.

23. Austen, 113.

24. Zunshine, "From the Social to the Literary."

25. Lee, "Measuring the Stomach of a Gentleman," 205. For in-depth discussion of "kinship sociality," see Lee's *Stranger*.

26. Lee, "Measuring the Stomach of a Gentleman," 209. Compare, too, to Breithaupt's analysis of Nietzsche's view of women: "In the world of Nietzsche's thought, women are masters at manipulating the way they are seen by others. They understand how they are observed but, unlike the objective person, they do not comport themselves purely receptively and projectively in the face of observation. Rather, they stake a claim to the observations of others by disguising, masking, beautifying, or withholding themselves" (*Dark Sides of Empathy*, 160).

27. Lee, "Measuring the Stomach of a Gentleman," 210.

28. Cao, *Story of the Stone*, vol. 1, 20.

29. For discussion, see Zunshine, "I Lie Therefore I Am."

30. Zunshine, "Think What You're Doing," 47.

31. *Tale of Frol Skobeev*; translation mine. Original: "Смотри, мой друг,—говорит Скобеев,—в каком она здравии: таков вот родительский гнев—они её за глаза бранят и клянут, оттого она и при смерти лежит."

32. Morris, *Literature of Roguery*, 51.

33. For a valuable cognitivist reading of the picaresque novel, see Simon, "Contextualizing Cognitive Approaches." As he puts it,

> The picaro, the hero of the genre, must live by his own wits to survive in a Spanish society in which vast social inequalities exist, in spite of the enormous fortune plundered from its American possessions. Throughout the story of his survival, the protagonist will rely on his own ability to read others' mental states and manipulate them. The resulting complexity of the interplay of minds, between the picaro and the other characters (picaros or otherwise) who are all trying to outsmart each other, is a central feature of this genre. In sum, as need breeds ruse and craftiness, the social-economic disparities of early modern Spain are an essential component of the genre and serve as an important contextual factor in the rise of the literary representation of intentionality. (19)

34. For a discussion of the master-manservant dialectic, see M. Gillespie, "From Beau Brummell to Lady Bracknell," 179.

35. Cusk, *Saving Agnes*, 157.

36. Ellison, introduction to *Invisible Man*, xxi.

37. Ellison, xix.

38. Ellison, xxii.

39. This applies not just to the Invisible Man's position as a character but also to his stance as a writer/narrator of his story. As John F. Callahan points out,

> Invisible Man's career as a failed orator teaches him that he must speak to us, his audience, in order to speak for us. And he returns to that condition of eloquence in the profound rhetorical question with which he ends: "Who knows but that on the lower frequencies I speak for you?" A writer's communication with his audience—citizens, some of which may also be other writers—may be an act of leadership. But, because of the nature of literature, narrative leadership is a symbolic act. Invisible Man, having set himself free, encourages his readers to take similar action. He does not attempt, as he has done presumptuously and blindly so many times, to lead his audience but to make contact on an equal individual basis. ("Frequencies of Eloquence," 87)

40. Ellison, *Invisible Man*, 508.

41. Ellison, 559.

42. For an analysis of the trickster/picaro references in *Invisible Man*, see Nadel, *Invisible Criticism*. Nadel complicates Norton Frye's comparison of Ellison's protagonist to such classic folklore and literary figures as Brer Bear, Brer Rabbit, and the "tricky

slave of Roman comedy," emphasizing "the specific conditions that created the Brer Bear and Brer Rabbit versions of the *eiron*," conditions that also obtain in the case of Invisible Man. For the latter "also manifest the specific marks of oppression and consequent encoding created by the racial caste system deeply embedded in the legal and extralegal institutions of the South. The slave in Roman comedy who wins his freedom ceases to be in an ironic position; he is not merely ostensibly free, not like the free black in the South, a slave without a master" (32–33).

43. As Valerie Smith observes, "Throughout the course of his life, the Invisible Man learns that he can never quite learn to be deceptive enough. No matter how devious he thinks he is, those who control him always manage to trick and betray him" ("Meaning of Narration," 209).

44. Ellison, *Invisible Man*, 257.

45. Ellison, 478.

46. Ellison, 477.

47. One wonders to what extent this dynamic is still at play in other literary contexts in which white characters are portrayed as not being able to appreciate the complex subjectivity of Black characters. Consider, for instance, Jennifer Riddle Harding's observation that in Charles Chesnutt's short story "Dave's Neckliss" (1889), the narratees of the story within the story, John and Annie, see the narrator, an old African American man named Julius, "largely as a childish, ham-loving old man who tells whimsical stories about slavery" and that neither John nor Annie are capable of appreciating "Julius's metaphors and his humor" ("Mind Enslaved?," 439).

48. Not only does Evelina signal to Mr. Smith her social superiority, but she also manages to do so without offending her grandmother, Mme. Duval, who is present and quite happy with Mr. Smith's courtship of her granddaughter. Here, as on many other occasions, Evelina's speech manifests the quality of what the Burney scholar Julia Epstein describes as "double-edgedness" (*Iron Pen*, 111), which is a particularly fascinating term if we consider the embedments that underlie it.

49. Burney, *Evelina*, 220.

50. Bakhtin, "Discourse in the Novel," 265.

51. Bakhtin, 308, 262.

52. Burney, *Evelina*, 327.

53. Burney, 330–331.

54. See Holquist, *Dialogism*, 154–155.

55. As Brian McCrea puts it, building on Michael McKeon's concept of "status inconsistency," Burney's "satire upon Mr. Smith doesn't imply an endorsement of characters like Coverley and Merton" (*Frances Burney*, 54).

56. Burney, *Evelina*, 221.

57. Burney, 146.

58. Burney, 177.

59. Burney, 188.

60. Burney, 206.

61. Burney, 275.

62. That Captain Mirvan seems to embed mental states on the second level more consistently than, say, Mme. Duval or the Branghtons (who tend to stay around the first level) may reflect his peculiar role in *Evelina*. As Ruth Bernard Yeazell observes, "though there is scarcely a character in the novel who seems more distant from Evelina than this crude ex-sailor, he nonetheless has a remarkable tendency to aim his practical jokes at targets whom she herself has strong motives to attack" (*Fictions of Modesty*, 141).

63. Burney, 141. See Francesca Saggini for a discussion of Burney's possible appropriation of the "long-established theatrical technique of employing particular speech patterns for characterization" (*Backstage in the Novel: Frances Burney and the Theater Arts* [University of Virginia Press, 2012], 78) in her representation of Captain Mirvan and Mme Duval.

64. Burney, 13, 19, 227, 355, 257.

65. Jane Spencer concurs: "on the whole the novel shows remarkably little sympathy for a grandmother deprived of her grandchild." Spencer sees this as part of the general pattern informing Burney's narrative: "With its strong emotional investment in the heroine's relationship to her father and to father figures, *Evelina* honours the patriline and is ambivalent about the matriline" ("Evelina and Cecilia," 27). While I agree with Spencer's analysis, my focus here is on specific rhetorical strategies (such as low-level emdedment of mental states associated with her) that make Mme. Duval a less sympathetic character than her personal losses might have entitled her to be. See also Kristina Straub's useful discussion of the novel's divided consciousness when it comes to the treatment of older women, such as Mme. Duval (*Divided Fictions*, 30).

66. Which may work, as it does more often than not, in Evelina's case, as a heightened awareness of one's own feelings (e.g., shame) in response to other people's perceptions of oneself, what Yeazell calls Evelina's "obsession with watching herself being watched" (*Fictions of Modesty*, 123).

67. As Epstein puts it, Mme. Duval's "roughhewn sensibility makes it impossible for her to empathize with others" (*Iron Pen*, 113).

68. At least this is what we encounter in *Evelina*. We can't assume that this is a general rule in fiction. Complexity does not imply moral goodness. High embedders

may come across as sensitive and intelligent people, or they may come across as peculiarly misguided, betrayed as it were by their sociocognitive complexity into ethically questionable or socially debilitating behavior. And do not forget about evil masterminds, whose hubristic Machiavellianism may render them abhorrent in the eyes of the reader. (Compare to Vermeule's important discussion of masterminds in *Why Do We Care about Literary Characters?*, 86.)

69. This happens when Tom Branghton uses his connection with Evelina to mooch a free ride out of Lord Orville (*Evelina*, 248–249).

70. Santos, Grossmann, and Varnum, "Class, Cognition and Cultural Change."

71. Santos, Grossmann, and Varnum.

72. Some literary critics feel very uncomfortable applying insights from contemporary psychology to the historical past. As Elfenbein explains,

> Such investigations open themselves to an easy charge of anachronism: since most psychological findings derive from participants who postdate [the past centuries], we cannot know if those findings apply to earlier periods. Yet literary scholars routinely apply approaches and insights honed in the twentieth- and twenty-first century academy to works written in earlier periods. Nervousness about the use of cognitive science is an arbitrary invocation of rigor that misrecognizes the field's enabling anachronisms. Also, there is no reason to decide a priori that contemporary psychological findings are irrelevant to the past. If it is wrong to assume that there is no difference between now and then, it is equally wrong to assume that there are no continuities either; assertions of historical difference do not guarantee truth any more than do ones of continuity. (*Gist*, 168)

73. Pocock, *Virtue, Commerce, and History*, 119. Of course, the dynamic of the relationship between social class and cognitive complexity can be reversed. For a compelling example of such a reversal, see Ellen Bayuk Rosenman's argument that the working poor can be presented as having a more "layered consciousness of social interactions," and thus "the most satisfying understanding" of a social situation, than their middle-class "betters" do ("Rudeness, Slang, and Obscenity," 58).

74. Doody, *Frances Burney*, 3.

75. Zunshine, "Why Jane Austen Was Different."

76. Shakespeare, *Measure for Measure*, 4.3.110–114.

77. Zunshine, *Getting Inside Your Head*, chap. 3.

78. Vineberg, "Problem Plays," 33.

79. For an important discussion of what it means for an author to display an intuitive awareness of various "cognitive" insights that couldn't have been known to the scientific (or natural-philosophic) thought of their day, see Hogan, *Sexual Identities*.

80. For a discussion of goals, methods, and the disciplinary trajectory of cognitive historicism, see Crane, "Cognitive Historicism."

81. Lee, "Measuring the Stomach of a Gentleman," 219. See also Lee, "Society Must Be Defended."

82. See also Emmerich, "GDR and Its Literature," 28.

83. For a recent discussion of functions of the unreliable narrator in the Russian novel of the 1920s–1930s, see Zhilicheva, "Функции 'Ненадежного.'"

84. For a discussion of a related dynamic in the GDR, i.e., when "some of the most significant works of GDR literature were either never published or only published years or even decades after the fact," see Emmerich, *GDR and Its Literature*, 20.

85. Borden, "Leonid Dobychin's *The Town of N*," viii.

86. Morris, "Russia," 213. See also Karin Leeder's observation about the literature produced in the GDR, which describes, to some extent, attitudes toward the official literary output of the Soviet Union. As she writes, "it is undoubtedly the case that for many years critical judgments in East and West were skewed more to political or moral considerations than aesthetic ones" (introduction to *Rereading East Germany*, 2).

87. See Günther, "Soviet Literary Criticism," 105. Elsewhere, Günther also points out that, "just as in the Soviet Union, in Germany in the 1930s, struggle for classicism meant the struggle against modernism" ("Zeleznaja Garmonija," 38; translation mine).

88. As Nancy Easterlin observes, "the history of literary theoretical approaches in the twentieth century . . . demonstrates the problem of assuming a correlation between aesthetic practices and the ideological implications of artworks. The vagaries of Marxist criticism serve as the most prominent example." In the three decades subsequent to the Russian Revolution in 2017, "arguments that unusual and non-representational techniques exposed class struggle kept pace with contrary arguments that mainstream realism served as the best mechanism for enlightenment" (*Biocultural Approach*, 98).

89. Hake, "Political Affects," 106–107. See also her discussion of "socialist modernism" (Hake, 108–109). See, however, Günther, "Zheleznaya Garmonija," for a different perspective. As he puts it, "on the whole, totalitarian culture is an enemy of modernism; it represents, rather, a form of enforced postmodernism. Everything associated with the avant-garde is subjected to a strictest selection; what is selected is useful from the point of view of the official ideology" (31; translation mine).

90. Compare to Günther: "In literature and in art, the danger arose above all from ambiguity and lack of transparency, which—not without foundation—were seen as the source of dissidence and the possibility of 'ideological contraband'" ("Soviet Literary Criticism," 96). See also Emmerich's discussion of GDR literature after 1965, in which, as he observes, poets, in particular, became aware "of literature's potential as subversive counter-discourse, in which they could practice diverse ways of speaking—heteroglossia, dialogism and intertextuality—and thus undermine the monosemy-affirming language environment" ("The *GDR and Its Literature*," 25).

91. Günther, "Soviet Literary Criticism," 106.

92. Günther, 97. See also David Brandenberger's discussion of the resolution of the Central Committee, of December 1928, "on state publishing," which mandated, among other things, that publishing industry provided "for mass literature's maximal accessibility (in both form and content) in order to find the broadest swath of readership" (Propaganda State in Crisis, 23).

93. Günther, 106.

94. Günther, "Zheleznaya Garmonia," 40; translation mine.

95. For a discussion of the official GDR literature after 1965, see Emmerich, "The GDR and Its Literature," 22. Particularly, as he observes, in that time period, writers "made use of unconventional, eccentric or dislocated narrative perspectives, which reached the standards set by modernism decades earlier (self-reflexivity, discontinuity and lack of plot)" (24).

96. Ellison, introduction to Invisible Man, xix.

97. Ellison, xxi, xxii.

98. Günther, "Zheleznaya Garmonia," 27; translation mine.

99. Lee, "Measuring the Stomach of a Gentleman," 219.

100. Gladkov, Цемент, 241; translation mine. Original:

> Глеб положил голову на колени Даши и увидел над собою ее лицо с огнистым пушком на щеках и глаза—пристальные, большие, встревоженные и любящие.
> —Здесь, под небом, чувствуешь себя другим, Дашок. Вот лежу у тебя на коленях . . Когда это было?.. И никогда я, кажется, не переживал ничего подобного. Я знаю только одно, что твоя любовь была больше и глубже моей, и я тебя недостоин. Я и сотой доли не пережил того, что пережила ты. Расскажи же мне сама о своих мытарствах . . . Может быть, я и себя тогда узнаю лучше.
> Воздух внезапно вспыхнул молнией: везде большими и маленькими звездами зароились огни. Волна восторга охватила Глеба; в волнении он поднялся на локоть.
> —Даша, голубка, гляди . . . как хорошо бороться и строить свою судьбу!.. Ведь это— все наше . . . мы!.. Наша сила и труд . . . Будто вздох чувствуешь . . . вздох перед первым ударом . . . когда хочется размахнуться . . .
> Даша опять положила руки на его грудь. Она сама волновалась, и Глеб слышал, как глухими толчками билось ее сердце.
> —Да, милый, хорошо бороться за свою судьбу . . . Пусть муки, пусть смерть . . Страшно это . . . и не всякий может вынести . . . Я вот вынесла, потому что люблю тебя сильнее страха . . . А потом и другое поняла, другое полюбила . . . может, даже больше тебя . . .
> —Говори, Дашок . . . что бы ни было—говори . . . Я уж научился не только слушать, но и . . . бороться с собой . . .

101. Of course, the socialist realist protagonists' tongue-tiedness also harks back to the literary convention, already well in place by the end of the nineteenth century according to which a strong emotional experience "was thought to elude expression in language" (Martens, "Corporeality," 237).

102. Gladkov, Цемент, 236. Original: "Внутренние Прослойки."

103. Gladkov, 220, 221, 222.

104. Emmerich, "The *GDR and Its Literature*," 19.

105. Paul, "Gender in GDR Literature," 108.

106. Claudius, *Menschen an unserer Seite*, 274–75; translated by and quoted in Paul, "Gender in GDR Literature," 109.

107. Claudius, 171–172; translated by and quoted in Paul, 108.

108. See also Lee's suggestive exploration of Chinese socialist realism in "When Nothing Is True."

109. Wolf, *They Divided the Sky*, 34–35.

110. On Wolf's subsequent commitment to "revise" socialist realism, expressed "in her comments in various interviews and essays," see Wiesehan, "Christa Wolf Reconsidered," 79.

111. "100 Must-Reads."

112. See A. Richardson, *Neural Sublime*; Crane, *Shakespeare's Brain*; Spolsky, *Satisfying Skepticism*; Phillips, *Distraction*; and Keen, *Thomas Hardy's Brains*.

Chapter 3

1. Over the years, literary scholars have suggested several other essential features. See, for instance, Miall, "Science in the Perspective of Literariness"; and A. Richardson, "Studies in Literature and Cognition."

2. As Nancy Easterlin puts it, some literary scholars

> point to the youth of neuroscience and the inferential nature of experimental psychology to confirm the prejudice that that science has nothing of interest to offer the student of literature. But drawing such a hasty conclusion is unwarranted in the light of the pragmatic process whereby ideas are tested, gain or lose force, and are rejected as invalid or accepted as legitimate knowledge by communities of learning. The more intellectually defensible conclusion to draw, given the experimental and provisional nature of so many psychological findings, is that our own uses of ideas from this new field will themselves be provisional and experimental. (*Biocultural Approach*, 154)

3. B. Schieffelin, "Found in Translation," 143.

4. Keane, *Ethical Life*, 131.

5. Consider, for instance, how easily Soviet writers ignored the long tradition of whimsical narrators, under socialist realism.

6. Astuti, "Some After Dinner Thoughts."

7. Astuti.

8. Duranti, "Further Reflections," 492. In fact, the etymology of the word "intention" testifies to the boundedness of mind reading with embodiment. As Duranti points out, "the contemporary understanding of intention comes from the Latin intentio originally understood as an embodied movement or 'tension'" (*Anthropology of Intention*, 72–73). But see also Gregory Hickok for an alternative explication of the role of embodiment in mindreading (*Myth of Mirror Neurons*, 171–172).

9. Boyer, *Minds Make Societies*, 222.

10. As Karin Kukkonen observes, "It seems quite likely that thought does not coincide with language, but this does not mean that language would not have an effect on the ways in which we think" ("Does Cognition Translate?," 251). Specifically, as Lisa Feldman Barrett proposes, "words invite the formation of emotion concepts, and . . . these emotion concepts, by providing the kind of attention-guiding and perception-shaping predictions mentioned above, then enable us to understand inner bodily states in terms of meaningful emotions" (quoted in Kukkonen, 249).

11. Wolf, *City of Angels*, 19. As Elfenbein puts it, "we are . . . battling the linearity of writing to capture the nonlinear, weblike experience of reading" (*Gist*, 36).

12. See Drucker, *Graphesis*, 64–137. For a related discussion of "pictorialism" and cognition, see Troscianko, *Kafka's Cognitive Realism*, 41–42, 54.

13. Sabbagh and Baldwin, "Understanding the Role," 171.

14. M. Johnson, "Embodiment of Language," 630. See also Martins and Fitch, "Do We Represent Intentional Action as Recursively Embedded?" As they point out in the conclusion of their essay, "the point is that models positing recursion are only relevant for human cognition if humans actually represent and use implicit knowledge of recursion in their activities: a demonstration of which requires hard empirical work" (20). But see also Tomasello, *Origins of Human Communication*, 336. Finally, for a discussion of enactivism as an alternative to representationalism ("according to the enactivist view, there is no representation of the world inside the brain") and for a useful alternative view of "functional" representations, see Troscianko, *Kafka's Cognitive Realism*, 114, 73–75.

15. I owe this insight to the anonymous reader of my manuscript enlisted by the MIT Press.

16. Mercier and Sperber, *Enigma of Reason*, 81.

17. Some cognitive literary scholars working with an enactive paradigm (e.g., Kukkonen, Polvinen, Caracciolo, and Troscianko) experiment with moving in that direction. To appreciate challenges that they face—that is, to see how their analyses may still depend on their awareness of the dynamic created by complex embedments—consider Troscianko's study of emotional appraisal as "an affective and enactive cognitive act" in Kafka's *Metamorphosis*. When Gregor Samsa overhears

his "mother worrying out loud to his sister that it would be a cruelty, not a kindness, to empty his room of furniture," he wonders about his response to her suggestion, while Kafka's readers are made aware that he may not fully realize what his response entails. As Kafka (in Troscianko's translation) puts it, "On hearing these words from his mother Gregor realised that the lack of all direct human speech, along with the monotony of life within the family, must have confused his mind over the past two months, for he could not account otherwise for the fact that he could quite earnestly have longed for his room to be emptied" (*Kafka's Cognitive Realism*, 167). As Troscianko explains,

> After the fact, Gregor's elaborative, experienced appraisal is that he'd longed for the room to be emptied because he'd been confused by the lack of verbal contact with other people and the monotony of family life. But the reader is also prompted here, precisely by the mention of confusion, to interpret things differently from Gregor: the obvious alternative is to understand what he glosses as confusion to be instead the natural cognitive changes resulting from his transformation into an insect and the simultaneous changes in his behavioural preferences. The antecedent appraisal would then be "this is a good thing, because I want to be able to crawl over the walls and ceiling more easily," making Gregor happily anticipatory. Then, by the time of the realisation quoted here, he has reappraised events as "I thought this was a good thing because I was confused by being so isolated and bored (so I am upset)," implicitly with a new antecedent appraisal of "this is a bad thing, because I'm scared of becoming more fully an animal." By highlighting the discrepancies between possible emotional appraisals, the text dramatises the ambivalences and fears involved in Gregor's gradual transition to a more fully insect state, as well as his attempts to deny them. . . . [The] potentially unsettling aspect of this kind of evocation consists in the way that Gregor's introspective capacities are so clearly yet subtly flagged as flawed: his access to the causes of his own mental states is presented as confused even as he supposedly identifies prior confusion in himself. The possibility of introspective insight, and in particular the transparency of emotion to itself, are called into question by the (somewhat messy) inseparability of thought and emotion in the form of appraisal. (167)

Note how Troscianko's analysis of emotional reappraisals experienced by Gregor directly depends on her (and our) processing of a series of complex embedments, which involve her/our awareness of Gregor's flawed understanding of the meaning of his introspective insights. For Troscianko, highlighting complex embedments present in the text serves as a means toward a particular interpretive end: it is a stepping stone toward her compelling exploration of antecedent and elaborative appraisals. What I propose here is that we slow down and become aware of that stepping stone as an underappreciated but crucial element of a critical analysis centering on enactive cognition.

18. For an overview of some alternatives to "cognitivism" (which provides the base for the current, metaphor-laden concept of theory of mind), such as, for instance, "antirepresentationalism" and "relationalism," see Baggs, "Book Review," 1947.

19. Vessel, Starr, and Rubin, "Brain on Art." As Randy L. Buckner and Daniel Carroll observe, "default modes of cognition are characterized by a shift from perceiving the external world to internal modes of cognition that simulate worlds that are separate from the one being directly experienced" ("Self-Projection and the Brain," 53).

20. Buckner, Andrews-Hanna, and Schacter, "Brain's Default Network," 1, quoted in Hickok, *Myth of Mirror Neurons*, 171.

21. Li, Mai, and Liu, "Default Mode Network."

22. Frith, "Social Mind?"

23. Dunbar, *How Many Friends Does One Person Need?*, 180. See also Whalen, Zunshine, and Holquist, "Increases in Perspective Embedding"; and Whalen et al., "Validating Judgments of Perspective Embedding."

24. Stiller and Dunbar, "Perspective-Taking and Memory Capacity," 95, 100.

25. For further discussion, see Stiller, Nettle, and Dunbar, "Small World of Shakespeare's Plays," 401.

26. Kanske et al., "Dissecting the Social Brain," 6.

27. For a review, see Apperly, *Cognitive Basis*, 11–34. See also Milligan, Astington, and Dack, "Language and Theory of Mind."

28. Astington, Pelletier, and Homer, "Theory of Mind and Epistemological Development," 133, 142. See also Deena Skolnick and Paul Bloom's study of children's perspective taking in the case of fictional characters. As they point out, there is "considerable evidence that children have difficulty with conceptual perspective taking before the age of five" ("What Does Batman Think about Sponge Bob?," B13.

29. See, for instance, an important volume, *Theory of Mind in the Pacific*, edited by Wassmann, Träuble, and Funke (2013). As Tanya Luhrmann summarizes in her review of the collection,

> [This] book presents a series of research projects in the Pacific by experimenters and ethnographers. That region has long been famous for its so-called opacity of mind—for the strong sense that it is inappropriate to ask about someone else's intentions, beliefs, and desires, or presume that one knows what they are. In what way might this strong cultural bias affect theory of mind? [What the researches found was that, although] theory of mind abilities develop universally among all human populations, [the] onset of mental state reasoning . . . varies across cultures as a consequence of different socialisation practices and ethnotheories concerning, for example, mental state talk. ("*Theory of Mind in the Pacific*," 443)

30. Astington, Pelletier, and Homer, "Theory of Mind and Epistemological Development," 133.

31. Onishi and Baillargeon, "Do 15-Months-Old Infants Understand." For a discussion, see Mercier and Sperber, *Enigma of Reason*, 94–96.

32. To quote Rebecca Saxe, there is the puzzling divergence between "recent advances in developmental psychology [that] suggest that children have some understanding of false beliefs much *earlier* than age 3 years, and initial neuroimaging studies of children's brains [that] suggest that key maturational changes in the

right temporo-parietal junction] occur much *later* than age 5 years" ("New Puzzle," 108). To account for this divergence, some scientists now propose a two-system theory of mindreading; that is, they suggest that infants' theory of mind is housed in a neural system distinct from that in which it will be housed later in development (see Saxe, 110; and Apperly, *Cognitive Basis*, 108–181). The first system encompasses "low-level' processes that are cognitively efficient but inflexible, and the second, 'high-level' processes that are highly flexible but cognitively demanding." In this view, when we "make explicit judgments about what others think or want" (or want us to think), we rely on the "high-level" processes, but what "gets us through our social day" is a "combination of low-level mindreading processes and the rich endowment of social knowledge that we gain through development" (Apperly, *Cognitive Basis*, 143, 155). For some responses to this proposal, see Mercier and Sperber, *Enigma of Reason*, 341; and Carruthers, "Two Systems for Mindreading?"

33. Carruthers, "Two Systems for Mindreading?," 159.

34. This may date to David G. Premack and Guy Woodruff's "Does the Chimpanzee Have a Theory of Mind?"

35. Miller, Kessel, and Flavell, "Thinking about People Thinking," 623.

36. Saxe and Kanwisher, "People Thinking about Thinking People," 1835.

37. See Saxe and Powell, "It's the Thought That Counts."

38. Skerry and Saxe, "Neural Representations of Emotion," 1951.

39. Saxe and Kanwisher, "People Thinking about Thinking People," 1841.

40. Kanske et al., "Dissecting the Social Brain," 17.

41. See Keen, *Empathy and the Novel*; Savarese, *See It Feelingly*; and Breithaupt, *Dark Sides of Empathy*. Also, see Joshua Landy for a witty critique of the view that good literature "simply gives us no choice but to be improved by it" (*How to Do Things with Fictions*, 30). For important studies of empathy and literature conducted by social psychologists, see Mar, Tackett, and Moore, "Exposure to Media"; and Djikic, Oatley, and Moldoveanu, "Reading Other Minds."

42. As Elfenbein puts is, "Literary scholars assume that characters are not real people and that the questions appropriate to ask about them are not the same ones that we might ask about real people. Yet no matter how often we stress such a point, both students in literature classes and many critics find that it never fully takes hold. For all our efforts, readers persist in treating literary characters as if they were people they had met" (*Gist*, 59).

43. Nettle, email communication, June 28, 2006. Also, see David Herman's suggestive argument about "thinking about thinking—or intelligence about intelligence" (*Storytelling and the Sciences of the Mind*, 278).

44. Boyer, *Minds Make Societies*, 131, 153. See also Tomasello, *Origins of Human Communication*, chap. 5 ("Phylogenetic Origins"). As Tomasello puts it, "The combination of helpfulness and recursive mindreading led to mutual expectations of helpfulness and the Gricean communicative intention as a guide to relevance inferences which could then come under social norms created by still another uniquely human propensity, in this case to be like and to be liked by others in this social group, as opposed to those other social groups" (218).

45. Kanske et al., "Dissecting the Social Brain," 17.

46. William James, *The Principles of Psychology*, quoted in Easterlin, *Biocultural Approach*, 156.

47. Easterlin, *Biocultural Approach*, 156.

48. See Palmer, *Fictional Minds*, for an exploration of this point.

49. As Auyoung puts it, the "sustained experience of reading between the lines intensifies the reader's consciousness of being uniquely able to comprehend the implied author" (*When Fiction Feels Real*, 55). For a discussion of such "nonreciprocal sense of intimacy" between the reader and the author, see Auyoung, 109; and Auyoung, "Unspoken Intimacy."

50. Compare to the religious studies scholar Paul C. Dilley's critique of "strict universalism of . . . [a] proposal for a 'cognitive historiography' dedicated to discovering universal historical 'rules' based on evolutionary trends." In its stead, similarly to literary scholars working with cognitive approaches to literature (e.g., see Crane, "Cognitive Historicism"), Dilley proposes a "'cognitive historicism,' on the model of New Historicism, with its emphasis on understanding texts in their ideological context, as well as uncovering the mechanisms of power in cultural representation." (*Monasteries and the Care of Souls*, 12).

Chapter 4

1. Luhrmann, "Toward an Anthropological Theory of Mind," 6.

2. Luhrmann, 7.

3. B. Schieffelin, "Speaking Only Your Own Mind," 433.

4. Robbins and Rumsey, "Introduction," 407–408.

5. Astuti, "Some After Dinner Thoughts."

6. Luhrmann, "Toward an Anthropological Theory of Mind," 11.

7. Luhrmann, 11.

8. Bambi Schieffelin, personal communication, March 1, 2019.

9. Steven Feld, email communication, March 1, 2019.

10. Luhrmann, "Toward an Anthropological Theory of Mind," 7.

11. Throop, "Suffering," 128.

12. Throop, 129.

13. Lepowsky, "Personhood," 48.

14. For a useful related analysis of comparative "perceptions of mental state access in the United States and Japan," see Wice at al. As they observe, the overall "results indicate that culturally variable norms specifying appropriate levels of mental state access play an important role in how individuals estimate their knowledge of other people's minds in close relationships" (n.p.).

15. Ochs and Schieffelin, "Language Acquisition and Socialization," 298.

16. B. Schieffelin, "Speaking Only Your Own Mind," 434.

17. B. Schieffelin, 433.

18. B. Schieffelin, 438. See also Bambi Schieffelin's more extensive treatment of this topic in her book *Give and Take of Everyday Life*, in which she observes that Kaluli say that "one cannot know what another thinks or feels" (72).

19. B. Schieffelin, *Give and Take*, 73. For a discussion, see Sabbagh and Baldwin, "Understanding the Role," 167. Also, compare to Ellen Dissanayake's argument about baby talk as a "proto-aesthetic device" ("Prelinguistic and Preliterate Substrates," 63).

20. B. Schieffelin, *Give and Take*, 71.

21. B. Schieffelin, 72–73.

22. Steven Feld, email communication, March 1, 2019.

23. Consider, for instance, premodern China, which, as Haiyan Lee argues, may have selectively drawn on the "opacity principle" in some of its "mainstream expressive culture," which prioritized "dramatic speech and action over interior monologues and embedded mental states" ("Measuring the Stomach of a Gentleman," 205).

24. Bambi Schieffelin, personal communication, March 1, 2019.

25. Caldwell-Harris, Kronrod, and Yang, "Do More, Say Less," 53.

26. Oforlea, "Dilemma of the African American Detective."

27. Ochs and Schieffelin, "Language Acquisition," 303–304.

28. Keane, *Ethical Life*, 127.

29. B. Schieffelin, "Found in Translation," 143.

30. Throop, "Suffering," 133.

31. Ochs and Schieffelin, "Language Acquisition," 299.

32. Throop, "Suffering," 134.

33. Trawalter et al., "Attending to Threat," 1325. See also Argyle and Cook, *Gaze and Mutual Gaze*; and Mason, Tatkow, and Macrae, "Look of Love."

34. As Lasse Hodne puts it, in studies of gaze orientation in Western portraits, conducted by experimental art historians, "the overall reaction" of subjects showed that "frontal faces with direct gaze . . . were regarded to be more caring, trustworthy, harmonic, inclusive and respectable than the corresponding images with averted gaze and face" ("Memling's Portraits of Christ," 254). Compare to Kayo Muira and Motoko Koike's discussion of gaze orientation in Japanese Ukiyo-e pictures, in "Judgment, Interpretation and Impression of Gaze Direction." As they point out, an "averted" gaze or an "ambiguous" gaze direction may give rise to "negative emotion," even in a culture known for "a tendency to avoid direct eye contact" (218). As they suggest, one reason for this negative response may be that the avoidant figure (in this case, a woman standing next to a young child) may be perceived as refusing to engage in a culturally "desirable" practice of "viewing together" or "side-by-side relation" (219, 220).

35. Landau, "11 Reasons a Child Cannot Look You in the Eyes." As Landau puts it, "Particularly [in] Asian cultures, eye contact can be seen as a sign of disrespect. To look at someone directly can be considered bold or defiant, so it is avoided."

36. Landau.

37. Bambi Schieffelin, personal communication, March 1, 2019.

38. Abbott, *Real Mysteries*, 146.

39. Abbott, 146–147.

40. See Savarese and Zunshine, "Critic as Neurocosmopolite," 21–26.

41. Hickok, *Myth*, 208.

42. Hickok, 224.

43. Blackman, "Reflections on Language," 149, 153. Quoted in Savarese and Zunshine, "Critic as Neurocosmopolite," 23.

44. Ochs and Schieffelin, "Language Acquisition," 299.

45. S. Richardson, *Clarissa*, 460.

46. Throop, "Suffering," 133.

47. Of course, direct eye contact can be experienced as intrusive and frightening in the context of the culture of transparency, too. See, for instance, Blakey Vermuele's

discussion of Big Brother's "enormous black staring eyes that seem to follow you everywhere" (*Why Do We Care about Literary Characters?*, 53).

48. Dilley, *Monasteries*, 14–15.

49. Dilley, 234.

50. Dilley, 295.

51. Dilley, 295.

52. Asztalos, "Faculty of Theology," 409.

53. Leff, "*Trivium* and the Three Philosophies," 308.

54. Robbe-Grillet, *Two Novels*, 51.

55. David Richter, email communication, April 23, 2018.

56. I use the word "naturalize" here the same way that Abbott does when he explains that "most readers of modernist texts . . . have learned to naturalize and thus accept without strain" certain "unique, even disturbing" textual features (*Real Mysteries*, 39; also 39n16). See also Monika Fludernik, *Towards a "Natural" Narratology*, 274.

57. Abbott, *Real Mysteries*, 145.

58. Lorant, *Melville*, 333. Quoted in Abbott, *Real Mysteries*, 129.

59. Abbott, *Real Mysteries*, 128–130.

60. Troscianko, *Kafka's Cognitive Realism*, 199. For the full description of the experiment, see Troscianko, 217–218.

61. See Zunshine, *Getting Inside Your Head*, 150.

62. Abbott, *Real Mysteries*, 152.

63. B. Schieffelin, "Found in Translation," 150.

64. Ochs and Schieffelin "Language Acquisition," 294.

65. B. Schieffelin, "Found in Translation," 150.

66. B. Schieffelin, "Two Dukula Sulo: & One Dog"; B. Schieffelin, email communication, January 23, 2019.

67. Rumsey, "Empathy and Anthropology," 222.

68. Compare to Dissanayake's argument that, although "literary scholars occasionally pay lip service to the existence of oral literature, it may not be fully realized that a minute proportion of all humans throughout history have been readers or writers, yet they nevertheless invented and responded to literary language" ("Prelinguistic and Preliterate Substrates," 70).

69. For a full discussion, see Zunshine, "Who Is He to Speak of My Sorrow?"

70. E. Schieffelin, *Sorrow of the Lonely*, 197.

71. E. Schieffelin, 184.

72. Feld, *Sound and Sentiment*, 223.

73. Duranti, "Further Reflections," 493.

74. Aristotle, *Poetics*, 71.

75. Booth, "Control of Distance," 574.

76. Spivak, "Women's Texts," 1095.

77. Sontag, "Against Interpretation," 406.

78. If it's indeed the case that readers who treat *The Plum* as pornography miss complex embedments involving mental states of the implied reader and the implied author, then it would be similar to the phenomenon described by Chris Gavaler and Dan Johnson, who found that, when readers assume a priori that what they are reading has "lower literary merit" (such as science fiction) they "exert less inference effort" in figuring out characters' motivations ("Genre Effect," 86, 91). Perhaps one can even test this hypothesis, telling one group of the first-time readers of *The Plum* that it's a pornographic novel and another that it's literary classic and seeing how deeply these respective groups would delve into the text's intentionality. Also, compare the history of the critical analysis of *The Plum* to that of another classical Chinese novel, Wu Cheng'en's *Journey to the West* (ca. 1592), as highlighted by Yuanfei Wang. Read as a religious text, *Journey to the West* offers up very different constellations of mental states than when read as a humorous, even nonsensical story ("Fantastic Jokes").

79. Plaks, The Four Masterworks of the Ming Novel, 122, 123, 128.

80. Ballaster, *Seductive Forms*, 35.

81. See, for example, Nussbaum, "Finely Aware"; Nussbaum, "Literary Imagination"; and Bruner, *Actual Minds*. In general, as John Guillory points out, recent attacks on the humanities "provoked a torrent of books, articles, reports, and blogs in [their] defense . . . , all attesting to the value of critical thinking and other skills produced by humanities study" ("Monuments and Documents," 10).

82. See Guillory for a critique of the notion that the sciences do not foster critical thinking ("Monuments and Documents," 15).

83. According to Robbins and Rumsey, in "Pacific societies where the opacity doctrine is present, . . . people tend to put little store in the veracity of what others say about their own thoughts"—but it does mean that the further open discussion would go against established daily practices and would be "regarded as extremely invasive and unethical" ("Introduction," 408, 416). When such a discussion does

ensue—when, for instance, others have good reasons to disbelieve the person's initial claim and are invested in bringing the matter to communal attention—they still carefully refrain from openly articulating their versions of the person's mental states. Instead, a social context may be arranged in which the person would eventually revise their earlier statement. For a description of such a situation, see B. Schieffelin, "Speaking Only Your Own Mind," 438.

34. Ochs, "Clarification and Culture," 335. See also McNamara et al., who point out in their study of how "cultural models of mind shape moral reasoning," that in "North American samples, judgments of wrongdoing are scaled almost exclusively by the 'did they mean to,' intent-oriented mental state reasoning process, while judgments about punish-worthiness are scaled by the degree of severity calculated by the more mind-blind 'whodunnit' process (though scope of punishment can be scaled by intent). Because these processes do not perfectly overlap, mis-matches in intent and outcome (i.e., an accident that results in a bad outcome despite a positive or neutral intent) can receive more severe reactions than would be expected in a strictly intent-focused system" (96; in-text quotations omitted).

35. See Rosenberg, *How History Gets Things Wrong*.

36. Howard, *First World War*, 3; emphasis added.

37. Tro, *Chemistry*, 4.

38. Note that we can't just say that the author added an extra bit of narrative to his list of basic SI units and be content with this general explanation. For, while it may be true that we now experience the text as having slightly more of a narrative arc to it—as in, "first we do this, and then we do that"—what has made the actual specific difference is the introduction of a mental state.

39. Charon, "Spoken Body," 264.

90. Charon, 270.

91. See Columbia University School of Professional Studies, "Narrative Medicine."

92. See Charon, *Narrative Medicine*; and Charon et al., *Principles and Practice*.

93. Hoekstra, "From Darwin to DNA," 278.

Chapter 5

1. As Karin Kukkonen puts it eloquently in her discussion of the "curse of realism," the "novel is the literary genre of realism. Its narratives are situated in a place and time that are recognisably real, its characters are subject to coincidence and contingencies that are conceivably every-day, and their feelings are expressed in a language that gets progressively better at capturing the subtleties of experience" (*4E Cognition*, 14).

2. See, for instance, Cravens, "Lyric and Narrative Consciousness." See also Petrone's *Life Has Become More Joyous* for a discussion of the appropriation of Pushkin's by Stalinist ideology. Starting from the 1930s, the official Party line was that the "traits of simplicity and realism linked Pushkin ot the prevailing literary style of the time, socialist realism" (117).

3. Pushkin, *Eugene Onegin*, 99.

4. Compare to Mark Turner's argument about compression in "Compression and Representation."

5. The general argument that art exaggerates familiar characteristics of reality is, of course, quite old. As Roman Jakobson wrote in 1922, "Exaggeration in art is unavoidable [according to] Dostoevskij; in order to show an object, it is necessary to deform the shape it used to have; it must be tinted, just as slides to be viewed under the microscope are tinted. You color your object in an original way and think that it has become more palpable, clearer, more real" ("On Realism in Art," 26).

6. For instance, when you listen to papers delivered by literary critics at scholarly conferences, such papers tend to embed complex mental states at an unusually high rate, for, after all, they often report and interpret complexly embedded mental states of literary characters.

7. Compare to Troscianko's observation that fiction "has the structural potential to prompt more reflexive instances than may occur in real life, resulting in an experience which is compelling, as we enactively engage, but may also be unsettling as moments of reflection accumulate, through perspectival shifts away from the primary focaliser" (*Kafka's Cognitive Realism*, 179).

8. As Jakobson puts it, "classicists, sentimentalists, the romanticists to a certain extent, even the 'realists' of the nineteenth century, the modernists to a large degree and, finally, the futurists, expressionists, and their like have more than once steadfastly proclaimed faithfulness to reality, maximum verisimilitude—in other words, realism—as the guiding motto of their artistic program" ("On Realism in Art," 20) See also McHale, "Revisiting Realisms."

9. Troscianko, *Kafka's Cognitive Realism*, 213.

10. Plaks, "Full-Length Hsiao-shuo," 173.

11. Plaks, 172.

12. On the "weighty seriousness" of *Patterns of Childhood*, see Olney, *Memory and Narrative*, 255.

13. Iliopoulou, *Because of You*, 83.

14. Plaks, "Full-Length Hsiao-shuo," 176; emphasis added.

15. Plaks, 175. This relation would also work, to some extent, for Russian literature; see Munro, "Finance and Credit," 552.

16. Japan of the Heian period (794 to 1185) did not experience anything comparable to the kind of economic development that could be associated, however broadly, with the rise of the novel in the eighteenth-century western Europe, China, or Russia. Moreover, to expand our range of historical contexts beyond economics, we can look at Plaks's argument about the incommensurable intellectual environments of seventeenth- and eighteenth-century China and eleventh-century Japan. As he puts it, while in "terms of intellectual history, at any rate, it does make some sense to see in the novel a manifestation of the need for some kind of a synthesis, a comprehensive reevaluation of the sum total of past cultural experience, in order to adapt that to the perception of emerging new directions, [such] speculations, however, cannot satisfactorily account for a work such as the *Tale of Genii*, which partakes of a number of the defining characteristics of the novel form enumerated above, yet appeared in the vastly more restrictive social and intellectual context of the Heian court in eleventh-century Japan" ("Full-Length Hsiao-shuo," 176).

17. Plaks, 176.

18. Keane, *Ethical Life*, 131.

19. See Feld, *Sound and Sentiment*; and Zunshine, "Who Is He to Speak of My Sorrow?"

20. Such a refutation will still be grist for the mill of "cognitive historicism" and thus a welcome addition to a cognitivist project.

21. There is a long tradition of critical publications on deception in ancient Greek and Roman poetics. For an example of a specifically "cognitive" perspective, see Minchin, "Cognition of Deception."

22. Saussy, "Comparative Literature and Translation," 79.

23. See Austin, *New Testaments*.

24. I say "one such shift" because it's possible that Russian literature had undergone another, earlier shift before the Mongol Yoke (1237–1480). This argument depends on how one reads the late twelfth-century epic poem *The Song of Igor's Campaign*. If one is willing to see third-level embedments in such lines as, for instance, "Let us, however, / begin this song / in keeping with the happenings / of these times / and not with the contriving of / Boyan" (31), then one would say, first, that third-level embedment of mental states not driven by deception was already available to the anonymous author of *The Song* and, second, that the literary context in which such an embedment had been possible must have been largely obliterated during the Yoke. Alternatively, one can adopt the view that "most old Russian literature

[including *The Song*] was not what we would consider fictional, or at least it presented itself as dealing with fact and reality" (Børtnes, "Literature of Old Russia," 1). This would mean that occasional third-level embedments encountered in such works as *The Song* should be read the same way in which we read occasional third-level embedments in contemporary literary nonfiction; that is, their presence does not impact the overall status of the text.

25. Kuritzyn, *Tale of Dracula*; translation mine. Original:

> И отправил царь к Дракуле посла, требуя от него дани. Дракула же воздал послу тому пышные почести, и показал ему свое богатство, и сказал ему: «Я не только готов платить дань царю, но со всем воинством своим и со всем богатством хочу идти к нему на службу, и как повелит мне, так ему служить буду. И ты передай царю, что, когда пойду к нему, пусть объявит он по всей своей земле, чтобы не чинили зла ни мне, ни людям моим, а я вскоре вслед за тобою пойду к царю, и дань принесу, и сам к нему прибуду». Царь же, услышав все это от посла своего, что хочет Дракула прийти к нему на службу, послу его честь воздал и одарил его богато. И рад был царь, ибо в то время вел войну на востоке. И тотчас послал объявить по всем городам и по всей земле, что, когда пойдет Дракула, никакого зла ему не причинять, а, напротив, встречать его с почетом. Дракула же, собрав все войско, двинулся в путь, и сопровождали его царские приставы, и воздавали ему повсюду почести. Он же, углубившись в Турецкую землю на пять дневных переходов, внезапно повернул назад, и начал разорять города и села, и людей множество пленил, перебил, одних—на колья сажал, других рассекал надвое или сжигал, не щадя и грудных младенцев. Ничего не оставил на пути своем, всю землю в пустыню превратил, а всех, что было там, христиан увел и расселил в своей земле. И возвратился восвояси, захватив несметные богатства, а приставов царских отпустил с почестями, напутствуя: «Идите и поведайте царю вашему обо всем, что видели. Сколько сил хватило, послужил ему. И если люба ему моя служба, готов и еще ему так же служить, сколько сил моих станет». Царь же ничего не смог с ним сделать, только себя опозорил.

26. Ermolay-Erazm, *Povest' o Petre and Fevronii*; translation mine. Original: "Это, брат, козни лукавого змея—тобою мне является, чтобы я не решился убить его, думая, что это ты—мой брат."

27. I am grateful to Denis Akhapkin for bringing this tale to my attention.

28. Zenkovsky, "Misery-Luckless-Plight," 497. Original:

> "Откажи ты, молодец, невесте своей любимой:
> быть тебе от невесты истравлену,
> еще быть тебе от тое жены удавлену,
> из злата и сребра бысть убитому!
> Ты пойди, молодец, на царев кабак,
> не жали ты, пропивай свои животы,
> а скинь ты платье гостиное,
> надежи [*] ты на себя гунку кабацкую,—
> кабаком то Горе избудетца,
> да то злое Горе-злочастие останетца:
> за нагим то Горе не погонитца,
> да никто к нагому не привяжетца,
> а нагому, босому шуметь розбой!"
> Тому сну молодець не поверовал (*Tale of Misery*)

29. Zenkovsky, "Misery-Luckless-Plight," 497.

30. Morris, *Literature of Roguery*, 51.

31. Serman, "Eighteenth Century," 69.

32. Emin, *Letters of Ernest and Doravra*; translation mine. Original: "Забудь вину мою и знай, что пожирающая меня любовь наказания, но не презрения достойна. На осужденного на смерть никто не гневается; все о нем сожалеют; и ты, небесная красота, последуя светскому правосудию, сожалей о несчастном, от которого последнее сие получаешь письмо, который тебя больше ничем огорчить не может и который идет на вечное заточение, неся с собою лютейшую о твоих приятностях память, коя бесконечно все его мысли, все чувства и всю природу мучить не перестанет."

33. Karamzin, "Bednaya Liza," 607; translation mine.

34. Karamzin, 612; translation mine, emphasis in the original. Original:

> Все жилки в ней забились, и, конечно, не от страха. Она встала, хотела идти, но не могла. Эраст выскочил на берег, подошел к Лизе и—мечта ее отчасти исполнилась: ибо он *взглянул на нее с видом ласковым, взял ее за руку* . . . Ах! Он поцеловал ее, поцеловал с таким жаром, что вся вселенная показалась ей в огне горящею! «Милая Лиза!—сказал Эраст.—Милая Лиза! Я люблю тебя!», и сии слова отозвались во глубине души ее, как небесная, восхитительная музыка; она едва смела верить ушам своим и . . . Но я бросаю кисть. Скажу только, что в сию минуту восторга исчезла Лизина робость—Эраст узнал, что он любим, любим страстно новым, чистым, открытым сердцем.

35. Being shaped by does not mean, of course, copying. As Boyer puts is, "creating a tradition does not really consist in imitation but includes the constant reconstruction and correction of input" (*Minds Make Societies*, 253).

36. Note how thinking of literature as seeking new ways of representing fictional consciousness shifts our focus from the *subject* of literary discourse to the *effect* this discourse may have on the mind of the reader. According to Porter Abbott, this shift relocates "the intention of the art to what it does to the mind of the reader or viewer: from what art is *about* to what it cognitively *is*" (*Real Mysteries*, 82).

37. Pushkin, *Novels, Tales, Journeys*, 39.

38. Pushkin, 41.

39. Gogol, "Overcoat," 394; emphasis in the original.

40. Nabokov, *Lectures on Russian Literature*, 60.

41. Nabokov, 54.

42. Wood, *Fun Stuff*, 233.

43. As Ann Gaylin puts it, "the eavesdropping scene [in *Wuthering Heights*] is crucial to the very existence of the narrative" (*Eavesdropping*, 26).

44. Nabokov, "Translator's Foreword," x.

45. Lee, "Measuring the Stomach," 205.

46. Note that this account focuses on the European novel and leaves out poetry, particularly the Romantics. A complementary line of inquiry may look, for instance, at the pattern of embedment of mental states in narrative poems that are known to have influenced Pushkin, such as Byron's *Childe Harold's Pilgrimage*.

47. Compare to Fritz Breithaupt's analysis of Nietzschean's view of shame: that is, the "awareness of being observed" (*Dark Sides of Empathy*, 43).

48. Pushkin, *Novels, Tales, Journeys*, 41.

49. Pushkin, 42.

50. Mersereau, "Nineteenth Century," 173.

51. Martinsen, *Surprised by Shame*, 1.

52. Martinsen, xv.

53. Ginzburg, Записные Книжки, 379; translation mine. Original: "Достоевщина как явление моральное и идейное мне в высокой степени противна, не потому, что чужда, но потому, что в какой-то мере свойственна."

54. I have explored this topic more fully in *Why We Read Fiction*.

55. Dostoevsky, *Idiot*; translation mine. Original:

> Он понимал также, что старик вышел в упоении от своего успеха но ему все-таки предчувствовалось, что это был один из того разряда лгунов, которые хотя и лгут до сладострастия и даже до самозабвения, но и на самой высшей точке своего упоения все-таки подозревают про себя, что ведь им не верят, да и не могут верить. В настоящем положении своем старик мог опомниться, не в меру устыдиться, заподозрить князя в безмерном сострадании к нему, оскорбиться.

56. Martinsen, *Surprised by Shame*, 31, 35. Also, as she reports, in "an 1873 *Diary of a Writer* article titled "Something about Lying" (*Nechto o vran'e*), Dostoevsky's narrative persona identifies shame as a critical motive for lying" (4).

57. Hegel, "Traditional Chinese Fiction," 395.

58. Lu, *Brief History*, 4.

59. This is the perspective taken by Y. W. Ma and Joseph S. M. Lau, the editors of *Traditional Chinese Stories*. Note that both their introductory notes (especially "Explanations," xx) and the subsequent reviews of this volume (see, for instance, Mair, "Review of *Traditional Chinese Stories*," 466; and DeWoskin, "Review of *Traditional Chinese Stories*," 774) provide a good example of what Hegel characterizes as the lack "of general agreement on criteria by which to identify" early Chinese fiction ("Traditional Chinese Fiction," 395).

60. Ma and Lau, "Explanations," in *Traditional Chinese Stories*, xx. As Ma and Lau observe,

> Judging from fragments of these works that have survived, these writings can hardly be called fiction in the modern sense of the word. Nor, in our opinion, can passages from early historical works be regarded as fiction, properly speaking. No matter how lively the portraits of historical figures through the use of direct speech, they are nevertheless historical figures meant to be eyewitnesses of history. Indeed, unless one draws a line between fiction and history, it might even be possible to find some of the earliest examples of Chinese story in the markings on oracle bones. It would be a great irony if the ingenuity of a historian in recording fact should be read as fiction. (xx)

61. I am lifting this phrase from Riftin, *От Мифа к Роману*, 78. Riftin uses it to describe a different kind of evolution in Chinese literary history, but his critique of the expectations of the linear development when it comes to literary history is relevant here.

62. Ma and Lau, *Traditional Chinese Stories*, 387. Original: 不能忍，夜伺其寢後，盜照視之，其腰已上生肉如人，腰下但有枯骨。

63. Ma and Lau, 414.

64. Owen, *Anthology of Chinese Literature*, 548.

65. Yu, "Story of Yingying," 184.

66. Yu, 185.

67. Quoted in Owen, *End of the Chinese "Middle Ages,"* 92.

68. Quoted in Owen, 96.

69. Owen, 101–102.

70. Note that the poetic convention of invoking present or future observers had preceded the poetry written in the Tang period. See, for instance, the ending of the famous "Preface to the Poems Collected from the Orchid Pavilion" ("Lantingji Xu," "蘭亭集序," fourth century AD) by Wang Xizhi, in which the speaker suggests that the future readers will empathize with the feelings expressed by the collection: "For the people who read this in future generations, perhaps you will likewise be moved by these words" ("後之覽者，亦將有感於斯文。").

71. Owen, *End of the Chinese "Middle Ages,"* 150.

72. Owen, *Anthology of Chinese Literature*, 518.

73. Owen, *End of the Chinese "Middle Ages,"* 168.

74. Owen, *Anthology of Chinese Literature*, 540.

75. Owen, *End of the Chinese "Middle Ages,"* 168.

76. I make this claim advisedly, taking into a consideration that (1) an expert reader of classical Chinese literature may see a broader variety of embedments in *Romance*

than I do and that (2) in principle, the subjective element unavoidably present in such claims serves to invite further research and discussion (and disagreement) rather than settling this question once and for all.

77. As Haiyan Lee puts it, "Martial tales are as much about the joust of brain as about the joust of brawn; and swashbuckling warriors invariably have to share the spotlight with shrewd strategists. In *The Romance of the Three Kingdoms* . . . , beloved warriors like Guan Yu and Zhao Yun at times are little more than pawns in the elaborate schemes cooked up by the master strategist Zhuge Liang—whose name is synonymous with strategic wisdom in Chinese culture" ("Measuring the Stomach of a Gentleman," 215).

78. Luo G., *Three Kingdoms*, 96. Original: "呂布入內問安，正值卓睡。貂蟬於床後探半身望布，以手指心，又以手指董卓，揮淚不止。布心如碎。"

79. On *The Western Wing* and literati culture, see M. Luo, *Literati Storytelling*, 179.

80. See, respectively, Tillman, "Selected Historical Sources"; and Shen, "Studies of *Three Kingdoms*," 163.

81. Hayden, "Beginning of the End," 43.

82. Shen, "Studies of *Three Kingdoms*," 156.

83. Tillman, "Selected Historical Sources," 53.

84. See Shen on "various artistic forms [that] aided in the dissemination of the story cycle" ("Studies of *Three Kingdoms*," 156).

85. Monaco, "Review of DiaoChan."

86. Schonebaum, introduction to *Approaches to Teaching*, 63. Scholars of Chinese literature have long been aware of the special role of *The Plum* for the course of Chinese literary history. Chen Dakang, for instance, saw 1590 "as the date at which the vernacular novel began to flourish" (Lu, *Brief History*, 101).

87. Please note that I use Wade-Giles in transcribing names of characters in *The Plum*, while using pinyin to transcribe names of characters in several other classical Chinese texts, such as *Dream of the Red Chamber*. For instance, I say Hsi-men Ch'ing instead of Ximen Qing, and Li P'ing-erh instead of Li Ping'er. The reason for this, potentially confusing, usage is that I want my English-speaking readers to be able to refer to David Tod Roy's translation of *The Plum*, which uses Wade-Giles, and to David Hawkes's translation of *Dream* (that is, *The Story of the Stone*), which uses pinyin. Generally, throughout this book, I use whichever system of Romanization the English translator of the cited text used.

88. Link, "Wonderfully Elusive Chinese Novel."

89. Link.

90. Scott, "*Story of the Stone* and Its Antecedents," 266.

91. Chang, "How to Read the *Chin P'ing Mei*," 204, 206.

92. Chang, 211.

93. Roy, *Plum in the Golden Vase*, vol. 2, 87.

94. Roy, 96.

95. Roy, 87.

96. T. Lu, "Interiority in *Jinpingmei cihua*." For a discussion of psychological interiority in classical Chinese texts, such as the *Analects* of Confucius, see Slingerland, "Cognitive Science and Religious Thought."

97. In fact, we can be made to feel sympathetic toward the liar. For instance, Utnapishtim does not judge Gilgamesh for attempting to deceive him: he sees that behavior as only too human. As he puts it to his wife, "Since the human race is duplicitous, he'll endeavor to dupe you" (Foster, *Epic of Gilgamesh*, 92).

98. Lee, "Response to the Panel."

99. Roy, *Plum in the Golden Vase*, 237. Original: "這個都是人氣不憤俺娘兒們，做作出這樣事來。爹，你也要個主張，好把醜名兒頂在頭上，傳出外邊去好聽?" (http://www.guoxue123.com/xiaosuo/jd/jpmch/025.htm).

100. See Zunshine, "Think What You're Doing."

101. Roy, *Plum in the Golden Vase*, vol. 2, 95.

102. Roy, 99.

103. Roy, 111. Original: "先不先只這個就不雅相，傳出去休說六鄰親戚笑話，只家中大小，把尔也不著在意裡。"

104. Roy, 121.

105. See Nabokov's commentary in Pushkin, *Eugene Onegin*.

106. "Но наш герой, кто б ни был он / Уж верно был не Грандисон" (Pushkin, *Eugene Onegin*, 154).

107. Roy, *Plum in the Golden Vase*, vol. 2, 96.

108. Roy, 494.

109. Roy, 494. See also Plaks, *The Four Masterworks of the Ming Novel*, 163–164.

110. Plaks, *The Four Masterworks of the Ming Novel*, 122, 123, 128.

111. Kidd and Castano, "Reading Literary Fiction and Theory of Mind," 8.

112. To appreciate how unwelcome a sustained argument about Pushkin's engagement with Western literature would have been in Soviet Russia, see chapter 5 ("A

Double-Edged Discourse on Freedom: The Pushkin Centennial of 1937") of Petrone's *Life Has Become More Joyous* (113–148).

113. Plaks, "The Four Masterworks of the Ming Novel, 121.

114. Wu, *Scholars*, 40. Original: "知縣心里想道: '這小斯那里害什么病! 想是翟家這奴 才, 走下鄉, 狐假虎威, 著實恐嚇了他一場; 他從來不曾見過官府的人, 害怕不 敢來了。老師既把這個人 托我, 我若不把他就叫了來見老師, 也惹得老師笑我 做事疲軟; 我不如竟自己下鄉去拜他。他看見賞 他臉面, 斷不是難為他的意思, 自然大著膽見我。我就順便帶了他來見老師, 卻不是辦事勤敏? ' 又 想道: '堂 堂一個縣令, 屈尊去拜一個鄉民, 惹得衙役們笑話。●●●'又想到: '老師前 日口气, 甚是敬 他; 老師敬他十分, 我就該敬他一百分。況且屈尊敬賢, 將來志 書上少不得稱贊一篇; 這是万古千年 不朽的勾當, 有甚么做不得?"

115. Cao, *Story of the Stone*, vol. 1, 193. Original: "姨媽不知道。幸虧是姨媽這裏, 倘或在別 人家, 人家豈不惱?好說就看得人家連個手爐也沒有, 巴巴的從家裏送個來。不說丫頭們太小心過餘, 還只當我素日是這等輕狂慣了呢。"

116. Cao, 193.

117. Cao, 436.

118. Cao, 437.

119. Cao, 438.

120. See Zunshine, "From the Social to the Literary."

121. Shang Wei points out that the literati novels "of the mid- and late Qianlong era . . . had so little in common with the earlier novels that their emergence in the mid-eighteenth century could well indicate the rise of a new narrative form" ("Literati Era," 269). The profound influence of *The Story of the Stone* on the subsequent development of Chinese literature is a well-explored topic in critical studies; see Plaks, "Novel in Premodern China"; and Schonebaum, introduction to *Approaches to Teaching*. We may further enrich our understanding of that influence if we retrace its history by looking specifically at the patterns of embedment associated with it. It would be interesting to see, for instance, to what extent numerous imitations and revisions of this novel (see Wei, "*Stone* Phenomenon") embed complex mental states in the same ambiguous, open-ended manner.

122. Cao, *Story of the Stone*, vol. 1, 205. Original: "黛玉忙又叫住, 問道: 「你怎麼不去辭辭 你寶姐姐呢?」寶玉笑而不答, 一逕同秦鐘上學去了。"

123. Talwar, Gordon, Lee, "Lying in the Elementary School Years," 804.

124. Ermer, Cosmides, and Tooby, "Cheater Detection."

125. Compare to William Flesch's suggestive exploration of punishing cheaters in fiction, in *Comeuppance*.

126. On the adaptive role of the emotion of sexual jealousy, see Cosmides and Tooby, "Evolutionary Psychology and the Emotions"; Daly, Wilson, and Weghorst, "Male Sexual Jealousy"; and Buss, *Evolution of Desire*.

127. Although the critical conversation about the relationship between form and context, both within and outside cognitivist contexts, has been rich and variegated (for some recent examples, see, for instance, Kramnick and Nersessian, "Form and Explanation"; and Levine, *Forms*), I focus deliberately on the earliest cases, such as Schiller and Vygotsky. For a recent engagement with Schiller's "conception of the liberating force of form," specifically from a cognitive perspective (i.e., that of probability designs), see Kukkonen, *Probability Designs*, 187.

128. "Darin also besteht das eigentliche Kunstgeheimnis des Meisters, dass er den Stoff durch die Form vertilgt" (Schiller, *Schillers Sämmtliche Werke*, 644; translation mine).

129. Vygotsky, *Psichologia Iskusstva*, 180, 186. Original:

С этой точки зрения становится совершенно ясным, что, если те два плана в басне, о которых мы все время говорим, поддержаны и изображены всей силой поэтического приема, т.е. существуют не только как противоречие логическое, но гораздо больше, как противоречие аффективное,—переживание читателя басни есть в основе своей переживание противоположных чувств, развивающихся с равной силой, но совершенно вместе. . . . Разве не то же самое разумел Шиллер, когда говорил о трагедии, что настоящий секрет художника заключается в том, чтобы формой уничтожить содержание? И разве поэт в басне не уничтожает художественной формой, построением своего материала того чувства, которое вызывает самым содержанием своей басни?

130. I speak of heavy lifting intentionally, building here on Vygotsky's useful metaphor of the airplane in *Psichologia Iskussstva*, 288; for translation, see Vygotsky, *Psychology of Art*, 227.

131. Mateo Alemán, *Life and Adventure* (1823).

132. Garrido Ardila, "Origins and Definition," 1.

133. Wood, "Reality Testing."

134. Lerner, *Leaving the Atocha Station*, 23.

135. Lerner, 119.

136. Lerner, 133.

137. On egocentricity and mindreading, see Riva et al., "Emotional Egocentricity Bias."

138. Lerner, *Leaving the Atocha Station*, 57.

139. Defoe, *Robinson Crusoe*, 97.

140. Wood, "Reality Testing."

Chapter 6

1. See Luhrmann, *"Theory of Mind in the Pacific,"* 443. See also McNamara et al., "Weighing Outcome," who point out that (in-text citations omitted),

> Developmentally, children living in more traditional and community-oriented groups often pass psychological tests that require them to use others' beliefs to predict their behavior (i.e., false belief measures) at later ages . . . This is particularly so for children from cultural contexts with Opacity of Mind norms . . . This suggests a weaker, more distant cognitive/semantic connection between thoughts and behaviors for children in these societies. Conversely, exposure to a larger lexicon of mental state terms and more formal Western education predicts children will perform these tasks at younger ages . . . This suggests a tighter linkage between mental states and behaviors, and further implies more emphasis on intention for judging behavior. (96)

2. Dyer, Shatz, and Wellman, "Young Children's Storybooks," 19.

3. Dyer, Shatz, and Wellman, 34. See also Wulandini, Kuntoro, and Handayani, "Effect of Literary Fiction."

4. See, for instance, Astington and Baird, *Why Language Matters.*

5. Peskin and Astington, "Effects of Adding Metacognitive Language," 256.

6. Peskin and Astington, 254.

7. Peskin and Astington, 255.

8. An earlier version of this section appeared in Zunshine, "From the Social to the Literary," under the heading "What Rosie Knew."

9. Peskin and Astington, "Effects of Adding Metacognitive Language," 253.

10. Harris, Rosnay, and Pons, "Language and Children's Understanding," 72.

11. Peskin and Astington, "Effects of Adding Metacognitive Language," 265. See also Zunshine, "Secret Life of Fiction."

12. Peskin and Astington, "Effects of Adding Metacognitive Language," 266.

13. Peskin and Astington, 267.

14. Of course, "the exact interpretation of [Peskin and Astington's] results needs more research" (Paul L. Harris, email communication, April 18, 2014). To begin with, the emphasis on the importance of reading fictional stories that make children work hard at deducing mental states does not mean to downplay the crucial role of talking to children about thoughts and feelings, and it may shed an interesting light on the underlying structure of those conversations. See, for instance, the study by Harris and his colleagues, who looked at children's attribution of emotions, attendant upon their attributions of false beliefs, and found that while the four- to six-year-olds may judge correctly that Red Riding Hood doesn't know that

the Wolf is waiting for her in her grandmother's cottage, they may still say that she is afraid rather than happy as she approaches the cottage. While thus positing a lag between "children's understanding of a protagonist's mistaken beliefs and their grasp of the emotions that flow from such beliefs," this study also found that "children with mothers who use more mental-state language make more correct attributions" of emotions (Harris, Rosnay, and Ronfard, "Mysterious Emotional Life," 107). Moreover, as Harris observes elsewhere, "a simple count of mental-state terms [used by mothers] may not be the most sensitive measure of effective maternal input even if it is a useful correlate. [It's possible] that it is the mother's pragmatic intent, notably her efforts to introduce varying points of view into a given conversation, that is the underlying and effective source of variation" (Harris, "Conversation," 77).

15. Rolston, *Traditional Chinese Fiction*, 217.

16. Dyer, Shatz, and Wellman, "Young Children's Storybooks," 22.

17. Burnett, *Annotated Secret Garden*,1.

18. Burnett, 6.

19. Milne, *Winnie-the-Pooh*, 14.

20. For a witty analysis of mindreading in the honey scene, see Cave, *Thinking with Literature*, 42–45.

21. Kinney, *Diary of a Wimpy Kid*, 15.

22. Kinney, 35.

23. Jansson, *Moomin Falls in Love*, 17.

24. Jansson, 18.

25. Jansson, 19; my emphases throughout.

26. Carroll, *Alice's Adventures*, 29.

27. Travers, *Mary Poppins Omnibus*, 26.

28. Travers, 56.

29. Travers, 133.

30. White, *Stuart Little*, 11.

31. White, 25.

32. Compare to Emer O'Sullivan's suggestion that a study in "comparative poetics" of children's literature may concern itself with such a question as "how (and why) the beginnings of the new, complex, literary children's literature, which embraces

techniques common to the psychological novel, can be traced back to the end of the 1950s in England, the 1960s in Sweden, and around 1970 in Germany [Nikolajeva, *Children's Literature Comes of Age*] and why this form has taken longer to be accepted and produced in other children's literatures" (O'Sullivan, "Comparative Children's Literature," 192). It would be interesting to see if the frequency of complex embedments in the books for children described by O'Sullivan as having embraced "techniques common to the psychological novel" would indeed approach that of the psychological novel.

33. Knoepflmacher and Myers, "From the Editors," viii.

34. Beckett, *Crossover Picturebooks*, 175.

35. Beckett, 176.

36. Zunshine, "Secret Life of Fiction," 729.

37. Lockington, *For Black Girls like Me*, 221.

38. Lockington, 219–220.

39. Clark, *Kiddie Lit*, 80.

40. Clemens and Howells, *Mark Twain–Howells Letters*, 196.

41. Clark, *Kiddie Lit*, 80.

42. Clemens and Howells, *Mark Twain–Howells Letters*, 122.

43. Twain, *Mississippi Writings*, 3.

44. Clark, *Kiddie Lit*, 84, 81.

45. Quoted in Clark, 89.

46. Clark, 101.

47. Phelan and Rabinowitz. "Narrative Values," 163.

48. Phelan and Rabinowitz, "Authors," 35.

49. Compare to Hogan's observation that literature can present us "with emotionally affective situations, where emotion systems interact in sometimes very subtle ways—unlike the artificially limited situations that are necessary for the control of variables in experimental research" (*Sexual Identities*, 20). Also, on the processing of high-level embedments, see Dunbar, *How Many Friends Does One Person Need?*, 180.

50. Zunshine, "Commotion of Souls," 139.

51. Miller, Kessel, and Flavell, "Thinking about People Thinking," 622.

52. Phelan and Rabinowitz, "Narrative Values," 163.

53. Phelan and Rabinowitz, "Authors," 35.

54. Phelan and Rabinowitz, "Narrative Values," 163.

55. Another fascinating outlier, similarly classed with children's literature, yet embedding complex mental states at a frequency we would associate with literature for adults, is J. M. Barrie's *Peter Pan*. For a useful discussion of the "labyrinth of subjectivities" of its various textual incarnations, see Butte, *Suture and Narrative*, particularly chap. 4, "The Wounds of Peter Pan" (quote on 133). Although Butte does not deal with embedded mental states per se, his analysis of Barrie's texts as speaking "to several audiences in several registers at the same time" (136) reveals proliferation of complex embedments.

56. As one anonymous Amazon purchaser puts it, "Although there are wonderful little snippets of family life, and a few hints of the conflicts between the feisty Laura and her more reserved and perfect sister Mary, the truth is, there isn't much of a plot here." slow-mamma, review of *Little House in the Big Woods*, Amazon, April 12, 2002, https://www.amazon.com/Little-House-Woods-Ingalls-Wilder/dp/0060581808/ref=sr_1_1?s=books&ie=UTF8&qid=1475604634&sr=1-1&keywords=little+house+in+the+big+woods.

57. Hill, "Introduction," xvi.

58. Quoted in Hill, xliii

59. Hill, xxxvi.

60. Hill, xxx.

61. Fellman, *Little House*, 6–7, 106.

62. Hill, "Introduction," lv.

63. Fellman, *Little House*, 85.

64. Wilder, *Little Town*, 12, 140.

65. Wilder, *These Happy Golden Years*, 49.

66. Wilder, 176, 184; my emphases throughout.

67. Fellman, *Little House*, 127.

68. For a broader discussion of "intermediality" of children's literature—that is, "a synergistic relation between stories and characters that originally appear in print and the forms into which they are subsequently transformed across media boundaries: film, video, DVD, audio adaptations," etc.—see O'Sullivan, "Comparative Children's Literature," 193. In one striking example she mentions, the Canadian classic *Anne of Green Gables*, by L. M. Montgomery, was not translated into German until the mid-1980s, and the translation was based on the film version" (O'Sullivan, 193).

69. Fellman, *Little House*, 123, 127–128.

70. Compare to Bettina Kümmerling-Meibauer's classic argument that "most of the key elements of sophisticated narratives are present in a simpler form in picture books" ("Metalinguistic Awareness," 177).

71. For a discussion of "pleasure" involved in children's interaction with twist endings, see Bellorín and Silva-Díaz, "Surprised Readers," 118.

72. On the gendering of the small fish and the big fish, see Drabble, "Jon Klassen."

73. On the gendering of Gruffalo, see Nick Miller, "Gruffalo Creator."

74. And, of course, as Alexandra Berlina helpfully reminds me, they also know that the mouse is surprised that the Gruffalo, the monster that the mouse thinks it has invented, turns out to exist!

75. For a valuable review of problems inherent in the issue of identification, see Keen, *Empathy and the Novel*.

76. Talwar, Gordon, and Lee, "Lying in the Elementary School Years," 804–810.

77. As Milligan et al. point out, age itself is not an "explanatory variable, but rather a proxy for various maturational factors that may explain variation, an important one of which is language ability" ("Language and Theory of Mind," 638).

78. For a review, see Ahrens, "Picturebooks."

79. See Kümmerling-Meibauer and Meibauer, "Early-Concept Books."

80. Rey, Curious George at the Zoo, n.p.

81. Miller, *Pooh's Honey Trouble*.

82. The number of reviews is growing, so by the time this book is in press, it will be higher.

83. The full review reads, "There is pretty much no story here, but if your little loves Winnie the Pooh, it'll be a hit anyway. However, I must disagree with the recommended age of 3 & up. This book is for toddlers" (Linklau, "Cute for Toddlers," Amazon, September 2015, https://www.amazon.com/gp/customer-reviews/R1YY7D6HAYZ6SX/ref=cm_cr_arp_d_rvw_ttl?ie=UTF8&ASIN=1423135792).

84. As Bettina Kümmerling-Meibauer observes, the "text in Hutchins's book merely informs the reader in a few words about Rosie's walk and is supplemented only by participial constructions with changing place names. . . . The completely dull [text relates] events with almost no mention of the emotional reactions of those who participate in them" ("Metalinguistic Awareness," 170).

85. Amazon, reviews of *Rosie's Walk*, https://www.amazon.com/Rosies-Walk-Pat-Hutchins/dp/0020437501/ref=sr_1_1?s=books&ie=UTF8&qid=1477417845&sr=1-1&keywords=rosie%27s+walk.

86. For a discussion of the role of adults in mediating children's relationships with books, see O'Sullivan, "Comparative Children's Literature." As she observes, "at every stage of literary communication we find adults acting for children" (191).

87. Richardson, *Literature*, 109; cf. Deppner, "Parallel Receptions of the Fundamental," 58–59. See also O'Sullivan, "Comparative Children's Literature," 190. Also, in my book *Strange Concepts* (chap. 14), I have looked at an eighteenth-century text specifically geared toward three- to five-year-olds, Anna Laetitia Barbauld's *Hymns in Prose for Children* (1781), but I don't want to conclude too much based on just one case study.

88. Harris, Rosnay, and Pons, "Language," 71–72. See also Rosnay et al., "Lag between Understanding"; and Hughes, White, and Ensor, "Talking about Thoughts."

89. On the relationship between cross-writing and crossovers, see Falconer, "Children's Literature."

90. For a useful discussion of the role of illustrations in "metafictional picturebooks" (355), see Lissi Athanasiou-Krikelis, "Mapping."

91. Lipson, *New York Times Parent's Guide*, 48; Amazon, *Hush Little Baby* page, accessed 06/08/2021. https://www.amazon.com/Hush-Little-Baby-Folk-Pictures/dp/0152058877/ref=sr_1_1?s=books&ie=UTF8&qid=1535232923&sr=1-1&keywords=hush+little+baby+marla+frazee.

92. J. Freeman, *Books Kids Will Sit Still For 3*, 236; Gillespie, *Best Books for Children*, 712; Scholastic, *Hush Little Baby* page, accessed 12/18/2018. As of 06/08/21, Scholastic doesn't seem to feature Frazee's *Hush Little Baby* anymore, so this reference more accurately pertains to their former characterization. https://www.scholastic.com/teachers/books/hush-little-baby-by-marla-frazee/.

93. Amazon, *Hush Little Baby* reviews, accessed 12/18/2018. https://www.amazon.com/Hush-Little-Baby-Folk-Pictures/product-reviews/0152058877/ref=cm_cr_getr_d_paging_btm_4?ie=UTF8&reviewerType=all_reviews&showViewpoints=1&sortBy=recent&pageNumber=4.

Conclusion

1. Wolf, *Patterns of Childhood*, 8. The translation is missing a part of one sentence. The full original (with the missing part italicized) reads, "Auffallend ist, daß wir in eigener Sache entweder romanhaft lügen oder stockend und mit belegter Stimme sprechen. Wir mögen wohl Grund haben, von uns nichts wissen zu wollen (oder doch nicht alles—was auf das gleiche hinausläuft). Aber selbst wenn die Hoffnung gering ist, sich allmählich freizusprechen *und so ein gewisses Recht auf den Gebrauch jenes Materials zu erwerben, das unlösbar mit lebenden Personen verbunden ist*—so wäre es doch nur diese geringfügige Hoffnung, die, falls sie durchhält, der Verführung zum Schweigen und Verschweigen trotzen könnte" (Wolf, *Kindheitsmuster*, 15).

2. See, for instance, Black and Barnes, "Fiction and Social Cognition."

Bibliography

Abbott, H. Porter. *Real Mysteries: Narrative and the Unknowable*. Columbus: Ohio State University Press, 2013.

Ahrens, Kathleen. "Picturebooks: Where Literature Appreciation Begins." In *Emergent Literary: Children's Books from 0 to 3*, edited by Bettina Kümmerling-Meibauer, 77–89. Amsterdam: John Benjamins, 2011.

Alemán, Mateo. *The Life and Adventure of Guzman D'Alfarache, or the Spanish Rogue*. Vol. 1. Translated from the French Edition of Mons. Le Sage by John Henry Brady. London: J. Nichols and Son, 1823. https://www.gutenberg.org/files/52806/52806-h/52806-h.htm.

Alemán, Mateo. *The Life and Adventures of Guzman D'Alfarache, or the Spanish Rogue*. Vol. 1. Translated by Alain-René Le Sage and John Henry Brady. London: Printed for Longman, Hurst, Rees, Orme, Brown, and Green, Paternoster-Row, 1823. https://www.gutenberg.org/files/52806/52806-h/52806-h.htm.

Al Harthi, Jokha. *Celestial Bodies*. Translated by Marilyn Booth. Croydon, UK: Sandstone, 2018.

Alakhverdov, Victor Michailovitch. *Psychology of Art* [Психология искусства]. St. Peterburg: Izdatelstvo DNK, 2001. http://mhp-journal.ru/upload/Library/Allakhverdov_VM_(2001)_Psychology_of_Art.pdf.

Athanasiou-Krikelis, Lissi. "Mapping the Metafictional Picturebook." *Narrative* 28 no. 3 (2020): 355–374.

Anonymous. *The Epic of Gilgamesh*. Translated by Nancy K. Sandars. Penguin, 1960.

Anonymous. "The Wanderer." Translated by Sean Miller. *Anglo-Saxons.net*. http://www.anglo-saxons.net/hwaet/?do=get&type=text&id=Wdr

Apperly, Ian. *The Cognitive Basis of "Theory of Mind."* New York: Psychology Press, 2011.

Apuleius. *The Golden Ass*. Translated by Sarah Ruden. New Heaven, CT: Yale University Press, 2013.

Argyle, Michael, and Mark Cook. *Gaze and Mutual Gaze*. Cambridge: Cambridge University Press, 1976.

Aristotle. *Poetics*. In *The Critical Tradition: Classic Texts and Contemporary Trends*, edited by David H. Richter, 50–72. Boston: Bedford/St. Martin's, 2016.

Asimov, Isaac. *The Bicentennial Man and Other Stories*. New York: Doubleday, 1976.

Astington, Janet Wilde, and Jodie A. Baird, eds. *Why Language Matters for Theory of Mind*. New York: Oxford University Press, 2005.

Astington, Janet Wilde, Janette Pelletier, and Bruce Homer. "Theory of Mind and Epistemological Development: The Relation between Children's Second-Order False Belief Understanding and Their Ability to Reason about Evidence." *New Ideas in Psychology* 20 (2002): 131–144.

Astuti, Rita. "Some After Dinner Thoughts on Theory of Mind." *Anthropology of This Century* 3 (2012): n.p. http://eprints.lse.ac.uk.

Asztalos, Monika. "The Faculty of Theology." In *A History of the University in Europe*, vol. 1, *Universities in the Middle Ages*, edited by Hilde de Ridder-Symoens, 409–441. Cambridge: Cambridge University Press, 1992.

Austen, Jane. *Mansfield Park*. Oxford: Oxford University Press, 1990.

Austen, Jane. *Pride and Prejudice*. New York: Dover, 1995.

Austin, Michael. *New Testaments: Cognition, Closure, and the Figural Logic of the Sequel, 1660–1740*. Newark: University of Delaware Press, 2011.

Auyoung, Elaine. "The Unspoken Intimacy of Aesthetic Experience: Hardy and Degas," *Poetics Today* 41, no. 2 (2020): 301–314.

Auyoung, Elaine. *When Fiction Feels Real: Representation and the Reading Mind*. New York: Oxford University Press, 2018.

Baggs, Ed. "Book Review: Evolving Enactivism: Basic Minds Meet Content." *Frontiers in Psychology* 8 (2017): 1947. doi: 10.3389/fpsyg.2017.01947.

Bakhtin, Mikhail. "Discourse in the Novel." In *The Dialogic Imagination: Four Essays*, edited and translated by Michael Holquist and Caryl Emerson, 259–422. Austin: University of Texas Press, 1982.

Ballaster, Ros. *Seductive Forms: Women's Amatory Fiction from 1684 to 1740*. Oxford, UK: Clarendon, 1992.

Barrie, J. M. *Peter Pan: The Original Tale of Neverland*. New York: Simon and Schuster, 2000.

Baum, Andrew, J. P. Garofalo, and Ann Marie Yali. "Socioeconomic Status and Chronic Stress: Does Stress Account for SES Effects on Health?" *Annals of the New York Academy of Sciences* 896, no. 1 (1999): 131–144.

Beckett, Sandra. *Crossover Picturebooks: A Genre for All Ages*. New York: Routledge, 2013.

Bellorín, Brenda, and Cecilia Silva-Díaz. "Surprised Readers: Twist Endings in Narrative Picturebooks." In *New Directions in Picturebook Research*, edited by Teresa Colomer, Bettina Kümmerling-Meibauer, and Cecilia Silva-Díaz, 113–127. New York: Routledge, 2010.

Black, Jessica, and Jennifer L. Barnes. "Fiction and Social Cognition: The Effect of Viewing Award-Winning Television Dramas on Theory of Mind." *Psychology of Aesthetics Creativity and the Arts* 9, no. 4 (2015): 423–429. doi: 10.1037/aca0000031.

Blackman, Lucy. "Reflections on Language." In *Autism and the Myth of Person alone*, edited by Douglas Biklen with Richard Attfield, Larry Bissonnette, Lucy Blackman, Jamie Burke, Alberto Frugone, Tito Rajarshi Mukhopadhyay, and Sue Rubin, 146–167. New York: New York University Press, 2005.

Bloom, Paul. *Against Empathy: The Case for Rational Compassion*. New York: Harper Collins, 2016.

Bolens, Guillemette. *Kinesic Humor: Literature, Embodied Cognition, and the Dynamics of Gesture*. New York: Oxford University Press, 2021.

Booth, Wayne C. "Control of Distance in Jane Austen's *Emma*." In *The Critical Tradition: Classic Texts and Contemporary Trends*, edited by David H. Richter, 570–582. Boston: Bedford/St. Martin's, 2016.

Borden, Richard C. "Leonid Dobychin's *The Town of N*: Introduction." In *The Town of N*, by Leonid Dobychin, translated from Russian by Richard C. Borden with Natalia Belova, i–xi. Evanston, IL: Northwestern University Press, 1998.

Børtnes, Jostein. "The Literature of Old Russia, 988–1730." In *The Cambridge History of Russian Literature*, edited by Charles A. Moser, 1–44. Cambridge: Cambridge University Press, 1989.

Bowes, Andrea, and Albert Katz. "Metaphor Creates Intimacy and Temporarily Enhances Theory of Mind." *Memory and Cognition* 43 (2015): 953–963. https://doi.org/10.3758/s13421-015-0508-4.

Boyer, Pascal. *Minds Make Societies: How Cognition Explains the World Humans Create*. New Haven, CT: Yale University Press, 2018.

Brandenberger, David. *Propaganda State in Crisis: Soviet Ideology, Indoctrination, and Terror under Stalin, 1927–1941*. New Haven: Yale University Press, 2011.

Breithaupt, Fritz. *The Dark Sides of Empathy*. Translated from German by Andrew B. B. Hamilton. Ithaca, NY: Cornell University Press, 2019.

Bruner, Jerome S. *Actual Minds, Possible Worlds*. Cambridge, MA: Harvard University Press, 1986.

Buckner, Randy L., Jessica R. Andrews-Hanna, and Daniel L. Schacter. "The Brain's Default Network: Anatomy, Function, and Relevance to Disease." *Annals of the New York Academy of Sciences* 1124 (2008): 1–38.

Buckner, Randy L., and Daniel C. Carroll. "Self-Projection and the Brain." *Trends in Cognitive Sciences* 11, no. 2 (February 2007): 49–57.

Burnett, Frances Hodgson. *The Annotated Secret Garden*. Edited by Gretchen Holbrook Gerzina. New York: Norton, 2007.

Burney, Frances. *Evelina; or, The History of a Young Lady's Entrance into the World*. Edited by Edward A. Bloom. Introduction and notes by Vivien Jones. Oxford: Oxford University Press, 2002.

Buss, David M. *The Evolution of Desire*. New York: Basic Books, 1994.

Butte, George. *Suture and Narrative: Deep Intersubjectivity in Fiction and Film*. Columbus: Ohio State University Press, 2017.

Caldwell-Harris, Catherine, Ann Kronrod, and Joyce Yang. "Do More, Say Less: Saying 'I Love You' in Chinese and American Cultures." *Intercultural Pragmatics* 10, no. 1 (2013): 41–69.

Callahan, John F. "Frequencies of Eloquence: The Performance and Composition of *Invisible Man*." In *New Essays on Invisible Man*, edited by Robert O'Meally, 55–94. Cambridge: Cambridge University Press, 1988.

Cao Xueqin. *The Story of the Stone*. Vol. 1, *The Golden Days*. Translated by David Hawkes. London: Penguin, 1973.

Cao Xueqin. *The Story of the Stone*. Vol. 2, *The Crab-Flower Club*. Translated by David Hawkes. London: Penguin, 1977.

Carney, James, Rafael Wlodarski, and Robin Dunbar. "Inference or Enaction? The Impact of Genre on the Narrative Processing of Other Minds." *PLoS ONE* 9, no. 12 (2014): e114172. https://doi.org/10.1371/journal.pone.0114172.

Carroll, Lewis. *Alice's Adventures in Wonderland*. London: Usborne, 2013.

Carruthers, Peter. "Two Systems for Mindreading?" *Review of Philosophy and Psychology* 7, no. 1 (2016): 141–162. doi: 10.1007/s13164-015-0259-y.

Castano, Emanuele. "Art Films Foster Theory of Mind." *Humanities and Social Sciences Communications* volume 8, Article number: 119 (2021). https://doi.org/10.105/s41599-021-00793-y

Castano, Emanuele, Alison Jane Martingano, and Pietro Perconti. "The Effect of Exposure to Fiction on Attributional Complexity, Egocentric Bias and Accuracy in Social Perception." *PLoS ONE* 15, no. 5 (2020). https://doi.org/10.1371/journal.pone.0233378.

Cave, Terence. *Thinking with Literature: Towards a Cognitive Criticism*. Oxford: Oxford University Press, 2016.

Cavell, Stanley. *Pursuits of Happiness: The Hollywood Comedy of Remarriage*. Cambridge, MA: Harvard University Press, 1981.

Chang Chu-p'o. "How to Read the *Chin P'ing Mei*." Translated and annotated by David T. Roy; additional annotation by David L. Rolston. In *How to Read the Chinese Novel*, edited by David L. Rolston, 202–243. Princeton, NJ: Princeton University Press, 1990.

Charon, Rita. *Narrative Medicine: Honoring the Stories of Illness*. Oxford: Oxford University Press, 2008.

Charon, Rita. "Spoken Body: An *Infinite Jest* of Life, Death, and the Medical Tongue," *Poetics Today* 41, no. 2 (2020): 261–280.

Charon, Rita. "To See the Suffering." *Academic Medicine* 92, no. 12 (December 2017): 1668–1670.

Charon, Rita, Sayantani DasGupta, Nellie Hermann, Craig Irvine, Eric R. Marcus, Edgar Rivera Colsn, Danielle Spencer, and Maura Spiegel. *The Principles and Practice of Narrative Medicine*. New York: Oxford University Press, 2016.

Chekhov, Anton P. "Skripka Rotshil'da." *Polnoie Sobranie Sochinenii I Pisem v 30-ti Tomach*. Tom 8. Moscow: Nauka, 1986. https://ilibrary.ru/text/978/p.1/index.html.

Clark, Beverly Lyon. *Kiddie Lit: The Cultural Construction of Children's Literature in America*. Baltimore: Johns Hopkins University Press, 2003.

Claudius, Eduard. *Menschen an unserer Seite*. Berlin: Volk und Welt, 1951.

Clemens, Samuel L., and William D. Howells. *Mark Twain–Howells Letters: The Correspondence of Samuel L. Clemens and William D. Howells, 1872–1910*. Edited by Henry Nash Smith and William M. Gibson. Cambridge, MA: Harvard University Press, 1960.

Columbia University School of Professional Studies. "Narrative Medicine." Accessed 06/08/21. http://sps.columbia.edu/narrative-medicine.

Cook, Amy. "4E Cognition and the Humanities." In *The Oxford Handbook of 4E Cognition*, edited by Albert Newen, Leon De Bruin, and Shaun Gallagher, 875–890. Oxford: Oxford University Press, 2020.

Cook, Amy. *Shakespearean Neuroplay: Reinvigorating the Study of Dramatic Texts and Performance through Cognitive Science*. New York: Palgrave Macmillan, 2010.

Cosmides, Leda, and John Tooby. "Evolutionary Psychology and the Emotions." In *Handbook of Emotions*, 2nd ed., edited by Michael Lewis and Jeannette M. Haviland-Jones. New York: Guilford, 2000. https://www.cep.ucsb.edu/emotion.html.

Crane, Mary Thomas. "Cognitive Historicism: Intuition in Early Modern Thought." In *The Oxford Handbook of Cognitive Literary Studies*, edited by Lisa Zunshine, 18–33. New York: Oxford University Press, 2015.

Crane, Mary Thomas. *Shakespeare's Brain: Reading with Cognitive Theory*. Princeton, NJ: Princeton University Press, 2001.

Cravens, Craig. "Lyric and Narrative Consciousness in Eugene Onegin." *Slavic and East European Journal* 46, no. 4 (Winter 2002): 683–709. http://www.jstor.com/stab e /3219907.

Culler, Jonathan. "The Closeness of Close Reading." *ADE Bulletin* 149 (2010): 20–23.

Currie, Gregory. "Framing Narratives," *Royal Institute of Philosophy Supplements* 60 (2007): 17–42.

Cusk, Rachel. *Saving Agnes*. New York: Picador, 1994.

Cusk, Rachel. *Transit*. New York: Farrar Straus and Giroux, 2017.

Daly, Martin, Margo Wilson, and Suzanne J. Weghorst. "Male Sexual Jealousy." *Ethology and Sociobiology* 3 (1982): 11–27.

Dangarembga, Tsitsi. *Nervous Conditions*. London: Women's Press, 1988.

Defoe, Daniel. *Robinson Crusoe*. New York: Oxford University Press, 2007.

De Jaegher, Hanne, and Ezequiel Di Paolo. "Participatory Sense-Making: An Enactive Approach to Social Cognition." *Phenomenology and the Cognitive Sciences* 6 (2007): 485–507.

Deppner, Martin Roman. "Parallel Receptions of the Fundamental: Basic Designs in Picturebooks and Modern Art." In *Emergent Literary: Children's Books from 0 to 3*, edited by Bettina Kümmerling-Meibauer, 55–74. Amsterdam: John Benjamins, 2011.

DeWoskin, Kenneth J. "Review of *Traditional Chinese Stories: Themes and Variations*, by Y. W. Ma, Joseph S. M. Lau." *Journal of Asian Studies* 38, no. 4 (August 1979): 773–775.

Dick, Philip K. *Do Androids Dream of Electric Sheep?* New York: Ballantine Books, 1996.

Dilley, Paul C. *Monasteries and the Care of Souls in the Late Antique Christianity: Cognition and Discipline*. Cambridge: Cambridge University Press, 2017.

Dissanayake, Ellen. "Prelinguistic and Preliterate Substrates of Poetic Narrative." *Poetics Today* 32, no. 1 (2011): 55–79.

Djikic, Maja, Keith Oatley, and Mihnea C. Moldoveanu. "Reading Other Minds: Effects of Literature on Empathy." *Scientific Study of Literature* 3, no. 1 (2013): 28–47.

Dodell-Feder, David, and Diana I. Tamir. "Fiction Reading Has a Small Positive Impact on Social Cognition: A Meta-Analysis." *Journal of Experimental Psychology General* 147, no. 11 (2018): 1713–1727.

Doody, Margaret Anne. *Frances Burney: The Life in the Works*. New Brunswick, NJ: Rutgers University Press, 1988.

Drabble, Emily. "Jon Klassen: Kate Greenaway Medal Winner 2014 – in Pictures." *The Guardian* 23 June, 2014. https://www.theguardian.com/childrens-books-site /gallery/2014/jun/23/jon-klassen-this-is-not-my-hat-kate-greenaway-in-pictures.

Dostoevsky, Fedor. *The Idiot*. Достоевский, Ф. М., *Идиот*. Том 2. Париж, YMCA-Press, 1900. https://babel.hathitrust.org/cgi/pt?id=pst.000006918101&view=1up&seq=1F.

Drucker, Johanna. *Graphesis: Visual Forms of Knowledge Production*. Cambridge, MA: Harvard University Press, 2014.

Dunbar, Robin I. M. *How Many Friends Does One Person Need? Dunbar's Number and Other Evolutionary Quirks*. London: Faber and Faber, 2010.

Dunbar, Robin I. M. "Why Are Good Writers So Rare? An Evolutionary Perspective on Literature." *Journal of Cultural and Evolutionary Psychology* 3, no. 1 (March 2005): 7–21. doi: 10.1556/JCEP.3.2005.1.1.

Duranti, Alessandro. *The Anthropology of Intention: Language in a World of Others*. Cambridge: Cambridge University Press, 2015.

Duranti, Alessandro. "Further Reflections on Reading Other Minds." *Anthropological Quarterly* 81, no. 2 (2008): 483–494.

Dyer, Jennifer R., Marilyn Shatz, and Henry M. Wellman. "Young Children's Storybooks as a Source of Mental State Information," *Cognitive Development* 15, no. 1 (January–March 2000): 17–37. doi: 10.1016/S0885-2014(00)00017-4.

Easterlin, Nancy. *A Biocultural Approach to Literary Theory and Interpretation*. Baltimore: Johns Hopkins University Press, 2012.

Easterlin, Nancy. "Voyages in the Verbal Universe: The Role of Speculation in Darwinian Literary Criticism." *Interdisciplinary Literary Studies* 2, no. 2 (2001): 59–73.

Elfenbein, Andrew. *The Gist of Reading*. Stanford, CA: Stanford University Press, 2018.

Elfenbein, Andrew. "Mental Representation." In *Further Reading*, edited by Matthew Rubery and Leah Price, 246–256. Oxford: Oxford University Press, 2020.

Eliot, George. *Middlemarch*. Edited by Rosemary Ashton. New York: Penguin Classics, 2003.

Ellison, Ralph. Introduction to *Invisible Man*, vii–xxiii. New York, Vintage, 1989.

Ellison, Ralph. *Invisible Man*. New York: Vintage Books, 1989.

Emin, Fedor. *Letters of Ernest and Doravra*. [1769] In *Русская литература XVIII века, 1770–1775. Хрестоматия*, edited by В. А Западов. Москва: Просвещение, 1979. http://az.lib.ru/e/emin_f_a/text_0020.shtml.

Emmerich, Wolfgang. "The GDR and Its Literature: An Overview." In *Rereading East Germany: The Literature and Film of the GDR*, edited by Karen Leeder, 8–34. Cambridge: Cambridge University Press, 2015.

Ender, Evelyne, and Deidre Shauna Lynch. "On 'Learning to Read.'" *PMLA* 130, no. 3 (2015): 539–545.

Epstein, Julia. *The Iron Pen: Frances Burney and the Politics of Women's Writing*. Madison: University of Wisconsin Press, 1989.

Ermer, Elsa, Leda Cosmides, and John Tooby. "Cheater Detection." *Iresearchnet .com: Psychology Research and Reference*. Accessed 06/10/2021. http://psychology .iresearchnet.com/social-psychology/antisocial-behavior/cheater-detection/.

Ermolay-Erazm. *Повесть о Петре и Февронии* [The Tale of Peter and Fevroniya] 1547. Translated by A. A. Alekseev. In *Памятники литературы Древней Руси Выпуск 06: Конец XV - Первая Половина XVI века*, edited by Лев Александрович Дмитриев and Дмитрий Сергеевич Лихачев. Москва: Художественная Литература. 1984. http://drevne-rus-lit.niv.ru/drevne-rus-lit/text/povest-o-petre-i-fevronii.htm.

Falconer, Rachel. "Children's Literature: Part II. Forms and Genres. 43. Crossover Literature." *Schoolbag*. Accessed September 20, 2020. http://schoolbag.info/literature /children/140.html.

Feld, Steven, *Sound and Sentiment: Birds, Weeping, Poetics, and Song in Kaluli Expression*. Philadelphia: University of Pennsylvania Press, 1990.

Fellman, Anita Clair. *Little House, Long Shadow: Laura Ingalls Wilder's Impact on American Culture*. Columbia: University of Missouri Press, 2008.

Fernyhough, Charles. "Metaphors of Mind." *Psychologist* 19 (2006): 356–358. https:/ thepsychologist.bps.org.uk/volume-19/edition-6/metaphors-mind.

Ferrante, Elena. *The Story of the Lost Child: Neapolitan Novels, Book Four*. Translated by Ann Goldstein. New York: Europa Editions, 2015.

Flesch, William. *Comeuppance: Costly Signaling, Altruistic Punishment, and Other Biological Components of Fiction*. Cambridge, MA: Harvard University Press, 2009.

Fletcher, Garth J. O., Paula Danilovics, Guadalupe Fernandez, Dena Peterson, and Glenn D. Reeder. "Attributional Complexity: An Individual Differences Measure." *Journal of Personality and Social Psychology* 51, no. 4 (1986): 875–884. https://doi.org/10 .1037/0022-3514.51.4.875.

Fludernik, Monika. "Collective Minds in Fact and Fiction: Intermental Thought and Group Consciousness in Early Modern Narrative." *Poetics Today* 35, no. 4 (Winter 2014): 689–730.

Fludernik, Monika. "The Many in Action and Thought: Towards a Poetics of the Collective in Narrative." *Narrative* 25, no. 2 (May 2017): 139–163.

Fludernik, Monika. *Towards a "Natural" Narratology*. New York: Routledge, 2010.

Forster, E. M. *Howards End*. New York: Vintage Books, 1954.

Frazee, Marla. *Hush, Little Baby: A Folk Song with Pictures Board Book*. Boston: Houghton Mifflin Harcourt, 2007.

Freeman, Charles. *The Greek Achievement: The Foundation of the Western World*. New York: Penguin, 1999.

Freeman, Judy. *Books Kids Will Sit Still For 3: A Read-Aloud Guide*. Children's and Young Adult Literature Reference Series 3. Westport, CT: Libraries Unlimited, 2006.

Frith, Christopher D. "The Social Mind?" *Philosophical Transactions of the Royal Society B*, April 29, 2007. doi: 10.1098/rstb.2006.2003 np.

Furlanetto, Tiziano, Andrea Cavallo, Valeria Manera, Barbara Tversky, and Cristina Becchio. "Through Your Eyes: Incongruence of Gaze and Action Increases Spontaneous Perspective Taking." *Frontiers in Human Neuroscience*, August 12, 2013. https://doi.org/10.3389/fnhum.2013.00455.

Garratt, Peter, ed. *The Cognitive Humanities: Embodied Mind in Literature and Culture*. New York: Palgrave Macmillan, 2016.

Garrido Ardila, J. A. "Origins and Definition of the Picaresque Genre." In *The Picaresque Novel in Western Literature: From the Sixteenth Century to the Neopicaresque*, edited by Garrido Ardila, 1–23. Cambridge: Cambridge University Press, 2015.

Gavaler, Chris, and Dan Johnson. "The Genre Effect: A Science Fiction (vs. Realism) Manipulation Decreases Inference Effort, Reading Comprehension, and Perceptions of Literary Merit." *Scientific Study of Literature* 7, no. 1 (2017): 79–108.

Gaylin, Ann. *Eavesdropping in the Novel from Austen to Proust*. Cambridge: Cambridge University Press, 2002.

Gillespie, John T. *Best Books for Children: Preschool through Grade Six*. Westport, CT: Libraries Unlimited, 2002.

Gillespie, Michael Patrick. "From Beau Brummell to Lady Bracknell: Re-viewing the Dandy in *The Importance of Being Earnest*." In *The Importance of Being Earnest*, by Oscar Wilde, edited by Michael Patrick Gillespie, 166–182. New York: Norton, 2006.

Ginzburg, Lidiya. Записные Книжки, Воспоминания, Эссе. Санкт-Петербург: Искусство-СПБ, 2011.

Gladkov, Fedor. Цемент. Москва: Художественная литература, 1925. https://www.e-reading.club/book.php?book=150460.

Gladkov, Fedor. *Cement*. Edited and translated by A. S. Arthur and C. Ashleigh. Evanston, IL: Northwestern University Press, 1994.

Gogol, Nikolai. "The Overcoat." In *The Collected Tales of Nikolai Gogol*, translated by Richard Pevear and Larissa Volokhonsky, 394–424. New York: Pantheon, 1998.

Graesser, Arthur C., Murray Singer, and Tom Trabasso. "Constructing Inferences during Narrative Text Comprehension." *Psychological Review* 101, no. 3 (1994): 371–395. https://doi.org/10.1037/0033-295X.101.3.371.

Guillory, John. "Monuments and Documents: Panofsky on the Object of Study in the Humanities." *History of Humanities* 1, no. 1 (2016): 9–30. http://dx.doi.org/10.1086/684635.

Gunn, Daniel P. "Free Indirect Discourse." In *Oxford Research Encyclopedias*, June 2019. doi: 10.1093/acrefore/9780190201098.013.1020m.

Günther, Hans. "Soviet Literary Criticism and the Formulation of the Aesthetics of Socialist Realism." In *A History of Russian Literary Theory and Criticism. The Soviet Age and Beyond*, edited by Evgeny Dobrenko and Galin Tihanov, 90–108. Pittsburgh: University of Pittsburgh Press, 2011.

Günther, Hans. "Zeleznaja Garmonija (Gosudarstvo Kak Total'noe Proizvedenie Iskusstva)" ["The Iron Harmony (a State as an All-encompassing Work of Art)"]. *Voprosy Literatury* 1 (1992): 27–41.

Hake, Sabina. "Political Affects: Antifascism and the Second World War in Frank Beyer and Konrad Wolf." In *Screening War: Perspectives on German Suffering*, edited by Paul Cooke and Marc Silberman, 102–122. Rochester, NY: Camden House, 2010.

Harding, Jennifer Riddle. "A Mind Enslaved? The Interaction of Metaphor, Cognitive Distance, and Narrative Framing in Chesnutt's 'Dave's Neckliss.'" *Style* 42, no. 4 (Winter 2008): 425–447.

Harris, Paul L. "Conversation, Pretense, and Theory of Mind." In *Why Language Matters for Theory of Mind*, edited by Janet Wilde Astington and Jodie A. Baird, 70–83. Oxford: Oxford University Press, 2005.

Harris, Paul L., Marc de Rosnay, and Francisco Pons. "Language and Children's Understanding of Mental States." *Current Directions in Psychological Science* 14, no. 2 (2005): 69–73.

Harris, Paul L., Marc de Rosnay, and Samuel Ronfard. "The Mysterious Emotional Life of Little Red Riding Hood." In *Children and Emotion: New Insights into Developmental Affective Sciences*, edited by Kristin H. Lagattuta, 106–118. Basel: Karger, 2014.

Hayden, George A. "The Beginning of the End: The Fall of the Han and the Opening of Three Kingdoms." In *Three Kingdoms and Chinese Culture*, edited by Kimberly Besio and Constantine Tung, 43–51. Albany: State University of New York Press, 2007.

Hegel, Robert E. "Traditional Chinese Fiction—The State of the Field." *Journal of Asian Studies* 53, no. 2 (May 1994): 394–426.

Heliodorus. *An Ethiopian Romance*. Translated by Moses Hadas. Philadelphia: University of Pennsylvania Press, 1957.

Henry, Nancy. *The Life of George Eliot: A Critical Biography*. Chichester, UK: Wiley-Blackwell, 2012.

Herman, David. "Multimodal Storytelling and Identity Construction in Graphic Narratives." In *Telling Stories: Language, Narrative, and Social Life*, edited by Deborah Schiffrin, Anna de Fina, and Anastasia Nylund, 195–208. Washington, DC: Georgetown University Press, 2010.

Herman, David. *Storytelling and the Sciences of the Mind*. Cambridge, MA: MIT Press, 2013.

Heyman, Gail D. "Children's Reasoning about Deception." In *Trust and Skepticism: Children's Selective Learning from Testimony*, edited by Elizabeth J. Robinson and Shiri Einav, 83–94. New York: Psychology Press, 2014.

Hickok, Gregory. *The Myth of Mirror Neurons: The Real Neuroscience of Communication and Cognition*. New York: Norton, 2014.

Highsmith, Patricia. Afterword to *Selected Novels and Short Stories*, edited by Joan Schenkar, 577–580. New York: Norton, 1993.

Highsmith, Patricia. *The Price of Salt*. In *Selected Novels and Short Stories*, edited by Joan Schenkar, 337–575. New York: Norton, 1993.

Hill, Pamela Smith. "Introduction: 'Will It Come to Anything?': The Story of *Pioneer Girl*." In *Pioneer Girl: The Annotated Autobiography*, edited by Pamela Smith Hill, xv–lix. Pierre: South Dakota Historical Society Press, 2014.

Hirschfeld, Lawrence A. "The Myth of Mentalizing and the Primacy of Folk Sociology." In *Navigating the Social World: What Infants and Other Species Can Teach Us*, edited by Mahzarin Banaji and Susan A. Gelman, 101–106. New York: Oxford University Press, 2014. doi: 10.1093/acprof:oso/9780199890712.003.0019.

Hodne, Lasse. "Memling's Portraits of Christ: A Cognitive Approach." *Acta ad archaeologiam et artium historiam pertinentia* 3, no. 17 (2019): 245–258. doi: 10.5617/acta.7810.

Hoekstra, Hopi E. "From Darwin to DNA: The Genetic Basis of Color Adaptation." In *In the Light of Evolution: Essays from the Laboratory and the Field*, edited by Jonathan B. Losos, 277–295. Greenville Village, CO: Roberts, 2010.

Hogan, Patrick Colm. *Sexual Identities*. New York: Oxford University Press, 2019.

Hogan, Patrick Colm. "Literary Universals." In *Introduction to Cognitive Cultural Studies*, edited by Lisa Zunshine, 37–60. Baltimore: The Johns Hopkins University Press, 2010.

Holquist, Michael. *Dialogism: Bakhtin and his World*. London: Routledge, 1990.

Homer. *The Odyssey. Rendered into English Prose for the Use by Those Who Cannot Read the Original by Samuel Butler.* London, New York, and Bombay: Longmans, Green, and Co, 1900.

Howard, Michael. *The First World War: A Very Short Introduction.* Oxford: Oxford University Press, 2002.

Hughes, Claire, Naomi White, and Rosie Ensor. "How Does Talking about Thoughts, Desires, and Feelings Foster Children's Socio-cognitive Development? Mediators, Moderators and Implications for Intervention." In *Children and Emotion: New Insights into Developmental Affective Science*, edited by Kristin Hansen Lagattuta, 94–105. Basel: Karger, 2014.

Iliopoulou, Evgenia. *Because of You: Understanding Second-Person Storytelling.* Bielefeld, Germany: Transcript-Verlag, 2019.

Ionescu, Andrei. "A Manifesto against Failures of Understanding: Ian McEwan's *Atonement.*" *Critique: Studies in Contemporary Fiction*, August 1, 2017, 1–19.

Irvine, Judith T. "Language Ideology." *Oxford Bibliographies.* New York: Oxford University Press, 2012. Accessed March 27, 2019. http://www.oxfordbibliographies.com /view/document/obo-9780199766567/obo-9780199766567-0012.xml.

Irving, John. *The 158-Pound Marriage.* New York: Ballantine, 1997.

Jackson, Shirley. "The Beautiful Stranger." In *Dark Tales*, 73–80. New York: Penguin, 2017.

Jakobson, Roman. "On Realism in Art." In *Language and Literature*, edited by Krystyna Pomorska and Steven Rudy, 19–27. Cambridge, MA: Harvard University Press, 1987.

Jansson, Tove. *Moomin Falls in Love.* New York: Farrar, Straus and Giroux, 2013.

Johnson, Barbara. "Teaching Deconstructively." In *Writing and Reading Differently*, edited by G. Douglas Atkins and Michael L. Johnson, 140–148. Lawrence: University Press of Kansas, 1986.

Johnson, Mark. *Embodied Mind, Meaning, and Reason: How Our Bodies Give Rise to Understanding.* Chicago: University of Chicago Press, 2017.

Johnson, Mark. "The Embodiment of Language." In *The Oxford Handbook of 4E Cognition*, edited by Albert Newen, Leon De Bruin, and Shaun Gallagher, 623–640. Oxford: Oxford University Press, 2020.

Kanske, Philipp, Anne Bockler, Fynn-Mathis Trautwein, and Tania Singer. "Dissecting the Social Brain: Introducing the EmpaToM to Reveal Distinct Neural Networks and Brain–Behavior Relations for Empathy and Theory of Mind." *NeuroImage* 122, no. 15 (November 2015): 6–19.

Karamzin, N. M. "Bednaya Liza." In *Izbrannye Sochinenia*, vol. 1, 605–621. Moscow: Khudozhestvennaya Literatura, 1964.

Keane, Webb. *Ethical Life: Its Natural and Social Histories*. Princeton, NJ: Princeton University Press, 2016.

Keane, Webb. "Others, Other Minds, and Others' Theories of Other Minds: An Afterword on the Psychology and Politics of Opacity Claims." *Anthropological Quarterly* 81, no. 2 (Spring 2008): 473–482.

Keen, Suzanne. *Empathy and the Novel*. New York: Oxford University Press, 2007.

Keen, Suzanne. *Thomas Hardy's Brains: Psychology, Neurology, and Hardy's Imagination*. Columbus: Ohio State University Press, 2014.

Kidd, David Comer, and Emanuele Castano. "Panero et al. (2016): Failure to Replicate Methods Caused the Failure to Replicate Results." *Journal of Personality and Social Psychology* 112 (2017): e1–e4. doi:10.1037/pspa0000072.

Kidd, David Comer, and Emanuele Castano. "Reading Literary Fiction and Theory of Mind: Three Preregistered Replications and Extensions of Kidd and Castano (2013)." *Social Psychological and Personality Science*, June 20, 2018, 1–10. doi: 10.1177/1948550618775410journals.sagepub.com/home/spp.

Kidd, David Comer, and Emanuele Castano. "Reading Literary Fiction Can Improve Theory of Mind." *Nature Human Behaviour* 2 (September 2018): 604. doi: https://doi.org/10.1038/s41562-018-0408-2 .

Kinney, Jeff. *Diary of a Wimpy Kid: Rodrick Rules*. New York: Amulet Books, 2008.

Knoepflmacher, Ulrich C., and Mitzi Myers. "From the Editors: 'Cross-Writing' and the Reconceptualizing of Children's Literary Studies." *Children's Literature* 25 (1997): vii–xvi.

Kompa, Nikola A. "Language and Embodiment—Or the Cognitive Benefits of Abstract Representations." *Mind and Language* 26 (2021): 27–47. doi: 10.1111/mila.12266

Kramnick, Jonathan, and Anahid Nersessian. "Form and Explanation." *Critical Inquiry* 43 (Spring 2017): 650–669.

Kukkonen, Karin. "Does Cognition Translate? Predictions, Plot, and World Literature." *Poetics Today* 41, no. 2 (2020): 243–259.

Kukkonen, Karin. *4E Cognition and Eighteenth-Century Fiction: How the Novel Found Its Feet*. Oxford: Oxford University Press, 2019.

Kukkonen, Karin. *Probability Designs: Literature and Predictive Processing*. Oxford: Oxford University Press, 2020.

Kulpa, Kathryn. Review of *Ninety-Nine Stories of God*, by Joy Williams. *Cleaver: Philadelphia's International Literary Magazine*, September 22, 2016. https://www.cleavermagazine.com/ninety-nine-stories-of-god-by-joy-williams-reviewed-by-kathryn-kulpa/.

Kümmerling-Meibauer, Bettina. "Emotional Connection: Representation of Emotions in Young Adult Literature." In *Contemporary Adolescent Literature and Culture:*

The Emergent Adult, edited by Mary Hilton and Maria Nikolajeva, 127–138. Abingdon, UK: Ashgate, 2012.

Kümmerling-Meibauer, Bettina. "Metalinguistic Awareness and the Child's Developing Concept of Irony: The Relationship between Pictures and Text in Ironic Picture Books." *Lion and the Unicorn* 23, no. 2 (April 1999): 157–183.

Kümmerling-Meibauer, Bettina, and Jörg Meibauer. "Early-Concept Books: Acquiring Nominal and Verbal Concepts." In *Emergent Literary: Children's Books from 0 to 3*, edited by Bettina Kümmerling-Meibauer, 91–114. Amsterdam: John Benjamins, 2011.

Kuritzyn, Fedor. *The Tale of Dracula*. ca. 1490. http://old-ru.ru/06-7.html.

Kuzmičová, Anezka. "Consciousness." In *Further Reading*, edited by Matthew Rubery and Leah Price, 271–281. Oxford: Oxford University Press, 2020.

Kuzmičová, Anezka and Katalin Bálint. "Personal Relevance in Story Reading: A Research Review." *Poetics Today* 40 no. 3 (2019): 429–451

Landau, Bonnie. "11 Reasons a Child Cannot Look You in the Eyes." *Special Mom Advocate*, March 2, 2018. https://www.specialmomadvocate.com/11-reasons-child-cannot-look-eyes/.

Landy, Joshua. *How to Do Things with Fictions*. Oxford: Oxford University Press, 2012.

Lantos, John D. "Reconsidering Action: Day-to-Day Ethics in the Work of Medicine." In *Stories Matter: The Role of Narrative in Medical Ethics*, edited by Rita Charon and Martha Montello, 154–159. London: Routledge, 2002.

Lee, Haiyan. "Chinese Feelings: Notes on a Ritual Theory of Emotion." *Wenshan Review of Literature and Culture* 9, no. 2 (June 2016): 1–37.

Lee, Haiyan. "Measuring the Stomach of a Gentleman with the Heart-Mind of a Pipsqueak: On the Ubiquity and Utility of Theory of Mind in Literature, Mostly." *Poetics Today* 41, no. 2 (2020): 205–222.

Lee, Haiyan. "Response to the Panel on Cognitive Approaches to Chinese Literature." Paper presented at the Modern Language Association, New York, January 2018.

Lee, Haiyan. "Society Must Be Defended: Chinese Spy Thrillers and the Enchantment of *Arcana Imperii*." Unpublished paper.

Lee, Haiyan. *The Stranger and the Chinese Moral Imagination*. Stanford, CA: Stanford University Press, 2014.

Lee, Haiyan. "When Nothing Is True, Everything Is Possible: On Truth and Power by Way of Socialist Realism." *PMLA* 134, no. 5 (2019): 1157–1164.

Leeder, Karen. Introduction to *Rereading East Germany: The Literature and Film of the GDR*, edited by Karen Leeder, 1–7. Cambridge: Cambridge University Press, 2015.

Leff, Gordon. "The *Trivium* and the Three Philosophies." In *A History of the University in Europe*, vol. 1, *Universities in the Middle Ages*, edited by Hilde de Ridder-Symoens, 307–336. Cambridge: Cambridge University Press, 1992.

Lepowsky, Maria. "Personhood, Empathy, and 'the Native's Point of View.'" In *The Anthropology of Empathy: Experiencing the Lives of Others in the Pacific Societies*, edited by Douglas W. Hollan and C. Jason Throop, 43–65. New York: Berghahn Books, 2011.

Lermontov, Mikhail. *A Hero of Our Time. A Novel.* translated by Vladimir Nabokov and Dmitri Nabokov. Garden City, N.Y.: Doubleday, 1958.

Lerner, Ben. *Leaving the Atocha Station.* Minneapolis: Coffee House, 2011.

Levinas, Emmanuel. *Totality and Infinity: An Essay on Exteriority.* Translated by Alphonso Lingis. Pittsburgh: Duquesne University Press, 1969.

Levine, Caroline. *Forms: Whole, Rhythm, Hierarchy, Network.* Princeton, NJ: Princeton University Press, 2017.

Li, Wanqing, Xiaoqin Mai, and Chao Liu. "The Default Mode Network and Social Understanding of Others: What Do Brain Connectivity Studies Tell Us." *Frontiers in Human Neuroscience* 8, no. 74 (2014). Published online February 24, 2014.

Link, Perry. "The Wonderfully Elusive Chinese Novel." *New York Review of Books*, April 23, 2015. https://www.nybooks.com/articles/2015/04/23/wonderfully-elusive -chinese-novel/.

Lipson, Eden Ross. *The New York Times Parent's Guide to the Best Books for Children.* New York: Three Rivers, 2000.

Lockington, Mariama J. *For Black Girls like Me.* New York: Farrar, Straus and Giroux, 2019.

Lorant, Laurie Robertson. *Melville: A Biography.* New York: Clarkson Potter, 1996.

Lu, Tina. "Interiority in *Jinpingmei cihua*." In *Approaches to Teaching "The Plum in the Golden Vase,"* edited by Andrew Schonebaum. New York: MLA, 2022.

Lu Xun [Lu Hsun]. *A Brief History of Chinese Fiction.* 1959. Translated by Yang Hsien-Yi and Gladys Yang. Introduction by Moss Roberts. Beijing: Foreign Language Press, 2014.

Lu Xun [Lu Hsun]. *A Madman's Diary: English and Chinese Bilingual Edition.* Translated by Paul Meighan. CreateSpace, 2014.

Luhrmann, Tanya Marie. "*Theory of Mind in the Pacific: Reasoning across Cultures*, edited by Jürg Wassmann, Birgit Träuble and Joachim Funke." *Anthropological Forum* 25, no. 4 (2015): 442–444.

Luhrmann, Tanya Marie. "Toward an Anthropological Theory of Mind: Overview." *Suomen Antropologi: Journal of the Finnish Anthropological Society* 36, no. 4 (2011): 5–13.

Luo Guanzhong. *The Three Kingdoms*. Vol. 1, *The Sacred Oath: The Epic Chinese Tale of Loyalty and War in a Dynamic New Translation*. Edited by Ronald C. Iverson. Translated by Yu Sumei. North Clarendon, VT: Tuttle, 2014.

Luo, Manling. *Literati Storytelling in Late Medieval China*. Seattle: University of Washington Press, 2015.

Ma, Y. W., and Joseph S. M. Lau, eds., *Traditional Chinese Stories: Themes and Variations*. New York: Columbia University Press, 1978.

Maiping, Chen. "The Intertextual Reading of Chinese Literature with Mo Yan's Works as Examples." *Chinese Literature Today* 5, no. 1 (2015): 34–36.

Mäkelä, Maria. "Possible Minds. Constructing – and Reading – Another Consciousness as Fiction." In *FREE Language INDIRECT Translation DISCOURSE Narratology*, edited by Pekka Tammi and Hannu Tommola, 231–260. Tampere: Tampere University Press, 2006.

Mair, Victor H. "Review of *Traditional Chinese Stories: Themes and Variations by Y. W. Ma, Joseph S. M. Lau*." *Harvard Journal of Asiatic Studies* 39, no. 2 (December 1979): 461–469.

Mar, Raymond A., Keith Oatley, Jacob Hirsh, Jennifer dela Paz, and Jordan B. Peterson. "Bookworms versus Nerds: Exposure to Fiction versus Non-fiction, Divergent Associations with Social Ability, and the Simulation of Fictional Social Worlds." *Journal of Research in Personality* 40 (2006): 694–712.

Mar, Raymond A., Keith Oatley, and Jordan B. Peterson. "Exploring the Link between Reading Fiction and Empathy: Ruling Out Individual Differences and Examining Outcomes." *Communications* 34, no. 4 (2009). doi: https://doi.org/10.1515/COMM.2009.025.

Mar, Raymond A., Jennifer L. Tackett, and Chris Moore. "Exposure to Media and Theory-of-Mind Development in Preschoolers." *Cognitive Development* 25, no. 1 (January–March 2010): 69–78.

Martens, Lorna. "Corporeality, Materiality, and Unnamed Emotions in Rilke's Dinggedichte." In *Feelings Materialized: Emotions, Bodies, and Things in Germany, 1500–1950*, edited by Heikki Lempa, Derek Hillard, and Russell Spinney, 235–251. New York: Berghahn Books, 2020.

Martins, Mauricio D., and W. Tecumseh Fitch. "Do We Represent Intentional Action as Recursively Embedded? The Answer Must Be Empirical. A Comment on Vicari and Adenzato (2014)." *Consciousness and Cognition* 38 (December 15, 2015): 16–21

Martinsen, Deborah A. *Surprised by Shame: Dostoevsky's Liars and Narrative Exposure*. Columbus: Ohio State University Press, 2003.

Mascaro, Olivier, and Olivier Molin. "Gullible's Travel: How Honest and Trustful Children Become Vigilant Communicators." In *Trust and Skepticism: Children's Selective Learning from Testimony*, edited by Elizabeth L. Robinson Shir Einav, 69–82. London: Psychology Press, 2014.

Mascaro, Olivier, and Dan Sperber. "The Moral, Epistemic, and Mindreading Components of Children's Vigilance towards Deception." *Cognition* 112 (2009): 367–380.

Mason, Malia F., Elizabeth P. Tatkow, and C. Neil Macrae. "The Look of Love: Gaze Shifts and Person Perception." *Psychological Science* 16 (2005): 236–239.

McCarthy, Cormac. *Blood Meridian: Or the Evening Redness in the West*. New York: Vintage, 1985.

McCrea, Brian. *Frances Burney and Narrative Prior to Ideology*. Newark: University of Delaware Press, 2013.

McHale, Brian. "Revisiting Realisms; or, WWJD (What Would Jakobson Do?)" *Journal of the Midwest Modern Language Association* 41, no. 2 (Fall 2008): 6–17. http://www.jstor.com/stable/20464270.

McKinnon, Margaret C., and Morris Moscovitch. "Domain-General Contributions to Social Reasoning: Theory of Mind and Deontic Reasoning Re-explored." *Cognition* 102 (2007): 179–218.

McNamara, Rita Anne, Aiyana K.Willard, Ara Norenzayan, and Joseph Henrich. "Weighing Outcome vs. Intent across Societies: How Cultural Models of Mind Shape Moral Reasoning." *Cognition* 182 (2019): 95–108. https://doi.org/10.1016/j.cognition.2018.09.008.

Mercier, Hugo, and Dan Sperber. *The Enigma of Reason*. Cambridge, MA: Harvard University Press, 2017.

Mersereau, John, Jr. "The Nineteenth Century: Romanticism, 1820–40." In *The Cambridge History of Russian Literature*, edited by Charles A. Moser, 136–188. Cambridge: Cambridge University Press, 1989.

Miall, David S. "Science in the Perspective of Literariness." *Scientific Study of Literature* 8 (2011): 8–14.

Miller, Nick. "Gruffalo Creator Finds Room for Girls—But They Don't Have to be Feisty." The Sydney Morning Herald. February 16, 2020. https://www.smh.com.au/culture/books/gruffalo-creator-finds-room-for-girls-but-they-don-t-have-to-be-feisty-20200214-p5411n.html.

Miller, Patricia H., Frank S. Kessel, and John H. Flavell. "Thinking about People Thinking about People Thinking about . . . : A Study of Social Cognitive Development." *Child Development* 41, no. 3 (September 1970): 613–623. doi: 10.2307/1127211.

Miller, Sara F. *Pooh's Honey Trouble*. Glendale, CA: Disney, 2012.

Miller, William Ian. *Losing It*. New Haven, CT: Yale University Press, 2011.

Milligan, Karen, Janet Wilde Astington, and Lisa Ain Dack. "Language and Theory of Mind: Meta-analysis of the Relation between Language Ability and False-Belief Understanding." *Child Development*, 78, no. 2 (2007): 622–646.

Milne, Alan Alexander. *The Complete Tales of Winnie-the-Pooh*. New York: Penguin, 1996.

Minchin, Elizabeth. "The Cognition of Deception: Falsehoods in Homer's Odyssey and Their Audiences." In *The Routledge Handbook of Classics and Cognitive Theory*, edited by Peter Meineck, William Michael Short, and Jennifer Devereaux, 109–138. London: Routledge, 2018.

Monaco, Loretta. "Review of DiaoChan: The Rise of the Courtesan." *LondonTheatre1* May 12, 2016. https://www.londontheatre1.com/reviews/review-of-diaochan-the -rise-of-the-courtesan/.

Morris, Marcia A. *The Literature of Roguery in Seventeenth- and Early Eighteenth-Century Russia*. Evanston, IL: Northwestern University Press, 2000.

Morris, Marcia A. "Russia: The Picaresque Repackaged." In *The Picaresque Novel in Western Literature: From the Sixteenth Century to the Neopicaresque*, edited by Garrido Ardila, 200–223. Cambridge: Cambridge University Press, 2015.

Moshfegh, Ottessa. Foreword to *Dark Tales*, by Shirley Jackson, edited by Ottessa Moshfegh, vii–x. New York: Penguin, 2016.

Muira, Kayo and Motoko Koike, "Judgment, Interpretation and Impression of Gaze Direction in an Ukiyo-e Picture." *Japanese Psychological Research* 45 no. 4 (2003): 209 – 220.

Munro, George E. "Finance and Credit in the Eighteenth-Century Russian Economy." *Jahrbücher für Geschichte Osteuropas* 45, no. 4 (1997): 552–560.

Murasaki Shikibu. *The Tale of Genji*. Translated by Edward G. Seidensticker. Gardners Books, 1992.

Murasaki Shikibu. *The Tale of Genji*. Translated by Royall Tyler. New York: Penguin, 2002.

Murasaki Shikibu. *The Tale of Genji*. Translated by Arthur Waley. North Clarendon, VT: Tuttle, 2010.

Nabokov, Vladimir. *Lectures on Russian Literature*. Edited by Fredson Bowers. New York: Harcourt Brace Jovanovich, 1981.

Nabokov, Vladimir. "Translator's Foreword." In *A Hero of Our Time*, by Mihail Lermontov, v–xv. New York: Doubleday Anchor Books, 1958.

Nadel, Alan. *Invisible Criticism: Ralph Ellison and the American Canon*. Iowa City: University of Iowa Press, 1988.

Newen, Albert, Leon De Bruin, and Shaun Gallagher. "4E Cognition: Historical Roots, Key Concepts, and Central Issues." In *The Oxford Handbook of 4E Cognition*, edited by Albert Newen, Leon De Bruin, and Shaun Gallagher, 3–15. Oxford: Oxford University Press, 2020.

Nielsen, Henrik Skov, James Phelan, and Richard Walsh. "Ten Theses about Fictionality." *Narrative* 23, no. 1 (2015): 61–73. doi: 10.1353/nar.2015.0005.

Nikolajeva, Maria. *Children's Literature Comes of Age: Toward a New Aesthetic*. New York: Garland, 1996.

Nikolajeva, Maria. "'The Penguin Looked Sad': Picturebooks, Empathy and Theory of Mind." In *Picturebooks: Representation and Narration*, edited by Bettina Kümmerling-Meibauer, 121–137. New York: Routledge, 2011.

Nizami. *The Story of Layla and Majnun*. Translated by Rudolf Gelpke. 2nd ed. Medford, OR: Omega, 1996.

Noel, Anne-Sophie. "What Do We Actually See on Stage? A Cognitive Approach to the Interaction of Visual and Aural Effects in the Performance of Greek Tragedy." In *The Routledge Handbook of Classics and Cognitive Theory*, edited by Peter Meineck, William Michael Short, and Jennifer Devereaux, 297–309. London: Routledge, 2018.

Nussbaum, Martha C. "'Finely Aware and Richly Responsible': Moral Attention and the Moral Task of Literature." *Journal of Philosophy* 82 (1985): 516–529. doi: 10.2307/2026358.

Nussbaum, Martha C. "Literary Imagination in Public Life." *New Literary History* 22 (1991): 877–910. doi: 10.2307/469070.

Ochs, Elinor. "Clarification and Culture." In *Meaning, Form, and Use in Context: Linguistic Applications*, edited by Deborah Schiffrin, 325–341. Washington, DC: Georgetown University Press, 1984.

Ochs, Elinor, and Bambi B. Schieffelin. "Language Acquisition and Socialization: Three Developmental Stories and Their Implications." In *Culture Theory: Essays on Mind, Self, and Emotion*, edited by Richard A. Shweder and Robert A. LeVine, 276–320. Cambridge: Cambridge University Press, 1984.

Oforlea, Aaron Ngozi. "The Dilemma of the African American Detective in Walter Mosley's Devil in a Blue Dress." Manuscript under review.

Olney, James. *Memory and Narrative: The Weave of Life-Writing*. Chicago: University of Chicago Press, 1998.

"100 Must-Reads." *DW*. Accessed 06/09/2021. https://www.dw.com/en/top-stories /100-must-reads/s-43415865.

Onishi, Kristine H., and Renée Baillargeon. "Do 15-Months-Old Infants Understand False Beliefs?" *Science* 308 (2005): 255–258.

O'Sullivan, Emer. "Comparative Children's Literature." *PMLA* 126, no. 1 (2011): 189–196.

Owen, Stephen. *An Anthology of Chinese Literature: Beginnings to 1911*. New York: Norton, 1996.

Owen, Stephen. *The End of the Chinese "Middle Ages": Essays in Mid-Tang Literary Culture*. Stanford, CA: Stanford University Press, 1996.

Palmer, Alan. *Fictional Minds*. Lincoln: University of Nebraska Press, 2004.

Palmer, Alan. *Social Minds in the Novel*. Columbus: Ohio State University Press, 2010.

Palmer, Alan. "Storyworlds and Groups." In *Introduction to Cognitive Cultural Studies*, edited by Lisa Zunshine, 176–192. Baltimore: Johns Hopkins University Press, 2010.

Panero, Maria Eugenia, Deena Skolnick Weisberg, Jessica Black, Thalia R. Goldstein, Jennifer L. Barnes, Hiram Brownell, and Ellen Winner. "Does Reading a Single Passage of Literary Fiction Really Improve Theory of Mind? An Attempt at Replication." *Journal of Personality and Social Psychology* 111, no. 5 (November 2016): e46–e54.

Parker-Pope, Tara. "Behind the 'Wimpy Kid' Phenomenon." *New York Times*, October 12, 2009. http://www.nytimes.com/2009/10/13/health/13well.html.

Pascal, Roy. *The Dual Voice: Free Indirect Speech and Its Functioning in the Nineteenth-Century European Novel*. Manchester: Manchester University Press, 1977.

Paul, Georgina. "Gender in GDR Literature." In *Rereading East Germany: The Literature and Film of the GDR*, edited by Karen Leeder, 106–125. Cambridge: Cambridge University Press, 2015.

Peskin, Joan, and Janet Wilde Astington. "The Effects of Adding Metacognitive Language to Story Texts." *Cognitive Development* 19 (2004): 253–273.

Petrone, Karen. *Life Has Become More Joyous, Comrades: Celebrations in the Time of Stalin*. Bloomington: Indiana University Press, 2000.

Petronius, Arbiter. *The Satyricon*. Translated by John Malcolm Mitchell. Edinburgh: Edinburgh University Press, 1923.

Phelan, James, and Peter J. Rabinowitz. "Authors, Narrators, Narration." In *Narrative Theory: Core Concepts and Critical Debates*, by David Herman, James Phelan, Peter J. Rabinowitz, Brian Richardson, and Robyn Warhol, 29–38. Columbus: Ohio State University Press, 2012.

Phelan, James, and Peter J. Rabinowitz. "Narrative Values, Aesthetic Values." In *Narrative Theory: Core Concepts and Critical Debates*, by David Herman, James Phelan,

Peter J. Rabinowitz, Brian Richardson, and Robyn Warhol, 160–164. Columbus: Ohio State University Press, 2012.

Phillips, Natalie M. *Distraction: Problems of Attention in Eighteenth-Century Literature*. Baltimore: Johns Hopkins University Press, 2016.

Phillips, Natalie M. "Literary Neuroscience and History of Mind: An Interdisciplinary fMRI Study of Attention and Jane Austen." In *The Oxford Handbook of Cognitive Literary Studies*, edited by Lisa Zunshine, 55–81. New York: Oxford University Press, 2015.

Pittard, Hannah. *Listen to Me*. New York: Houghton Mifflin Harcourt, 2016.

Plaks, Andrew H. "Full-Length Hsiao-shuo and the Western Novel: A Generic Reappraisal." *New Asia Academic Bulletin* 1 (1978): 163–176.

Plaks, Andrew H. "The Novel in Premodern China." In *The Novel*, vol. 1, *History, Geography, and Culture*, edited by Franco Moretti, 181–213. Princeton, NJ: Princeton University Press, 2006.

Plaks, Andrew H. *The Four Masterworks of the Ming Novel: Ssu ta ch'i-shu*. Princeton, NJ: Princeton University Press, 2016.

Pocock, J. G. A. *Virtue, Commerce, and History: Essays on Political Thought and History, Chiefly in the Eighteenth Century*. Cambridge: Cambridge University Press, 1985.

Polvinen, Merja. "Enactive Perception and Fictional Worlds." In *The Cognitive Humanities: Embodied Mind in Literature and Culture*, edited by Peter Garratt, 19–34. London: Palgrave Macmillan, 2016.

Polvinen, Merja. "Sense-Making and Wonder: An Enactive Approach to Narrative Form in Speculative Fiction." In *The Edinburgh Companion to Contemporary Narrative Theories*, edited by Zara Dinnen and Robyn Warhol, 67–80. Edinburgh: Edinburgh University Press, 2018.

Popova, Yanna B. *Stories, Meaning, and Experience*. New York: Routledge, 2015.

Premack, David G., and Guy Woodruff. "Does the Chimpanzee Have a Theory of Mind?" *Behavioral and Brain Sciences* 1, no. 4 (1978): 515–526. doi: 10.1017/SC140525X00076512.

Proust, Marcel. *Remembrance of Things Past*. Vol. 1. Translated by C. K. Scott Moncrieff. Hertfordshire, UK: Wordsworth Editions, 2006.

Pushkin, Alexander. *Eugene Onegin, Revised Edition: A Novel in Verse by Alexandr Pushkin, Translated from Russian, with a Commentary, by Vladimir Nabokov*. Vol. 1. Bollingen Series 77. Princeton, NJ: Princeton University Press, 1975.

Pushkin, Alexander. *Novels, Tales, Journeys: The Complete Prose of Alexander Pushkin*. Translated by Richard Pevear and Larissa Volokhonsky. New York: Knopf, 2016.

Rey, H. A. *Curious George at the Zoo*. Boston: Houghton Mifflin Harcourt, 2007.

Richardson, Alan. *Literature, Education, and Romanticism: Reading as Social Practice, 1780–1832*. Cambridge: Cambridge University Press, 1994.

Richardson, Alan. *The Neural Sublime: Cognitive Theories and Romantic Texts*. Baltimore: Johns Hopkins University Press, 2010.

Richardson, Alan. "Studies in Literature and Cognition: A Field Map." In *The Work of Fiction: Cognition, Culture, and Complexity*, edited by Alan Richardson and Ellen Spolsky, 1–29. Aldershot, UK: Ashgate, 2004.

Richardson, Samuel. *Clarissa; or, The History of a Young Lady*. Edited by Angus Ross. Penguin Books, 1985.

Riftin, Boris. *От Мифа к Роману: Эволюция Изображения Персонажа в Китайской Литературе [From myth to the serial novel: The evolution of the image of character in Chinese literature]*. Moscow: Наука, 1979.

Riva, Federica, Chantal Triscoli, Claus Lamm, Andrea Carnaghi, and Giorgia Silani. "Emotional Egocentricity Bias across the Life-Span." *Frontiers in Aging Neuroscience* 8, no. 74 (2016). doi: 10.3389/fnagi.2016.00074.

Robbe-Grillet, Alain. *Two Novels by Robbe-Grillet: Jealousy and In the Labyrinth*. Translated by Richard Howard. New York: Black Cat Books by Grove Press, 1965.

Robbins, Joel, and Alan Rumsey. "Introduction: Cultural and Linguistic Anthropology and the Opacity of Other Minds," *Anthropological Quarterly* 81, no. 2 (2008): 407–420.

Rolston, David L. *Traditional Chinese Fiction and Fiction Commentary: Reading and Writing between the Lines*. Stanford, CA: Stanford University Press, 1997.

Rooney, Sally. *Conversations with Friends*. London: Hogarth, 2017.

Rosenberg, Alex. *How History Gets Things Wrong: The Neuroscience of Our Addiction to Stories*. Cambridge, MA: MIT Press, 2018.

Rosenman, Ellen Bayuk. "Rudeness, Slang, and Obscenity: Working-Class Politics in London Labour and the London Poor." In *Victorian Vulgarity: Taste in Verbal and Visual Culture*, edited by Susan David Bernstein and Elsie B. Michie, 55–70. Farnham, UK: Ashgate, 2009.

Rosnay, Marc de, Francisco Pons, Paul L. Harris, and Julian M. B. Morrell. "A Lag between Understanding False Belief and Emotion Attribution in Young Children: Relationships with Linguistic Ability and Mothers' Mental State Language." *British Journal of Developmental Psychology* 22 (2004): 197–218.

Roy, David Tod, trans. *The Plum in the Golden Vase, or, Chin P'ing Mei*. Vol. 1, *The Gathering*. Princeton, NJ: Princeton University Press, 1993.

Roy, David Tod, trans. *The Plum in the Golden Vase, or, Chin P'ing Mei*. Vol. 2, *The Rival*. Princeton, NJ: Princeton University Press, 2001.

Ruden, Sarah. "Translator's Preface." In *The Golden Ass*, by Apuleius, translated by Sarah Ruden, ix–xvi. New Haven, CT: Yale University Press, 2013.

Rumsey, Alan, "Empathy and Anthropology: An Afterword." In *The Anthropology of Empathy: Experiencing the Lives of Others in the Pacific Societies*, edited by Douglas W. Hollan and C. Jason Throop, 215–224. New York: Berghahn Books, 2011.

Sabbagh, Mark A., and Dare Baldwin. "Understanding the Role of Communicative Intentions in Word Learning." In *Joint Attention: Communication and Other Minds*, edited by Naomi Eilan, Christoph Hoerl, Teresa McCormack, and Johannes Roessler, 165–184. Oxford: Oxford University Press, 2005.

Saggini, Francesca. *Backstage in the Novel: Frances Burney and the Theater Arts*. Charlottesville: University of Virginia Press, 2012.

Santos, Henri Carlo, Igor Grossmann, and Michael E. W. Varnum. "Class, Cognition and Cultural Change in Social Class." *PsyArxiv Preprints*, July 2, 2018. https://psyarxiv.com/92smf/.

Sartre, Jean-Paul. *Being and Nothingness*. Translated by Hazel E. Barnes. New York: Philosophical Library, 1956.

Saussy, Haun. "Comparative Literature and Translation." In *Introducing Comparative Literature: New Trends and Applications*, by Cesar Domingues, Haun Saussy, and Dario Villanueva, 78–87. London: Routledge, 2015.

Savarese, Ralph James. *See It Feelingly: Classic Novels, Autistic Readers, and the Schooling of a No-Good English Professor*. Durham, NC: Durham University Press, 2018.

Savarese, Ralph James, and Lisa Zunshine. "The Critic as Neurocosmopolite; or, What Cognitive Approaches to Literature Can Learn from Disability Studies: Lisa Zunshine in Conversation with Ralph James Savarese." *Narrative* 22, no. 1 (2014): 17–44.

Saxe, Rebecca. "The New Puzzle of Theory of Mind Development." In *Navigating the Social World: What Infants, Children, and Other Species Can Teach Us*, edited by Mahzarin R. Banaji and Susan A. Gelman, 107–112. New York: Oxford University Press, 2013.

Saxe, Rebecca. "The Right Temporo-Parietal Junction: A Specific Brain Region for Thinking about Thoughts." In *Handbook of Theory of Mind*, edited by Alan Leslie and Tamsin German. London: Psychology Press, 2010. https://saxelab.mit.edu/sites/default/files/publications/Saxe_RTPJChapter.pdf.

Saxe, Rebecca, and Nancy Kanwisher. "People Thinking about Thinking People: The Role of the Temporo-parietal Junction in 'Theory of Mind.'" *NeuroImage* 19 (2003): 1335–1842.

Saxe, Rebecca, and Lindsey J. Powell. "It's the Thought That Counts: Specific Brain Regions for One Component of Theory of Mind." *Psychology Science* 17 (2006): 692–699.

Scarry, Elaine. *Dreaming by the Book*. New York: Farrar, Straus and Giroux, 2013.

Scarry, Elaine. *Naming Thy Name: Cross Talk in Shakespeare's Sonnets*. New York: Farrar, Straus and Giroux, 2016.

Schacter, Daniel L. *Searching for Memory: The Brain, the Mind, and the Past*. New York: Basic Books, 1996.

Schenkar, Joan. *The Talented Miss Highsmith: The Secret Life and Serious Art of Patricia Highsmith*. New York: St. Martin's, 2009.

Schieffelin, Bambi B. "Found in Translation: Reflexive Language across Time and Texts in Bosavi, Papua New Guinea." In *Consequences of Contact: Language Ideologies and Sociocultural Transformation in Pacific Societies*, edited by Miki Makihara and Bambi B. Schieffelin, 140–165. New York: Oxford University Press, 2007.

Schieffelin, Bambi B. *The Give and Take of Everyday Life: Language Socialization of Kaluli Children*. Cambridge: Cambridge University Press, 1990.

Schieffelin, Bambi B. "Speaking Only Your Own Mind: Reflections on Confession, Gossip, and Intentionality in Bosavi (PNG)." *Anthropological Quarterly* 81, no. 2 (2008): 431–441.

Schieffelin, Bambi B. "Two Dukula Sulo: & One Dog." Told by Heina (Kaluli, Papua New Guinea, 1984). Unpublished manuscript, 2019.

Schieffelin, Edward L. *The Sorrow of the Lonely and the Burning of the Dancers*. New York: St. Martin, 1976.

Schiller, J. C. Friedrich von. *Schillers Sämmtliche Werke: Vollständige Ausgabe in Zwe Bänden*. Vol. 2. Philadelphia: Ferlag von J. Robler, 1880.

Schneider, Ralf. "The Cognitive Theory of Character Reception: An Updated Proposal." *Anglistik* 24, no. 2 (2013): 117–134.

Schonebaum, Andrew. Introduction to *Approaches to Teaching "The Story of the Stone,"* edited by Andrew Schonebaum and Tina Lu, 5–69. New York: Modern Language Association, 2012.

Scott, Mary. "*The Story of the Stone* and Its Antecedents." In *Approaches to Teaching "The Story of the Stone,"* edited by Andrew Schonebaum and Tina Lu, 266. New York: Modern Language Association, 2012.

Sedgwick, Eve Kosofsky. "Privilege of Unknowing: Diderot's *The Nun*." In *Tendencie*, 23–56. Durham, NC: Duke University Press, 1993.

Serman, Ilya. "The Eighteenth Century: Neoclassicism and the Enlightenmen:, 1730–90." In *The Cambridge History of Russian Literature*, edited by Charles A. Mose:, 45–91. Cambridge: Cambridge University Press, 1989.

Shakespeare, William. *Measure for Measure*. In *The Complete Works*, edited by G. B. Harrison. Fort Worth, TX: Harcourt Brace, 1980.

Shakespeare, William. *Twelfth Night*. In *The Complete Works*, edited by G. B. Harrison. Fort Worth, TX: Harcourt Brace, 1980.

Shen, Bojun. "Studies of *Three Kingdoms* in the New Century." In *Three Kingdoms and Chinese Culture*, edited by Kimberly Besio and Constantine Tung, 153–165. Albany: State University of New York Press, 2007.

Simon, Julien S., "Contextualizing Cognitive Approaches to Early Modern Spanish Literature." In *Cognitive Approaches to Early Modern Spanish Literature*, edited by Isabel Jaén and Julien J. Simon, 13–33. New York: Oxford University Press, 2016.

Sinnott, Susan, Michelle Jones, Yvette Lapierre, Jodi Evert, Chris Duke (Illustrator), Connie Russell (Illustrator), and Jamie Young (Illustrator). *Welcome to Kirsten's World, 1854: Growing Up in Pioneer America*. American Girl Publishing, 1999.

Skerry, Amy E., and Rebecca Saxe. "Neural Representations of Emotion Are Organized around Abstract Event Features." *Current Biology* 25, no. 15 (August 2015): 1945–1954.

Skolnick, Deena, and Paul Bloom. "What Does Batman Think about Sponge Bob? Children's Understanding of the Fantasy/Fantasy Distinction." *Cognition* 101 (2006): B9–B18.

Slingerland, Edward. "Cognitive Science and Religious Thought: The Case of Psychological Interiority in the *Analects*." In *Mental Culture: Towards a Cognitive Science of Religion*, edited by Dimitris Xygalatas and Lee McCorkle, 197–212. London: Acumen, 2013.

Smith, Valery. "The Meaning of Narration in Invisible Man." In *Ralph Ellison's Invisible Man: A Casebook*, edited by John F. Callahan, 189–220. Oxford: Oxford University Press, 2004.

Smith, Zadie. *On Beauty*. New York: Penguin, 2005.

Snodgrass, Sara E. "'Women's Intuition': The Effect of Subordinate Role on Interpersonal Sensitivity." *Journal of Personality and Social Psychology* 49 (1985): 146–155.

Solnit, Rebecca. "Nobody Knows." *Harper's Magazine*, March 2018. https://harpers.org/archive/2018/03/nobody-knows-3/.

Song of Igor's Campaign, The. Ca 1200 [*Слово о Полку Игореве*] Translated by Vladimir Nabokov. New York: Ardis Publishers, 2003.

Sontag, Susan. "Against Interpretation." In *The Critical Tradition: Classic Texts and Contemporary Trends*, edited by David H. Richter, 403–408. Boston: Bedford/St. Martin's, 2016.

Spark, Muriel. *The Girls of Slender Means*. New York: New Directions, 1998.

Spencer, Jane. "Evelina and Cecilia." In *The Cambridge Companion to Frances Burney* edited by Peter Sabor, 23–38. Cambridge: Cambridge University Press, 2007.

Sperber, Dan. *Rethinking Symbolism*. Translated by Alice L. Morton. Cambridge: Cambridge University Press, 1975.

Spivak, Gayatri Chakravorty. "The Women's Texts and a Critique of Imperialism." In *The Critical Tradition: Classic Texts and Contemporary Trends*, edited by David H. Richter, 1086–1099. Boston: Bedford/St. Martin's, 2016.

Spolsky, Ellen. *The Contracts of Fiction: Cognition, Culture, Community*. New York: Oxford University Press, 2015.

Spolsky, Ellen. *Satisfying Skepticism: Embodied Knowledge in the Early Modern World*. Aldershot, UK: Ashgate, 2001.

Stiller, James, and Robin I. M. Dunbar. "Perspective-Taking and Memory Capacity Predict Social Network Size." *Social Networks* 29 (2007): 93–104.

Stiller, James, Daniel Nettle, and Robin I. M. Dunbar. "The Small World of Shakespeare's Plays." *Human Nature* 14 (2003): 397–408.

Straub, Kristina. *Divided Fictions: Fanny Burney and Feminine Strategy*. Lexington: University Press of Kentucky, 1987.

Tale of Frol Skobeev, The. 1680–1720. [*Повесть о Фроле Скобееве*] *Древнерусская Литература*, при поддержке кафедры русской литературы и фольклора КемГУ [Ancient Russian Literature: site supported by the Department of Russian Literature and Folklore of Kemerovo State University (Kemerovo, Russia)], 2002–2003. Accessed 06/11/2021. http://www.drevne.ru/lib/frol.htm.

Tale of Misery-Luckless-Plight, The. 1600–1700. *Повесть о Горе и Злочастии*] *Древнерусская Литература*, при поддержке кафедры русской литературы и фольклора КемГУ [Ancient Russian Literature: site supported by the Department of Russian Literature and Folklore of Kemerovo State University (Kemerovo, Russia)], 2002-2003. Accessed 06/11/2021.http://www.drevne.ru/lib/zlos.htm.

Talwar, Victoria, Heidi Gordon, and Kang Lee. "Lying in the Elementary School Years: Verbal Deception and Its Relation to Second-Order Belief Understanding." *Developmental Psychology* 43, no. 3 (2007): 804–810. doi: 10.1037/0012-1649.43.3.804.

Throop, C. Jason. "Suffering, Empathy, and Ethical Modalities of Being in Yap." In *The Anthropology of Empathy: Experiencing the Lives of Others in the Pacific Societies*, edited by Douglas W. Hollan and C. Jason Throop, 119–149. New York: Berghahn Books, 2011.

Tillman, Hoyt Cleveland. "Selected Historical Sources for *Three Kingdoms*." In *Three Kingdoms and Chinese Culture*, edited by Kimberly Besio and Constantine Tung, 53–69. Albany: State University of New York Press, 2007.

Tobar, Héctor. "The Assassin Next Door." *New Yorker*, July 29, 2019. https://www.newyorker.com/magazine/2019/07/29/the-assassin-next-door.

Tolstoy, Lev. *Anna Karenina*. Translated by Richard Pevear and Larissa Volokhonsky. New York: Viking, 2000.

Tolstoy, Lev. Война и Мир. 1869. [*War and Peace*]. Том 3 [Volume 3] Moscow: Molodaia Gvardia, 1978.

Tomasello, Michael. *Origins of Human Communication*. Cambridge, MA: MIT Press, 2008.

Travers, P. L. *The Mary Poppins Omnibus*. London: Lions, 1994.

Trawalter, Sophie, Andrew R. Todd, Abigail A. Baird, and Jennifer A. Richeson. "Attending to Threat: Race-Based Patterns of Selective Attention." *Journal of Experimental Social Psychology* 44, no. 5 (September 2008): 1322–1327. doi: 10.1016/j.jesp 2008.03.006.

Tribble, Evelyn B. *Cognition in the Globe: Attention and Memory in Shakespeare's Theatre*. London: Palgrave Macmillan, 2011.

Tribble, Evelyn B., and John Sutton. "Cognitive Ecology as a Framework for Shakespearean Studies." *Shakespeare Studies* 39 (2011): 94–103.

Tro, Nivaldo J. *Chemistry: Structure and Properties*. Boston: Pearson, 2017.

Troscianko, Emily. *Kafka's Cognitive Realism*. London: Routledge, 2014.

Turner, Mark. "Compression and Representation." *Language and Literature* 15, no.1 2006): 17–27. doi: 10.1177/0963947006060550.

Twain, Mark. *Mississippi Writings: "Tom Sawyer," "Life on the Mississippi," "Huckleberry Finn," "Pudd'nhead Wilson."* New York: Library of America, 1982.

Van Duijn, Max J., Ineke Sluiter, and Arie Verhagen. "When Narrative Takes Over: The Representation of Embedded Mindstates in Shakespeare's *Othello*." *Language and Literature* 24, no. 2 (2015): 148–166.

van Kuijk, Iris, Peter Verkoeijen, Katinka Dijkstra, and Rolf A. Zwaan. "The Effect of Reading a Short Passage of Literary Fiction on Theory of Mind: A Replication of Kidd and Castano (2013)." *Collabra: Psychology* 4, no. 7 (2018). doi: 10.1525/collabra.117.

Vapnyar, Lara. *Still Here*. New York: Hogarth, 2016.

Vermeule, Blakey. *Why Do We Care about Literary Characters?* Baltimore: Johns Hopkins University Press, 2010.

Vessel, Edward A., G. Gabrielle Starr, and Nava Rubin. "The Brain on Art: Intense Aesthetic Experience Activates the Default Mode Network." *Frontiers in Human Neuroscience* 6, no. 66 (2012). Published online April 20, 2012. doi: 10.3389/fnhum.2012.00066.

Vignemont, Frédérique de. "Frames of Reference in Social Cognition." *Quarterly Journal of Experimental Psychology*, 2007, 1–27.

Vineberg, Steve. "Problem Plays." *Threepenny Review* 52 (1993): 32–34.

Vygotsky, Lev S. *Psichologia Iskusstva*. Moscow: Iskusstvo, 1968.

Vygotsky, Lev S. *The Psychology of Art*. Translated by Scripta Technica, Inc. Introduction by A. N. Leontiev. Commentary by V. V. Ivanov. Cambridge, MA: MIT Press, 1974.

Wang, Xiaojue. "*Stone* in Modern China: Literature, Politics, and Culture." In *Approaches to Teaching "The Story of the Stone,"* edited by Andrew Schonebaum and Tina Lu, 413–426. New York: Modern Language Association, 2012.

Wang Xizhi. "Preface to the Poems Composed at the Orchid Pavilion." https://en .wikisource.org/wiki/Translation:Preface_to_the_Poems_Composed_at_the_Orchid_ Pavilion.

Wang, Yuanfei. "Fantastic Jokes: Allegory, Humor, and Narrative in the Early Modern Chinese Novel *Journey to the West.*" Paper presented at the Modern Language Association, New York, January 2018.

Wassmann, Jürg, Birgit Träuble, and Joachim Funke, eds. *Theory of Mind in the Pacific: Reasoning across Cultures*. Heidelberg: Universitätsverlag Winter, 2013.

Wei, Shang. "The Literati Era and Its Demise (1723–1840)." In *The Cambridge History of Chinese Literature*, edited by Kang-o Sun Chang and Stephen Owen, 245–342. Cambridge: Cambridge University Press, 2010.

Wei, Shang. "The *Stone* Phenomenon and Its Transformation from 1791 to 1919." In *Approaches to Teaching "The Story of the Stone,"* edited by Andrew Schonebaum and Tina Lu, 390–412. New York: Modern Language Association, 2012.

Whalen, Douglas H., Lisa Zunshine, Evelyne Ender, Jason Tougaw, Robert F. Barsky, Peter Steiner, Eugenia Kelbert, and Michael Holquist. "Validating Judgments of Perspective Embedding: Further Explorations of a New Tool for Literary Analysis." *Scientific Study of Literature* 6, no. 2 (2016): 278–298.

Whalen, Douglas H., Lisa Zunshine, and Michael Holquist. "Increases in Perspective Embedding Increase Reading Time Even with Typical Text Presentation: Implications for the Reading of Literature." *Frontiers in Psychology*, November 2015. doi: 10.3389/fpsyg.2015.01778.

Whalen, Douglas H., Lisa Zunshine, and Michael Holquist. "Theory of Mind and Embedding of Perspective: A Psychological Test of a Literary 'Sweet Spot.'" *Scientific Study of Literature* 22, no. 2 (2012): 301–315.

Wharton, Edith. "Xingu." In *"Roman Fever" and Other Stories*, ed. Cynthia Griffin Wolff, 23–58. New York: Scribner, 1997.

White, E. B. *Charlotte's Web*. New York: HarperCollins, 1952.

White, E. B. *Stuart Little*. New York: Harper and Row, 1945.

Wice, Matthew, Minoru Karasawa, Tomoko Matsui, and Joan G. Miller, "Knowing Minds: Culture and Perceptions of Mental State Access." *Asian Journal of Social Psychology* 31 March 2020. https://doi.org/10.1111/ajsp.12404.

Wiesehan, Gretchen. "Christa Wolf Reconsidered: National Stereotypes in 'Kindheitsmuster.'" *Germanic Review* 68, no. 2 (Spring 1993): 79–87.

Wilder, Laura Ingalls. *Little House in the Big Woods*. New York: Scholastic, 1963.

Wilder, Laura Ingalls. *Little House on the Prairie*. New York: Scholastic, 1963.

Wilder, Laura Ingalls. *Little Town on the Prairie*. New York: Scholastic, 1941.

Wilder, Laura Ingalls. *These Happy Golden Years*. New York: Scholastic, 1943.

Williams, Joy. *Ninety-Nine Stories of God*. New York: Tin House Books, 2016.

Wolf, Christa. *City of Angels; or, The Overcoat of Dr. Freud*. Translated from German by Damion Searls. New York: Farrar, Straus and Giroux, 2013. Published in German in 2010.

Wolf, Christa. *Kindheitsmuster*. Berlin: Aufbau-Verlag, 1976.

Wolf, Christa. *Patterns of Childhood*. Translated by Ursule Molinaro and Hedwig Rappolt. New York: Farrar, Straus and Giroux, 1980.

Wolf, Christa. *They Divided the Sky*. Translated by Luise von Flotow. Ottawa: University of Ottawa Press, 2013.

Wood, James. *The Fun Stuff: And Other Essays*. New York: Farrar, Straus and Giroux, 2012.

Wood, James. "Reality Testing: A First Novel about Poetry and Imposture in Spain." *New Yorker*, October 24, 2011. https://www.newyorker.com/magazine/2011/10/31/reality-testing.

Wordsworth, William. "Lines Composed a Few Miles above Tintern Abbey, on Revisiting the Banks of the Wye during a Tour, July 13, 1798." *The Norton Anthology of English Literature, Eighth Edition: The Major Authors*, edited by Stephen Greenblatt and M. H. Abrams, 1491–1495. New York and London: Norton, 1990.

Wu Ching-Tzu. *The Scholars*. Beijing: Foreign Languages Press, 2000.

Wulandini, Gouvara, I. A. Kuntoro, and E. Handayani. "The Effect of Literary Fiction on School-Aged Children's Theory of Mind (ToM)." In *Diversity in Unity: Perspectives from Psychology and Behavioral Sciences*, edited by Amarina Ashar Ariyanto, Hamdi Maluk, Peter Newcombe, Fred P. Piercy, Elizabeth Kristi Poerwandari, and Sri Hartati R. Suradijono, 159–166. London: Taylor and Francis, 2018.

Yeazell, Ruth Bernard. *Fictions of Modesty: Women and Courtship in the English Novel*. Chicago: University of Chicago Press, 1991.

Yu, Pauline. "The Story of Yingying." In *Ways with Words: Writing about Reading Texts from Early China*, edited by Pauline Yu, Peter Bol, Stephen Owen, and Willard Peterson, 182–185. Berkeley: University of California Press, 2000.

Zamiatin, Evgenij. Мы [We]. Moscow: ACT, 2008.

Zenkovsky, Serge A., ed. and trans. "Misery-Luckless-Plight." In *Medieval Russia's Epics, Chronicles, and Tales*, 489–501. New York: Dutton, 1974.

Zhilicheva, G. A. "Функции «Ненадежного» Нарратора в Русском Романе 192(–1930-х Годов" [Functions of "unreliable narrator" in the Russian novel of the 1920s–1930s]. *Вестник ТГПУ* [*TSPU bulletin*] 11, no. 139 (2013): 32–38.

Zunshine, Lisa. "Bakhtin, Theory of Mind, and Pedagogy: Cognitive Construction of Social Class." *Eighteenth-Century Fiction* 30, no. 1 (2017): 109–126.

Zunshine, Lisa. "The Commotion of Souls." *SubStance* #140, 45, no. 2 (2016): 118–142.

Zunshine, Lisa. "Embodied Social Cognition and Comparative Literature: an Introduction." *Poetics Today* 41, no. 2 (2020): 171–186.

Zunshine, Lisa. "From the Social to the Literary: Approaching Cao Xueqin's *The Story of the Stone* (*Honglou meng*) from a Cognitive Perspective." In *The Oxford Handbook of Cognitive Literary Studies*, edited by Lisa Zunshine, 176–196. New York: Oxford University Press, 2015.

Zunshine, Lisa. *Getting Inside Your Head: What Cognitive Science Can Tell Us about Popular Culture*. Baltimore: Johns Hopkins University Press, 2012.

Zunshine, Lisa. "I Lie Therefore I Am." In *Approaches to Teaching "The Plum in the Golden Vase,"* edited by Andrew Schonebaum. New York: MLA, forthcoming.

Zunshine, Lisa. "Introduction to Cognitive Literary Studies." In *The Oxford Handbook of Cognitive Literary Studies*, edited by Lisa Zunshine, 1–9. New York: Oxford University Press, 2015.

Zunshine, Lisa. "May 2020 Bibliography for Cognitive Literary Studies." May 2020. https://mla.hcommons.org/deposits/item/hc:30131/.

Zunshine, Lisa. "The Secret Life of Fiction." *PMLA* 130, no. 3 (2015): 724–731.

Zunshine, Lisa. *Strange Concepts and the Stories They Make Possible: Cognition, Culture, Narrative*. Baltimore: The Johns Hopkins University Press, 2008.

Zunshine, Lisa. "Style Brings in Mental States: A Response to Alan Palmer's 'Social Minds.'" *Style* 45, no. 2 (2011): 349–356.

Zunshine, Lisa. "'Think What You're Doing, or You'll Only Make an Ugly Reputation for Yourself': *Chin P'ing Mei* (金瓶梅), Lying, and Literary History," in 认知诗学 [*Cognitive poetics*] (December 2017): 44–62.

Zunshine, Lisa. "Who Is He to Speak of My Sorrow?" *Poetics Today* 41, no. 2 (2020): 223–241.

Zunshine, Lisa. "Why Jane Austen Was Different, and Why We May Need Cognitive Science to See It." *Style* 41, no. 3 (2007): 287–290.

Zunshine, Lisa. *Why We Read Fiction: Theory of Mind and the Novel*. Columbus: Ohio State University Press, 2006.

Zwaan, Rolf A. "Effect of Genre Expectations on Text Comprehension." *Journal of Experimental Psychology: Learning, Memory, and Cognition* 20, no. 4 (1994): 920–933.

Index